PROOF AND COMPUTATION

Digitization in Mathematics, Computer Science, and Philosophy

PROOF AND COMPUTATION

Digitization in Mathematics, Computer Science, and Philosophy

Editors

Klaus Mainzer
Technische Universität München, Germany

Peter Schuster
Università degli Studi di Verona, Italy

Helmut Schwichtenberg
Ludwig-Maximilians-Universität München, Germany

W **World Scientific**

NEW JERSEY · LONDON · SINGAPORE · BEIJING · SHANGHAI · HONG KONG · TAIPEI · CHENNAI · TOKYO

Published by

World Scientific Publishing Co. Pte. Ltd.

5 Toh Tuck Link, Singapore 596224

USA office: 27 Warren Street, Suite 401-402, Hackensack, NJ 07601

UK office: 57 Shelton Street, Covent Garden, London WC2H 9HE

Library of Congress Cataloging-in-Publication Data

Names: Mainzer, Klaus, editor. | Schuster, Peter, 1966–   editor. |
    Schwichtenberg, Helmut, 1942–   editor.
Title: Proof and computation : digitization in mathematics, computer science and philosophy /
    edited by Klaus Mainzer (Technische Universität München, Germany),
    Peter Schuster (Università degli Studi di Verona, Italy),
    Helmut Schwichtenberg (Ludwig-Maximilians-Universität München, Germany).
Other titles: Proof and computation (World Scientific Publishing Company)
Description: [Hackensack] New Jersey : World Scientific, 2018. |
    Includes bibliographical references and index.
Identifiers: LCCN 2018014609 | ISBN 9789813270930 (hardcover : alk. paper)
Subjects: LCSH: Mathematics--Data processing. | Mathematics--Philosophy.
Classification: LCC QA76.95 .P766 2018 | DDC 004.01/51--dc23
LC record available at https://lccn.loc.gov/2018014609

British Library Cataloguing-in-Publication Data
A catalogue record for this book is available from the British Library.

For any available supplementary material, please visit
http://www.worldscientific.com/worldscibooks/10.1142/11005#t=suppl

Desk Editor: Benny Lim

Printed in Singapore

Preface

Today's debate about safety and reliability of data and algorithms is very similar to the logical-mathematical debate on foundations in the early 20th century — today, however, with grave consequences for technology and economy. The key to the solution of these problems is to be found in the logical foundations of mathematics and computer science themselves, the current development of which leads to new joint research methods and perspectives in mathematics and computer science.

In this vein, an international autumn school "Proof and Computation" was held from 3rd to 8th October 2016 at Aurachhof in Fischbachau near Munich. Its aim was to gather together young researchers active in the foundations of mathematics, computer science, and philosophy. There was the opportunity to form ad-hoc groups working on specific projects, but also to discuss in more general terms the vision of extracting correct programs from proofs. The participants worked on predicative foundations, constructive mathematics and type theory, computation in higher types, extraction of programs from proofs, and algorithmic aspects in financial mathematics. Lectures were given by Laura Crosilla on predicativity, Hajime Ishihara on constructive analysis, Stan Wainer on computability, Masahiko Sato on lambda calculus, Kenji Miyamoto on program extraction from proofs, Andreas Abel on Agda, and Josef Berger on algorithmic aspects of financial mathematics.

Acknowledgements

The autumn school was made possible by the Udo Keller Stiftung (Hamburg), the CORCON project of the European Commission (PIRSES-GA-2013-612638, Correctness by Construction), and a JSPS Core-to-Core project. Parts of the editing process took place within the John Templeton Foundation's project "A New Dawn of Intuitionism: Mathematical and Philosophical Advances" (ID: 60842) and the CATLOC project (Ricerca di

Base, Categorical Localisation: Methods and Foundations) of the University of Verona. The editors are thankful for the great patience exhibited by World Scientific, and would like to express their gratitude to Daniel Wessel, who most diligently prepared the camera-ready version of this volume. The editors especially thank the Udo Keller Stiftung, which sponsors interdisciplinary projects of science and philosophy in the spirit of the German physicist and philosopher Carl Friedrich von Weizsäcker (1912–2007).

Munich and Verona, February 2018

Klaus Mainzer
Peter Schuster
Helmut Schwichtenberg

Contents

Chapter 1

Proof and Computation

Perspectives beyond Turing Computability

Klaus Mainzer

Lehrstuhl für Philosophie und Wissenschaftstheorie
Technische Universität München (TUM)
Arcisstraße 21, 80333 München, Germany
mainzer@tum.de

Contents

1. Introduction

From a digital point of view, constructive proofs should be realized by digital systems. If mathematics is restricted to computable functions of natural numbers, Turing machines will do the job. But, in higher mathematics and physics (e.g., functional analysis), we have to consider functionals and spaces of higher types. Therefore, a general concept of a *digital information system* is necessary to compute finite approximations of functionals of higher types. Contrary to the machine-orientation of computational mathematics, *intuitionistic mathematics* is rooted in the philosophy of human creativity. Mathematics is understood as a human activity of constructing

1

and proving step-by-step. In the rigorous sense of Brouwer's intuitionism, we even get a concept of real continuum and infinity which differs from the ordinary understanding of these concepts in classical mathematics. In the foundational research programs of proof mining and reverse mathematics, we offer degrees of constructivity and provability to classify problem solving and axiomatic strength for different mathematical applications.

Proof mining aims to extract effective procedures by purely logical analyses of mathematical proofs and principles [52]. Sometimes it is even sufficient to find bounds for search procedures of problem solving. At least, proof mining tells us how far away a proof is from being constructive. The research program of proof mining goes back to Georg Kreisel's proof-theoretic research. He illustrates the extraction of effective procedures from proofs as "unwinding proofs" [25].

From a theoretical point of view, proof mining is an important link between logic, mathematics, and computer science. Logic is no longer only a formal activity beside and separated from mathematics. Actually, metatheorems of proof mining deliver *logical tools to solve mathematical problems* more effectively and to obtain new mathematical information: Proofs are more than verifications of theorems!

From a practical point of view, proof theory also has practical consequences in applied computer science. In this case, instead of formal theories and proofs, we consider formal models and computer programs of processes, e.g., in industry. Formal derivations of formulas correspond to, e.g., steps of industrial production. In order to avoid additional costs of mistakes and biases, we should test and prove that *computer programs* are correct and sound before applying them. *Automated theorem proving* is applied to integrated design and verification. Software and hardware design must be verified to prevent costly and life threatening errors. Dangerous examples in the past were flaws in the microchips of space rockets and medical equipment. Network security of the Internet is a challenge for banks, governments, and commercial organizations. In the age of Big Data and increasing complexity of our living conditions, rigorous proofs and tests are urgently demanded to avoid a dangerous future with overwhelming computational power running out of control.

But, in real mathematics, proofs cannot all be reduced to effective procedures. Actually, there are *degrees of constructivity, computability, and provability.* How strong must a theory be in order to prove a certain theorem? We try to determine which axioms are required to prove a theorem. How constructive and computational are these assumptions?

Instead of going forward from axioms to the theorems in usual proofs, we prove in the reverse way (backward) from a given theorem to the assumed axioms. Therefore, this research program is called *reverse mathematics* [29]. If an axiom system S proves a theorem T and theorem T together with an axiom system S' (the reversal) prove axiom system S, then S is called equivalent to theorem T over S'. In order to determine the degrees of constructivity and computability, one tries to characterize mathematical theorems by equivalent subsystems of arithmetic. The formulas of these arithmetical subsystems can be distinguished by different degrees of complexity.

In (second-order) arithmetic, all objects are represented as natural numbers or sets of natural numbers. Therefore, proofs about real numbers must use Cauchy sequences of rational numbers, which can be represented as sets of natural numbers. In reverse mathematics, theorems and principles of real mathematics are characterized by equivalent subsystems of arithmetic with different degrees of constructivity and computability [45]. Some of them are computable in Turing's sense, but others are definitely not and need stronger tools beyond Turing-computability. Therefore, it is proved that only parts of real mathematics can be reduced to the digital paradigm. Reverse mathematics and proof mining are important research programs aiming to connect mathematics, computer science, and logic in a rigorous way.

Philosophically, reverse mathematics and proof mining are deeply rooted in the epistemic tradition of efficient reasoning with the *principle of parsimony* in explanations, proofs, and theories. It was the medieval logician and philosopher William of Ockham (1285–1347) who first demanded that one should prefer explanations and proofs with the fewest number of assumed abstract concepts and principles. The reason is that, according to Ockham, universals (e.g., mathematical sets) are only abstractions from individuals (the elements of a set) without real existence. Ockham's principle of parsimony became later popular as "Ockham's razor" which should reduce abstract principles as far as possible.

The question arises of how far reduction is possible without loss of essential information. In order to analyze real computation in mathematics, we need an extension of *digital computability beyond Turing-computability*. *Real computing machines* can be considered idealized analog computers accepting real numbers as given infinite entities [8]. They are not only theoretically interesting but also practically inspiring with respect to their applications in scientific computing and technology (e.g., sensor technology) [62].

2. Basics of Computability

The early historical roots of computer science stem back to the age of classical mechanics. The mechanization of thoughts begins with the invention of mechanical devices for performing elementary arithmetic operations automatically. A mechanical calculation machine executes serial instructions step by step. In general, the traditional design of a mechanical calculation machine contains the following components.

First, there is an input mechanism by which a number is entered into the machine. A selector mechanism selects and provides the mechanical motion to cause the addition or subtraction of values on the register mechanism. The register mechanism is necessary to indicate the value of a number stored within the machine, technically realized by a series of wheels or disks. If a carry is generated because one of the digits in the result register advances from 9 to 0, then that carry must be propagated by a carry mechanism to the next digit or even across the entire result register. A control mechanism ensures that all gears are properly positioned at the end of each addition cycle to avoid false results or jamming the machine. An erasing mechanism has to reset the register mechanism to store a value of zero.

It was Leibniz's mechanical calculating machine for the first four rules of arithmetic which contained each of the mechanical devices from the input, selector and register mechanisms to the carry, control, and erasing mechanisms. The Leibniz machine became the prototype of a hand calculating machine. If we abstract from the technical details and particular mechanical constructions of Leibniz's machine, then we get a model of an ideal calculating machine.

Leibniz's design of a calculating machine is part of a general research program [59]. Leibniz's mathesis universalis [86] intended to simulate human thinking by calculation procedures ("algorithms") and to implement them on mechanical calculating machines. Leibniz proclaimed two basic subdisciplines:

- An *ars iudicandi* should allow every scientific problem to be decided by an appropriate arithmetic algorithm after its codification into numeric symbols.
- An *ars inveniendi* should allow scientists to seek and enumerate possible solutions of scientific problems.

Leibniz' mathesis universalis seems already to foreshadow the famous Hilbert program in the 20th century with its demands for formalization and axiomatization of mathematical knowledge. Actually, Leibniz developed some procedures to formalize and codify languages. He was deeply convinced that there are universal algorithms to decide all problems in the world by mechanical devices.

The modern formal logic of Frege and Russell and the mathematical proof theory of Hilbert and Gödel have been mainly influenced by Leibniz' program of mathesis universalis. The hand calculating machine which was abstracted from the Leibniz machine can easily be generalized to Marvin Minsky's register machine [74, 92]. It allows the general concept of computability to be defined in modern computer science.

A register machine with program F is defined to compute a function f with n arguments if for arbitrary $x_1, \ldots, x_n$ in the registers $1, \ldots, n$ (and zero in all other ones) the program F is executed and stops after a finite number of steps with the arguments of the function in the register $1, \ldots, n$ and the function value $f(x_1, \ldots, x_n)$ in register $n + 1$.

The program

$$\langle 1 \rangle := x_1; \ldots; \langle n \rangle := x_n$$

$$\downarrow$$

$$F$$

$$\downarrow$$

$$\langle n + 1 \rangle := f(x_1, \ldots, x_n)$$

works according to a corresponding matrix of instructions. A function f is called computable by a register machine RM (RM-computable) if there is a program F computing f.

Obviously, Minsky's register machine is an intuitive generalization of a hand calculating machine à la Leibniz. But, historically, some other equivalent formulations of machines were at first introduced independently by Alan Turing [104] and Emil Post [83] in 1936. A Turing machine (Fig. 1) can carry out any effective procedure. It consists of

a) a control box in which a finite program is placed,
b) a potentially infinite tape, divided lengthwise into squares,
c) a device for scanning, printing on one square of the tape at a time, and for moving along the tape or stopping, all under the command of the control box.

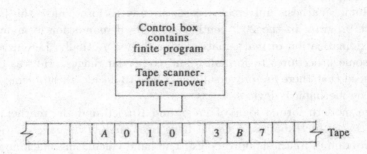

Fig. 1.   Turing machine

If the symbols used by a Turing machine are restricted to a stroke / and a blank *, then an RM-computable function can be proved to be computable by a Turing machine and vice versa. We must remember that every natural number x can be represented by a sequence of x strokes (for instance 3 by ///), each stroke on a square of the Turing tape. The blank * is used to denote that the square is empty (or the corresponding number is zero). In particular, a blank is necessary to separate sequences of strokes representing numbers. Thus, a Turing machine computing a function f with arguments $x_1, \ldots, x_n$ starts with tape $\cdots * x_1 * x_2 * \cdots x_n * \cdots$ and stops with $\cdots * x_1 * x_2 * \cdots x_n * f(x_1, \ldots, x_n) * \cdots$ on the tape.

From a logical point of view, a general purpose computer — as constructed by associates of John von Neumann in America and independently by Konrad Zuse in Germany — is a technical realization of a universal Turing machine which can simulate any kind of Turing program [37]. Analogously, we can define a universal register machine which can execute any kind of register program. Actually, the general design of a von Neumann computer consists of a central processor (program controller), memory, an arithmetic unit, and input-output devices. It operates step by step in a largely serial fashion. A present-day computer à la von Neumann is really a generalized Turing machine.

Besides Turing and register machines, there are many other mathematically equivalent procedures for defining computable functions. Recursive functions are defined by unbounded search, procedures of functional substitution and iteration, beginning with some elementary functions, for instance substitution and iteration, beginning with some elementary functions (for instance, the successor function $n(x) = x + 1$) which are obviously computable. All these definitions of computability by Turing machines, register machines, recursive functions, etc., have been proven to be mathematically equivalent. Obviously, each of these precise concepts defines a procedure which is intuitively effective (computable). Thus, Alonzo Church postulated his famous thesis that the informal intuitive notion of an effective procedure is identical with one of these equivalent precise concepts, such as that of a Turing machine:

> Every computational procedure (algorithm) can be implemented by a Turing machine.

Church's thesis cannot be proved, of course, because it compares mathematically precise concepts with an informal intuitive notion. Nevertheless, the mathematical equivalence of several precise concepts of computability which are intuitively effective supports the evidence of Church's thesis. Consequently, we can speak about computability, effectiveness, and computable functions without referring to particular effective procedures ("algorithms") like Turing machines, register machines, recursive functions, etc. According to Church's thesis, we may in particular say that every computational procedure (algorithm) can be implemented by a Turing machine. A general purpose computer is an (approximately) technical realization of a (universal) Turing machine. In that sense every recursive function, as a kind of machine program, can be calculated by a general purpose computer.

To be more precise, a class of functions which was accepted as intuitively computable from the very beginning [47, 93] concerns the primitive recursive functions with the zero function, successor function, and projection functions which is closed under composition and recursion. The class of primitive recursive functions can be extended by an intuitively computable search procedure for minimal numbers satisfying certain computable conditions. Therefore, the class $\mathcal{F}_r$ of total recursive functions is defined as the least class of functions satisfying the conditions of the class $\mathcal{F}_{pr}$ of primitive recursive functions and the additional demand that $\mathcal{F}_r$ is closed under the application of the μ-operator. In the following, $\vec{y}$ stands for a tupel $y_1, \ldots, y_n$:

If for all $\vec{y}$ there exist an x with $g(x, \vec{y}) = 0$ (formally $\forall \vec{y} \, \exists x \, g(x, \vec{y}) = 0$), then the μ-operator is defined by $f(\vec{y}) = \mu x (g(x, \vec{y}) = 0)$, i.e., the least x satisfying $g(x, \vec{y}) = 0$.

Sometimes it is not clear that functions are totally defined. $f(x) \simeq$

$g(x)$ means that function f is defined iff function g is defined, and if they are defined, then their values are equal. Now, we can extend the class of total recursive functions to the class of partial recursive functions (with corresponding definition of the μ-operator for partial recursive functions). The equivalence of partial recursive functions with RM (register machine)-computable functions supports Church's thesis.

Now, we are able to define effective procedures of decision and enumerability, which were already demanded by Leibniz' program of a *mathesis universalis* [16]. The characteristic function χ_M of a subset M of natural numbers is defined by $\chi_M(x) = 1$ if x is an element of M and as $\chi_M(x) = 0$ otherwise.

A set M is defined as decidable if its characteristic function saying whether or not a number belongs to M is computable.

Programs and computations consist of lists of symbols which can be coded by natural numbers. Then, we can define a decidable predicate $T(x, \vec{y}, z)$ (Kleene's T-predicate) with the meaning that "z codes a (terminating) computation according to program x for arguments $\vec{y}$". The total computable result-extracting function U extracts the result from the code for a terminating computation. It can be proven that each computable function f with code x of its computer program can be represented in Kleene's normal form:

$$f(\vec{y}) \simeq U(\mu z \chi_T(x, \vec{y}, z) = 1) \simeq U(\mu z T(x, \vec{y}, z)).$$

The partially defined expression $[x](\vec{y})$ denotes the result of applying program x to $\vec{y}$. Actually, x is the code number of a machine program (e.g., register machine) which is also called "machine number" (index) of the computable function f. Obviously, $[x](\vec{y})$ is defined if there is a (terminating) computation z of program x for input arguments $\vec{y}$:

$$[x](\vec{y}) \text{ is defined} \leftrightarrow \exists z \, T(x, \vec{y}, z).$$

A set M is defined to be effectively enumerable if there exists an effective (computable) procedure f for generating its elements, one after another (formally $f(1) = x_1, f(2) = x_2, \ldots$ for all elements $x_1, x_2, \ldots$ from M). The definition of recursive enumerability can be generalized from sets to predicates. In short:

> A predicate P is recursively enumerable, if P is empty or there is a recursive function f such that $\forall x\,(P(x) \leftrightarrow \exists y\, f(y) = x)$.

Recursive enumerability can be interpreted as a formal definition of Leibniz' *ars inveniendi*. It can be proven [38] that for all recursively enumerable predicates P there is a recursively decidable predicate Q with $P(x) \leftrightarrow \exists y\, Q(x, y)$ for all x.

Decidability can be characterized by enumerability: A predicate is recursively decidable if and only if the predicate and its complementary predicate is recursively enumerable (Post's theorem). The complementary set $\overline{M}$ of M contains all elements which do not belong to M. If M and $\overline{M}$ are recursively enumerable, then we can enumerate their elements step by step, in order to decide if a given number does belong to M or not. Thus, M is recursively decidable. By definition, it follows that every recursively decidable set is recursively enumerable. But there are recursively enumerable sets which are not decidable. These are the first hints that there are limits to Leibniz's original program of a "mathesis universalis", based on the belief in universal decision procedures.

At this point, Turing's famous halting problem comes in: Is there a universal decision procedure to determine whether an arbitrary computer program stops after finite steps for an arbitrary input? Turing proved that the halting problem is in principle unsolvable. Then Gödel's (first) incompleteness [34] is only a corollary of Turing's proof [14].

The unsolvability of the halting problem resolves Hilbert's Entscheidungsproblem. If there is a complete formal axiomatic system from which all mathematical truth follows, then it would give us a procedure to decide if a computer program will ever halt. That is impossible by Turing's proof.

The unsolvability of Turing's halting problem is not only fundamental for computability theory, but it has deep consequences for mathematical foundations. An example is Hilbert's 10th problem which is also proved to be in principle unsolvable due to Turing's Halting problem. In 1900, David Hilbert asked for an algorithm which will decide whether a Diophantine equation has a solution [40]. This was the 10th problem of his famous list of 23 mathematical problems which were still unsolved at the beginning of the 20th century. Algebraic equations which involve only multiplication, addition and exponentiation of whole numbers were named after the third-century Greek mathematician Diophantos of Alexandria.

In 1970, J.V. Matijasevic from the V.A. Steklov Institute in former Leningrad (St. Petersburg) proved that Hilbert's 10th problem is equivalent to Turing's Halting problem and, consequently, not decidable.

Matijasevic used results of [17]. (The Russian mathematician G.V. Cudnovskij was also involved.) According to Lagrange's representation of natural numbers as sum of four quadratic whole numbers, Hilbert's 10th problem can be reduced to the existence of solutions in natural numbers.

A predicate D is called Diophantine if it is definable by predicates $x+y = z, x \cdot y = z, x^y = z$ and logical operations $\vee$ (or), $\wedge$ (and), and $\exists$ (existence quantifier):

$$D(x_1, \ldots, x_n) \leftrightarrow \exists y_1, \ldots, y_r \, P(x_1, \ldots, x_n, y_1, \ldots, y_r) \text{ with } P \text{ recursive}$$
$$\leftrightarrow \exists y_1, \ldots, y_r \, \chi_P(x_1, \ldots, x_n, y_1, \ldots, y_r) = 1$$

with computable characteristic function χ_P.

According to our theorem of recursive enumerability, it follows that every Diophantine predicate is recursively enumerable. Vice versa, it can be proven that every enumerable predicate is Diophantine. Matijasevic used the Fibonacci sequence to define an appropriate Diophantine predicate. The Halting problem can be represented by an enumerable, but not decidable predicate. Therefore, the corresponding Diophantine predicate is also not decidable.

3. Hierarchies of Computability

According to Church's thesis, Turing computability is a representative definition of computability in general. In the following we want to consider problems with degrees of complexity below and beyond this limit. Below this limit there are many practical problems concerning certain limitations on how much the speed of an algorithm can be increased. Especially among mathematical problems there are some classes of problems that are intrinsically more difficult to solve algorithmically than others. Thus, there are degrees of computability for Turing machines which are made precise in complexity theory in computer science.

Complexity classes of problems (or corresponding functions) can be characterized by complexity degrees, which give the order of functions describing the computational time (or number of elementary computational steps) of algorithms (or computational programs) depending on the length of their inputs. Complexity theory delivers degrees for the algorithmic power of Turing machines or Turing-type computers. The theory has practical consequences for scientific and industrial applications. But does it imply limitations for mathematics and the human mind? The fundamental questions of complexity theory (for example P versus NP) refer to the measurement of the speed, computational time, storage capacity, and so on, of algorithms. What about the complexity of mathematical problems beyond Turing computability?

Enumerability is only the first step on a ladder of increasing computational complexity. By adding unrestricted quantifiers to recursive predicates, degrees of computability can be extended in the arithmetical hierarchy beyond Turing computability [41, 50, 93]. A predicate P is arithmetical iff it has an explicit definition

$$P(x) \leftrightarrow Q_1 \ldots Q_n R(x, x_1, \ldots, x_n) \qquad (*)$$

for all numbers x with R recursive and each number quantifier Q_i for either $\exists x_i$ or $\forall x_i$ with number variables x_i. Two quantifiers are of the same kind if they are both existential or both universal. Two adjacent quantifiers of the same kind can be replaced by a single quantifier (contraction of quantifiers). A predicate is $\Sigma_n^0 (\Pi_n^0)$ for $n \geq 1$ iff it has an explicit definition $(*)$ with no two adjacent quantifiers of the same kind and the first quantifier existential (universal). Every arithmetical predicate is equivalent to some Σ_n^0 predicate or to some Π_n^0 predicate for some $n \geq 1$. This classification of arithmetical predicates is called the arithmetical hierarchy. A predicate is Δ_n^0 iff it is both Σ_n^0 and Π_n^0. Kleene's normal form of recursive functions allows us to enumerate all recursive functions and recursively enumerable predicates (Section 2). It can be generalized to the arithmetical hierarchy.

Computations can depend on external agencies supplying answers to questions about a set (or property) M by an unknown procedure. According to Turing, the external agency of an unknown procedure is called an oracle [105]. In Fig. 2, the computation of a function value $f(x)$ depends on an oracle deciding if certain values satisfy a property M [84]. Most of our calculations in everyday life depend on assumptions we rely in without knowing a final confirmation. Thus, oracles lead to an important extension of relativized computation.

A function f is said to be computable relative to (recursive in) M iff the program of f needs an oracle M. An oracle can also be a machine operation (total function) ϕ supplying values $\phi(x)$ by an unknown procedure. Random oracles depend on random procedures.

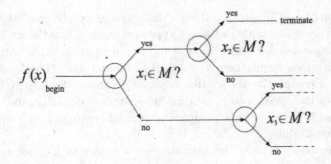

Fig. 2.   Relative computability with an oracle

With relativized computability, we can determine degrees of computational solvability in problem solving. How far can problem solving be reduced to Turing-compatibility?

A predicate A is Turing reducible to a predicate B (notation: $A \leq_T B$) iff A is computable in B.

The relation $\leq_T$ is transitive, symmetric, and reflexive. Thus, $\leq_T$ induces an equivalence relation $\equiv_T$ defining that two predicates are equivalent iff each is recursive in the other, i.e.,

$$A \equiv_T B \quad \text{iff} \quad A \leq_T B \text{ and } B \leq_T A.$$

Intuitively speaking, two predicates are equivalent if they are equally difficult to decide. Turing reducibility can also be applied to relativized computability of functions. The corresponding equivalence classes are called Turing degrees of unsolvability. They are denoted by $\mathbf{A}, \mathbf{B}, \ldots$ with predicates $A, B, \ldots$ as representatives of the corresponding equivalence classes. $\mathbf{A} \leq_T \mathbf{B}$ means that a predicate of Turing degree $\mathbf{A}$ is recursive in a predicate of Turing degree $\mathbf{B}$. Then, by transitivity, any predicate of Turing degree $\mathbf{A}$ is recursive in any predicate of Turing degree $\mathbf{B}$. A Turing degree $\mathbf{A}$ is said to be recursively enumerable in Turing degree $\mathbf{B}$ if a predicate of $\mathbf{A}$ is recursively enumerable in a predicate of $\mathbf{B}$. Then, by transitivity, that predicate of $\mathbf{A}$ is recursively enumerable in every predicate of $\mathbf{B}$.

It can be proven that among the degrees recursively enumerable in $\mathbf{A}$, there is a largest one. The largest degree which is recursively enumerable in $\mathbf{A}$ is called the (Turing) jump $\mathbf{A}'$ of $\mathbf{A}$. With Turing jump, we can prove that there is no largest Turing degree.

But before we consider remarkable applications of Turing jumps in the foundations of mathematics, we remind the reader that the degrees of arithmetical hierarchy can be generalized from arithmetic to analysis. In this case, the quantifiers refer to function variables [48, 49]. Every analytical predicate is equivalent either to some arithmetical predicate or to some Σ_n^1 predicate or to some Π_n^1 predicate for some $n \geq 1$. This classification of analytical predicates is called the analytical hierarchy. A predicate is Δ_n^1 iff it is both Σ_n^1 and Π_n^1. The analytical hierarchy can also be relativized, and corresponding properties can be applied to the relativized case.

In the arithmetical hierarchy, Δ_1^0 is a distinguished layer which contains exactly the recursive predicates. What about Δ_1^1 predicates in the analytical hierarchy? The arithmetical relations are only a proper subclass of layer Δ_1^1 in the analytical hierarchy. A complete characterization of Δ_1^1 relations needs an extension of the arithmetical hierarchy which is called the hyperarithmetical hierarchy.

The hyperarithmetical hierarchy starts in a first layer with the recursive predicates. In the following layers, they are relativized to sets of indices which are generated step by step in a sequence of Turing jumps and finalized by computable unions of these sets [15]:

1) The system $\mathcal{O}$ of indices is partially ordered with bifurcating branches by relation $<_{\mathcal{O}}$:

$$1 <_{\mathcal{O}} 2$$
$$x <_{\mathcal{O}} y \to x <_{\mathcal{O}} 2^y, y <_{\mathcal{O}} 1^y$$
$$\forall n \left([y](n) <_{\mathcal{O}} [y](n+1)\right), \exists n \left(x <_{\mathcal{O}} [y](n)\right), \lambda n[y](n) \text{ total} \to x <_{\mathcal{O}} 3 \cdot 5^y$$

$$1 <_{\mathcal{O}} 2^1 <_{\mathcal{O}} 2^2 <_{\mathcal{O}} \dots \begin{cases} 3 \cdot 5^{y_1} <_{\mathcal{O}} 2^{3 \cdot 5^{y_1}} \begin{cases} \dots \\ \dots \end{cases} \\ 3 \cdot 5^{y_2} <_{\mathcal{O}} 2^{3 \cdot 5^{y_2}} \begin{cases} \dots \\ \dots \end{cases} \\ \dots \\ \dots \end{cases}$$

2) The Turing jump M' of a set M is the effective disjoint union of all sets recursively enumerable in M, i.e., the largest Turing degree which is recursively enumerable in the degree of M.

H-sets are defined by recursion on $<_{\mathcal{O}}$:

$$H_1 = \emptyset$$
$$H_{2^y} = H'_y$$
$$H_{3 \cdot 5^y} = \{\langle a, b \rangle \mid b <_{\mathcal{O}} 3 \cdot 5^y, a \in H_b\} \text{ with } \langle a, b \rangle = 2^a \cdot 3^b$$

3) A set or predicate M is hyperarithmetical iff M is recursive in some H-set. The Turing degree of H-set H_y is determined by index number y.

The class HYP of all hyperarithmetical sets is proven to be equal to the class of Δ_1^1-sets in the analytical hierarchy, i.e., HYP $= \Delta_1^1$. At a first glance, Δ_1^1-sets seem to be far away from recursive and decidable sets and predicates. But, nevertheless, hyperarithmetic turns out to be an important part of predicative and constructive mathematics which we will consider in proof theory and foundations of mathematics in the next section [57, 69]. In general, the (relativized) analytical hierarchy is the bridge from the theory of computability to set theory as basis of mathematics, in order to classify mathematical sets and structures according to their degrees of complexity.

4. Constructive Ordinal Proof Theory

Before coming to predicative mathematics and hyperarithmetic, we consider Turing's historical approach to overcoming Gödel's incompleteness [105]. According to Gödel's incompleteness, each formal axiomatic theory T (containing arithmetic) is associated with a true but unprovable proposition A_T. Turing started with statements of the Π_1^0 form $\forall x\, R(x)$ expressing that a certain (primitive) recursive property R holds for all integers x. He tried to overcome Gödel's incompleteness theorem by constructing degrees of incompleteness. Therefore, he extended the incomplete theories by the corresponding unprovable propositions (as axioms) in a progression along a construction of ordinals, i.e., beginning with the initial theory T_1 and its unprovable A_1,

$$T_2 = T_1 \cup \{A_1\}, \ldots, T_n = T_1 \cup \{A_1, \ldots, A_{n-1}\}, \ldots$$
$$T_\omega = T_1 \cup \{A_1, \ldots, A_{n-1}, \ldots\}, \ldots$$

In order to maintain an effective generation, one must pass to a transfinite limit ordinal α and theory T_α only when α is the limit of an effectively presented sequence $\alpha_1, \ldots, \alpha_n, \ldots$ and the theories T_{α_n} are already obtained. Effectiveness is made precise by an effective representation of ordinals in the integers which can be realized in different ways. In 1936, Church and Kleene defined the index system $\mathcal{O}$ which we already considered in the previous section. The index set $\mathcal{O}$ is partially ordered by $<_{\mathcal{O}}$. It can be used to introduce constructive ordinals. Constructivity comes in by machine numbers representing effective procedures [36]: An ordinal α is constructive iff it has an effective notation $a \in \mathcal{O}$ with $\alpha = |a|$ with

$$|1| = 0$$
$$|2^a| = |a| + 1$$
$$|3 \cdot 5^e| = \sup_{n < \omega}(|[e](n)|).$$

It follows for ordering $<_{\mathcal{O}}$:

$$b <_{\mathcal{O}} a \rightarrow |b| < |a|,$$
$$\forall n \left([e](n) <_{\mathcal{O}} [e](n+1)\right) \text{ for limit ordinals.}$$

Addition and multiplication can be defined as partial recursive operations:

$$|a +_{\mathcal{O}} b| = |a| + |b|$$
$$|a \cdot_{\mathcal{O}} b| = |a| \cdot |b|.$$

The Church-Kleene ordinal ω_1^{CK} denotes the least non-constructive ordinal.

In his Ph.D. thesis 1939, Turing wanted to distinguish degrees of completeness with sequences $\Lambda = \{T_a \mid a \in \mathcal{O}\}$ of logical theories which are associated with indices a of constructive ordinals. The ordinal progression $\Lambda_T = \{T_a \mid a \in \mathcal{O}\}$ of an initially incomplete theory T (e.g., Peano arithmetic) is defined on the indices of constructive ordinals:

$$T_a = T$$
$$T_{2^a} = T_a \cup \{A_a\}$$
$$T_{3 \cdot 5^e} = \bigcup_n T_{[e](n)}.$$

A theorem A is said to be provable in the ordinal progression Λ_T iff there is a a $a \in \mathcal{O}$ with A is provable in T_a. A theorem A is deeper than theorem B iff the least constructive ordinal $\alpha = |a|$ of theory T_a for proving A is greater than the least constructive ordinal $\beta = |b|$ of T_b for proving B.

Indices a increase with increasing ordinals of a, thus overcoming incompleteness at least gradually at each state a. One of Turing's main results was that Λ is complete for true Π_1^0 statements. He intended to strengthen this completeness result to Π_2^0 statements. These sentences have the form $\forall x \exists y\, R(x, y)$ with a primitive recursive predicate R. The motivation of his interest in Π_2^0 predicates is the fact that this layer of arithmetical hierarchy includes very interesting mathematical statements (e.g., the Riemann hypothesis Turing was strongly interested in).

According to Gödel, terms and statements can be coded by numbers, e.g., statement A by Gödel number $\lceil A \rceil$. Turing defined a logical theory very generally as any recursively enumerable set T of (code numbers of) Π_2^0 sentences A such that $\lceil A \rceil \in T$ implies that A is true.

A logical derivation of A in T (abbreviation $A \vdash T$) means $\lceil A \rceil \in T$. A theory S of this kind is said to be at least as complete as T iff $T \subseteq S$. S is said to be more complete than T if T is a proper subset with $T \subset S$. In general, Turing expected to get a classification of Π_2^0 theorems according to their logical depth. A theorem which required a (constructive) ordinal α to prove it would be deeper than one which could be proved by the use of a (constructive) ordinal β less than α [21].

But, in 1962, Solomon Feferman proved that Λ is incomplete for Π_2^0 sentences. Anyway, besides Gentzen, Turing's Ph.D. thesis was the second initiative of ordinal proof theory which has analyzed measures of proof complexity with ordinals. An important extension of ordinal proof theory aims at the analysis of predicative mathematics which is closely connected with the Δ_1^1-layer of analytical hierarchy and hyperarithmetic beyond Turing-computability [26–28].

Predicativity dates back to historical debates on paradoxes with circular arguments. Famous mathematicians and philosophers were involved in these debates. In 1905, the French mathematician Jules Richard described the following example of a semantic paradox:

Real numbers which are defined by English phrases can be arranged in an infinite list of dictionary order. This ordering yields an infinite list of corresponding real numbers $r_1, r_2, r_3, \ldots, r_n, \ldots$ Then, we can define a real number r^* which is different to each number of this list by a diagonal construction. (For example, the integer part of r^* is 0, the n-th decimal place of r^* is 1 if the n-th decimal place of r_n is not 1, and the n-th decimal place of r^* is 2 if the n-th decimal place of r_n is 1.) This is a contradiction.

Henri Poincaré argued that there is a circular argument, because the definition of number r^* is referred to the supposed total class of decimal numbers which can be defined by English phrases. In short, r^* is an element of the class of objects which is used to define r^*. In 1908, Bertrand Russell suggested a syntactical version of the circular argument: Whatever contains an apparent variable must not be a possible value of that variable. A famous example is Russell's paradox of the set of all sets which are not element of themselves. Cantor's (unrestricted) comprehension principle states

$$\forall P \, \exists y \, \forall x \, (x \in y \leftrightarrow P(x)). \tag{$*$}$$

As example, we define $P(x) :\leftrightarrow \neg x \in x$. Then, by $(*)$, there is a set M with

$$\forall x \, (x \in M \leftrightarrow \neg x \in x).$$

For instance $x := M$, we get the contradiction

$$M \in M \leftrightarrow \neg M \in M.$$

Obviously, the definition of M refers to M itself.

Russell avoided impredicative concepts with circular arguments by his type theory. In 1918, Hermann Weyl [110] (and later Paul Lorenzen [60, 61]) criticized set-theoretical foundations of mathematics and suggested the introduction of mathematical analysis through a theory of real numbers which is only based on the construction of natural numbers, integers, and rational numbers step by step in an inductive manner without circular arguments. We will come back to this approach later on. Hierarchies of recursion theory and degrees of computability can be used to analyze predicative definitions and proofs. Actually, ordinals can be determined, up to which degree of complexity predicative definitions and proofs are possible.

It was Kleene et al. who started to refer definitions of sets of natural numbers to arithmetic and analytic hierarchies. The sets which are obtained by finite iteration of Turing jumps are, up to relative recursiveness, all the arithmetically definable sets. In a next step, Kleene extended this hierarchy by the index system $\mathcal{O}$ of Church-Kleene notations for constructive ordinals. The least ordinal α which cannot be indicated by an index $a \in \mathcal{O}$ as $\alpha = |a|$ is denoted ω_1^{CK}. It can also be proven that recursive well-orderings on the natural numbers [98] have order types as ordinals which are less than ω_1^{CK}. We already introduced the hyperarithmetical hierarchy along increasing ordinals $< \omega_1^{\mathrm{CK}}$: A set is called hyperarithmetical iff it is recursive in a set H_a for some $\mathcal{O}$. The class HYP of all hyperarithmetical sets is exactly the Δ_1^1-layer of the analytical hierarchy, i.e., HYP $= \Delta_1^1$.

Why is HYP predicative, i.e., without self-referencing sets? HYP is related to predicative definitions by the ramified type theory which is applied to the analytical hierarchy restricted to constructive ordinals [69]. At first, we introduce the hierarchy $(\mathcal{A}_k)_{k < \omega}$ generating predicatively definable sets step by step along the natural numbers starting with the class $\mathcal{A}$ of arithmetic predicates:

$\mathcal{A}_0 := \mathcal{A}$,

$\mathcal{A}_{k+1} := \mathcal{A}_k^*$ with predicates the 2nd order quantifiers of which are

relativized to $\mathcal{A}_k$,

$\omega := \bigcup_{k < \omega} \mathcal{A}_k$.

The ramified hierarchy is extended for constructive ordinals:

$$\forall y \in o \quad \mathcal{A}_y := \begin{cases} \mathcal{A}, & \text{if } y = 1 \\ \mathcal{A}_{(y)_0}^*, & \text{if } y = 2^{(y)_0} \\ \bigcup_n \mathcal{A}_{[(y)_2](n)}, & \text{if } y = 3 \cdot 5^{(y)_2}. \end{cases}$$

In 1955, Kleene proved that

$$\text{HYP} = \bigcup_{y \in o} \mathcal{A}_y = \mathcal{A}_{\omega_1^{\text{CK}}}.$$

In 1960, Kreisel suggested to identify the predicatively definable sets with hyperarithmetical sets from HYP. An ordinal is called predictively definable if it is the order type of a predicatively definable well ordering $\prec$ of the natural numbers. Obviously, all constructive ordinals are predicative. Consequently, a set is considered predictively definable iff it belongs to $\mathcal{A}_y$ for some $y \in o$. Because of HYP $= \mathcal{A}_{\omega_1^{\text{CK}}}$, all hyperarithmetical sets are accepted to be predicative. For the converse, the predicatively definable sets do not go beyond HYP, because they do not go beyond the recursive well-ordering with order type (ordinal) less than ω_1^{CK}.

In the case of predicatively definable sets, we considered the hierarchy of hyperarithmetical sets. In order to determine predicative provability, we start with a formal theory $\mathcal{H}$ of 2nd order arithmetic the comprehension axiom of which is restricted to Δ_1^1-sets:

$$\forall P \in \Delta_1^1 \, \exists X \, \forall x \, (x \in X \leftrightarrow P(x)),$$

where $\forall x \, (P(x) \leftrightarrow Q(x))$ with any Π_1^1-formula $P(x)$ and and Σ_1^1-formula $Q(x)$.

In analogy to index system o for predicatively definable sets, an index system N of ordinal numbers was introduced to represent all predicatively provable well-orderings. A recursive progression of theories $\mathcal{H}_a$ (with $a \in N$) starts with $\mathcal{H}$ and is extended step by step: If formula $F(x)$ is provable in System $\mathcal{H}_a$ for all x, then $\forall x \, F(x)$ is provable in $\mathcal{H}_{a \oplus 1}$ (with N-addition

$\oplus$). In formal theories, numbers (including ordinals) are represented by their formal counterparts. The enumerable predicate $pr(a, \lceil F(x) \rceil)$ means, that formula F (with Gödel number $\lceil F(x) \rceil$) is provable in $\mathcal{H}_a$ [26, 28]. Thus, we can define a hyperarithmetical theory progression:

$$\mathcal{H}_0 := \mathcal{H}$$

$$\mathcal{H}_{a \oplus 1} := \mathcal{H}_a \cup \{\forall x \, pr(a, \lceil F(x) \rceil) \to \forall x \, F(x) \,|\, F \text{ formula with free}$$
$$\text{first-order variable}\}$$

$$\mathcal{H}_a := \bigcup_{b < a} \mathcal{H}_b \text{ with } a \text{ numeral of limit ordinal.}$$

In order to guarantee predicative provability, the ordinal indices are restricted to those which are autonomous to $(\mathcal{H}_a)_a$, i.e.,

(i) 0 is an autonomous ordinal index.
(ii) If b is an autonomous ordinal index and a codes an ordinal provable in $\mathcal{H}_b$, then a is an autonomous ordinal index.
(iii) $\mathrm{Aut}(\mathcal{H}) :=$ the least ordinal α with $|c| < \alpha$ for all c autonomous to $(\mathcal{H}_a)_a$.

$(\mathcal{H}_a)_a$ starts with a predicatively justified theory $\mathcal{H}$ with predicatively definable predicates and a hyperarithmetical comprehension principle of sets. The justification of predicative provability is continued by the condition of autonomous ordinal index along the theory progression $(\mathcal{H}_a)_a$. In this sense, proofs in $(\mathcal{H}_a)_a$ are predicative. A well-ordering can be predicatively proven if its ordinal type is smaller than $\mathrm{Aut}(\mathcal{H})$.

But, actually, we can argue for more: All predicative proofs can be characterized by $(\mathcal{H}_a)_a$. This argument reminds us of Church's thesis: All Turing-computable functions are computable in an intuitive sense. But, why are intuitively computable functions Turing-computable? Church argued with many different approaches of computability which are all intuitively computable. The decisive point is that these approaches are mathematically equivalent with Turing-computability in a rigorous sense. Analogously, several approaches to predicative probability were suggested, which are equivalent to the hyperarithmetical approach in a rigorous sense. Therefore, predicative proofs are assumed to be characterized by the hyperarithmetical theory progression $(\mathcal{H}_a)_a$ [26, 27, 69].

An example of another approach is a progression of ramified theories the ordinal indices of which are also restricted by a condition of autonomy. It can be proven that these approaches have the same ordinal bound, i.e., the least ordinal α with $|c| < \alpha$ for all c autonomous to the considered theory progression. Further on, under certain conditions, the proofs in one of these theory progressions can be translated into appropriate proofs to another equivalent theory progression. All these arguments support the assumption that predicative provability is exactly realized in the approaches equivalent to hyperarithmetical theory progression $(\mathcal{H}_a)_a$.

The limit ordinal of the predicatively provable ordinals can independently be characterized by a hierarchy $(\chi_\alpha)_\alpha$ of critical functions of ordinals α. For $\alpha \neq 0$, critical function χ_α enumerates the set of fixed points of all χ_β for each $\beta < \alpha$ in order of size [26, 87, 88, 107]. We get the Veblen hierarchy of critical functions:

$\chi_\alpha(\xi) := \omega^\xi$, if $\alpha = 0$

χ_α enumerates the common fixed points of χ_β with $\beta < \alpha$, if $\alpha \neq 0$

$\Gamma_\alpha := \alpha$th fixed point ξ of $\chi_\xi(0) = \xi$

Γ_0 is the smallest ξ such that $\chi_\xi(0) = \xi$. It is exactly the ordinal bound of predicatively provable well-orderings which is sometimes called the Feferman-Schütte ordinal [24, 82]. Transfinite inductions up to ordinals smaller than Γ_0 can be proven in one of the equivalent theory progressions, e.g., in $(\mathcal{H}_a)_a$. But transfinite induction up to Γ_0 is excluded. Therefore, Γ_0 can also be interpreted as the smallest only impredicatively provable well-ordering. Predicativity and impredicativity beyond Γ_0 is discussed by Wolfram Pohlers in [82].

5. Computability in Higher Types

Mathematics cannot be completely reduced to sets and functions of numbers. Applied mathematics in natural and engineering sciences use functionals with number-theoretic functions as arguments (e.g., observables in quantum physics) or even spaces of functionals (e.g., differential geometry in the theory of relativity). In order to evaluate these abstract mathematical objects by computers, the concept of computability must be generalized to functionals in higher types. The question arises how to compute these abstract entities with a huge amount of data which are obviously needed in ordinary mathematics and scientific applications. Therefore, in the following section, a general concept of digital information system is introduced to solve these problems. The original idea stems back to Dana Scott in [91]. In the following, we refer to a presentation by Helmut Schwichtenberg in [89].

In physical computers, computations must be finite. Therefore, in any evaluation of functionals $\Phi(\varphi)$, the argument φ can be called only finitely many times, i.e., the value of $\Phi(\varphi)$ (if defined) must be determined by some finite subfunction of φ. Let $\mathcal{G}$ be a partial functional of type 3, mapping type-2 functionals Φ to natural numbers. How can $\mathcal{G}$ be computable? If Φ is given and $\mathcal{G}(\Phi)$ evaluates to a defined value, evaluation must be finite. Hence the argument Φ can only be called on finitely many functions φ. Each φ must be presented to Φ in a finite form (i.e., as a set of ordered pairs) [89, Chapter 6]. A finite approximation of a functional Φ is a finite set X of pairs (φ_0, n) such that

  (i) φ_0 is a finite function.
  (ii) $\Phi(\varphi_0)$ is defined with value n.
  (iii) If (φ_0, n) and (φ_0', n') belong to X (where φ_0 and φ_0' are consistent), then $n = n'$.

A functional Φ is viewed as the union of all its finite approximations. It is a type-2 argument of type-3 functional $\mathcal{G}$ which should satisfy the principle of finite support: If $\mathcal{G}(\Phi)$ is defined with value n, then there is a finite approximation Φ_0 of Φ such that $\mathcal{G}(\Phi_0)$ is defined with value n.

If $\mathcal{G}(\Phi)$ is evaluated, then we obtain the same value independent of any extension of argument Φ. Functional Φ' extends Φ if for any piece of data (φ_0, n) in Φ there exists another (φ_0', n) in Φ' such that ϕ_0 extends ϕ_0'. With the notion of functional extension, we can formalize the principle of monotonicity: If $\mathcal{G}(\Phi)$ is defined with value n and Φ' extends Φ, then also $\mathcal{G}(\Phi')$ is defined with value n. From these principles, we can conclude that any functional is determined by its set of finite approximations.

The concept of finite approximation motivates a first definition of generalized computability:

A generalized mathematical object (e.g., functional) is computable iff its set of finite approximations is (primitive) recursively enumerable (Σ_1^0-definable).

The computability of number-theoretic functions was defined by Turing machines. Turing machines (and their technical realizations as computers) are examples of information systems. We introduce a general concept of information system to compute finite approximations of generalized mathematical objects like functionals and spaces [91]. From a practical point of view, information systems seem to be universal in a digital world. They are needed everywhere to process the increasing mass of data in bits and bytes. Especially, general concepts of mathematical objects like functionals contain a huge amount of data. These considerations motivate the following mathematical definition of an information system.

In order to describe approximations of abstract objects like functionals by finite ones, we use an information system with a countable set A of bits of data ("tokens"). Approximations need finite sets U of data which are consistent with each other. An "entailment relation" expresses the fact that the information of a consistent set U of data is sufficient to compute a bit of information ("token"). An information system is a structure $\mathcal{A} :=$ $(A, \mathrm{Con}, \vdash)$ with a countable set A ("tokens"), non-empty set Con of finite ("consistent") subsets of A and subset $\vdash$ of $\mathrm{Con} \times A$ ("entailment relation") with

   i. $U \subseteq V \in \mathrm{Con} \implies U \in \mathrm{Con}$
   ii. $\{a\} \in \mathrm{Con}$
  iii. $U \vdash a \implies U \cup \{a\} \in \mathrm{Con}$
   iv. $a \in U \in \mathrm{Con} \implies U \vdash a$
   v. $U, V \in \mathrm{Con} \implies \forall a \in V \, (U \vdash a) \implies (V \vdash b \implies U \vdash b)$.

The ideals ("objects") of an information system $\mathcal{A} := (A, \mathrm{Con}, \vdash)$ are defined as subsets x of A with

   i. $U \subseteq x \implies U \in \mathrm{Con}$ (x is consistent)
   ii. $x \supseteq U \vdash a \implies a \in x$ (x is deductively closed).

Example. The deductive closure $\overline{U} := \{a \in A \mid U \vdash a\}$ of $U \in \mathrm{Con}$ is an ideal. The set of all ideals of information system $\mathcal{A}$ is denoted by $|\mathcal{A}|$.

Finally, we want to define the computability of functionals with information systems. In a first step, we introduce the class of approximable maps between two information systems which can intuitively be considered as a kind of "function space". A "function space" between information systems is also an information system which can be used to characterize (continuous) functionals. In the sense of information systems, these functionals are sets of tokens. Therefore, they can be called computable iff they (i.e., sets of tokens) are recursively enumerable. The approximable maps between two information systems can exactly be identified as the ideals of the function space [89, p. 255].

There are functions and functionals of different degrees of complexity which can be distinguished by the introduction of types. Especially in computer languages, types are essential to recognize the different domains of variables. With respect to our characterization of functionals with information systems, we also can introduce types of information systems. Types are built from base types by the formation of function types $\rho \to \sigma$. For every type ρ, the information system $\mathcal{C}_\rho := (C_\rho, \mathrm{Con}_\rho, \vdash_\rho)$ can be defined. The ideals $x \in |\mathcal{C}_\rho|$ are the partial continuous functionals of type ρ. Since $\mathcal{C}_{\rho \to \sigma} = \mathcal{C}_\rho \to \mathcal{C}_\sigma$, the partial continuous functionals of type $\rho \to \sigma$ correspond to the continuous functions from $|\mathcal{C}_\rho|$ to $|\mathcal{C}_\sigma|$.

A partial continuous functional $x \in |\mathcal{C}_\rho|$ of type ρ can be defined as computable iff it is recursively enumerable as set of tokens. For example, Fig. 3 illustrates tokens and entailments for the algebra $\mathbf{N}$ of natural numbers. For tokens a, b, an entailment $\{a\} \vdash b$ means that there is a path from a (up) to b (down).

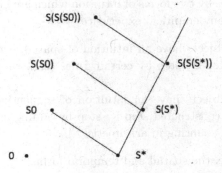

Fig. 3. Tokens and Entailments for algebra $\mathbf{N}$

Partial continuous functionals of type ρ can be used as semantics of a formal functional programming language (in the style of [81]): Every closed term of type ρ in the programming language denotes a computable partial continuous functional of type ρ, i.e., a recursive enumerable consistent and deductively closed set of tokens. Another approach uses recursive equations to define computable functionals [4].

6. Intuitionistic Mathematics and Human Creativity

In computational mathematics, proving and computing is reduced to digital machines evaluating mathematical data and information. Contrary to machine-oriented foundations, intuitionistic mathematics is deeply rooted in a philosophy of human creativity. The Dutch mathematician L.E.J. Bouwer (1881–1966) started with human intuitions which enable mathematical activities and constructions. But his philosophy of the human subject does not only exclude non-constructive abstractions in classical mathematics. His radical principles of construction even lead to theorems which are false in classical mathematics [109]. A crucial point of Brouwer's analysis is his intuitionistic understanding of the continuum and infinity as basis of ordinary mathematics. This discussion culminates in his fan theorem which plays a dominant role in intuitionistic mathematics. How far can it be accepted in computational and constructive mathematics? We begin with the intuitionistic philosophy of a creative subject.

According to Brouwer, mathematical truth is founded by constructions of a "creative subject" [10, 12]. He followed Kant's explanation of human understanding by human subjects. Kant assumed that human understanding is made possible by two forms of intuition which are given to the subject before ('a priori') any empirical experience [46]:

(1) Human subjects have an intuition of spatial forms which are constructed step by step by certain rules (e.g., ruler and compass in geometry).
(2) Human subjects have an intuition of sequential temporal points which are constructed step by step by adding a unit according to the rule of counting in arithmetic.

On the one side, the spatial and temporal forms of intuition enable the human subject to understand empirical objects and events in space and time. For Kant, human subjects are born with an intuitive understanding of space and time which makes possible their orientation in the world. On the other side, the spatial and temporal forms of intuition provide the schemes of geometric and arithmetic constructions and, by that, the foundations of mathematics [69, 71]. Kant illustrates the temporal form of intuition by an unlimited sequence of points $\cdot, \cdot\cdot, \cdot\cdot\cdot, \ldots$, which are extended step by step by a unique point. In our temporal intuition, these points represent a sequence of present moments ('now') which are passing in a linear order. Formally, this process corresponds to the construction of the natural numbers $1, 1 +$

$1, 1+1+1, \ldots$ or in the usual abbreviation of decimal numbers $1, 2, 3, \ldots$ according to the rule of counting.

After Kant, the mathematician Leopold Kronecker claimed that the natural numbers are "made by God", but "all the rest" is made by the human being (according to a talk of Kronecker quoted in [108]). Independent of Kronecker's reference to God, Kant argues on the same line: The natural numbers are given to the subject by the temporal form of intuition and its arithmetical scheme of counting. "All the rest" must be reduced to the fundamental form of arithmetical construction. Thus, for Kant and Kronecker, infinity is not given, but only a "facon de parler" (Poincaré) or "regular idea" (Kant) for the unlimited process of counting. Brouwer derived radical consequences for the truth of theorems and their proofs: In temporal intuition, only finite sequences of natural numbers can be constructed. Thus, for Brouwer, mathematical truth depends on finite stages of realization in time by a creative subject. Georg Kreisel suggested a formal definition [56, 99]. A creative subject has a proof of proposition A at stage m (abbreviation: $\Sigma \vdash_m A$) iff

(CS1) for any proposition A, $\Sigma \vdash_m A$ is a decidable function of A, i.e.,
$$\forall x \in \mathbb{N} \, (\Sigma \vdash_x A \vee \neg \Sigma \vdash_x A).$$
(CS2) $\forall x, y \in \mathbb{N} \, (\Sigma \vdash_x A \to (\Sigma \vdash_{x+y} A))$
(CS3) $\exists x \in \mathbb{N} \, (\Sigma \vdash_x A) \leftrightarrow A$.

A weaker version of CS3 is Kreisel's "Axiom of Christian Charity" (1967):

(CC) $\neg \exists x \in \mathbb{N} \, (\Sigma \vdash_x A) \to \neg A$.

The idea that only finite initial segments of infinite sequences are given leads to Brouwer's concept of choice sequences. A choice sequence means a process which is not necessarily predetermined by some law or algorithm [19, 103]:

(i) α lawless sequence iff at any stage of $\alpha 0, \alpha 1, \alpha 2, \ldots$ only finitely many values of α are known
(ii) α lawlike sequence iff all values of α are known by a law (algorithm).

Lawless sequences can be illustrated by sequences of casts with a die after some already realized casts in the beginning. The die can be thrown in arbitrarily many times. Nevertheless, at any stage, only finitely many casts are known and the following cast is unknown. In lawlike sequences, all stages are predetermined by a law. An example is the sequence of even natural numbers with law $\alpha n = 2n$ or the sequence of decimal places of real number π.

The radical difference to classical mathematics came in, when choice sequences were allowed in real analysis. If real numbers are defined by fundamental (Cauchy) sequences to be given by choice sequences, then the statement "all total functions from $\mathbb{R}$ to $\mathbb{R}$ are continuous" can be proven intuitionistically. At a first glance, this statement seems to be false in classical mathematics. But, the meaning of the condition "function f from $\mathbb{R}$ to $\mathbb{R}$ is total" is obviously much stronger if $\mathbb{R}$ is extended to reals defined via fundamental sequences as choice sequences [11]. Therefore, we get a different intuitionistic meaning of classical concepts. We will come back to this point in the Section 9 on "real computing".

In intuitionist mathematics, infinite objects are considered as ever growing and never finished. Therefore, sets need a new foundation. The intuitionistic analogue of a set is a spread which is defined as a countably branching tree labelled with natural numbers or other finite objects and containing only infinite paths. A fan is a finitely branching spread. A branch is an intuitionistic choice sequence, i.e. an infinite sequence of numbers (or finite objects) created step by step by a law (algorithm) or without law (e.g., coin). A lawless sequence is ever unfinished. The only available information about a lawless sequence at any stage is the initial segment of the sequence created thus far. In order to formalize these concepts, basic concepts of graph theory are needed [100, p. 186 f.].

Brouwer's intuitionistic mathematics is motivated by two fundamental intuitions: The first intuition is the scheme of counting as principle of natural numbers, in order to found proofs and theorems of arithmetic. The second intuition is the concept of choice sequences, in order to capture the idea of the continuum and to found the proofs and theorems of analysis. Especially, he wanted to guarantee the theorem that a continuous function from a closed interval to is uniformly continuous [3]. Historically, Brouwer derived this theorem from his bar-theorem [9, 10]. Following [100, 101], one starts with the so-called fan theorem (which was considered by Brouwer as a corollary of his bar theorem) as axiom. Actually, a fan as finitary spread is more convenient and all known important applications of the bar theorem can also be derived from the fan theorem.

Brouwer's famous bar induction (for decidable predicates P) can be understood as an induction principle over trees [10, 68]. Decidability of P with $\forall n\,(P(n) \vee \neg P(n))$ and $\forall\alpha\,\exists x\,P(\overline{\alpha}x)$ (i.e., for all branches α there is an initial segment $\overline{\alpha}x$ such that $P(\overline{\alpha}x)$) imply that $\{n \mid \forall m \prec n\,\neg P(m)\}$

is a regular well-founded tree (with ordering $m \prec n$ on the tree, i.e., m predecessor of n). The bar of this tree is the set $\{n \mid P(n) \wedge \forall m \prec n \neg P(m)\}$. The implications $\forall n \, (P(n) \to Q(n))$ and $\forall n \, (\forall y \, Q(n * \langle y \rangle) \to Q(n))$ express that Q is a property which holds on a bar and is transferred upwards. If Q holds for all immediate successors of a node n, then it also holds for n. At that point, Brouwer's bar induction concludes that Q holds for the empty node $\langle \rangle$.

Intuitionism draws rigorous consequences for proofs on infinitely growing trees which are not only important entities for mathematicians, but also for computer scientists. Obviously, bounds ("bars") in search trees spare us time and costs in the sense of Ockham's principle of parsimony. Intuitionistic mathematics yields controllable and reliable tools to approximate infinity by finitude (cf. lawless choice sequences). But, the question arises whether Brouwer's belief in human intuitions is necessary. The digital information system which we introduced in the previous section also approximated the infinity by finite means. In short: Humans seem only to be biological examples of an information system to process data and information. Information systems are the reference systems of (higher) computability [62].

7. Constructive Proof Mining

Since Antiquity, proofs are not only understood as verifications of truth, but also constructions. Applied mathematicians, natural scientists, and engineers have been interested in effective solutions of problems with geometric constructions in the past and computations with computers in modern times. Mathematical proofs sometimes seem to be abstract, ineffective, and beyond computational solutions. But, often, they include "hidden" effective information such as computable bounds which only must be extracted from the given proof. In this section, we consider the foundational research program of "proof mining" which aims at general logical tools to find ("unwind") effective procedures (algorithms) in mathematical proofs of ordinary and practical mathematics. Are there even logical metatheorems to solve these tasks in general for whole classes of theorems and theories? In this case, logic would no longer be an isolated enterprise of only theoretical and philosophical interest separated from mathematics and computer science. Proof mining would actually bridge logic, mathematics, and computer science.

A historically nice example of "proof mining" is yielded by a famous proof of Euclid that there are infinitely many prime numbers. His indirect proof is published in Book IX as proposition 20 of his "Elements" [22, p. 63], [1, p. 3]. A careful analysis of Euclid's proof reveals a computable bounding function of prime numbers. In the history of mathematics, even much better bounds were delivered (e.g., Euler's proof) [80]. All these examples underline that proofs can include more information than the verification of truth. They also show that effective procedures can be extracted with different degrees of computability. Obviously, the extraction of effective procedures from given proofs is deeply rooted in the practice of mathematics.

The logician and mathematician Georg Kreisel brought it to the point with his claim for "unwinding of proofs" [25]: "What more do we know if we have proved a theorem by restricted means than if we merely know that it is true?" Later on, Kreisel's unwinding of proofs was developed further and called "proof mining" by Ulrich Kohlenbach in [52, p. 13]. In the past, the extraction of effective procedures from given proofs only depends on ordinary mathematics (e.g., number-theoretic knowledge in the case of Euclid's proof). What is new and exciting in Kreisel's suggestion? He favored a research strategy which uses logical and proof-theoretical tools to extract effective procedures from given proofs.

Finally, the research strategy of proof mining can be linked with an automated extraction of effective procedures which can be realized by computers. These ideas have far-reaching consequences if computer programs are considered as formalized proofs. In this case, proof mining leads to program verification which is a crucial issue in a world with increasing automation: How can security and reliance of algorithms and computer programs be guaranteed? Anyway, logic and proof theory are no longer an isolated research of some specialists with no influence to the rest of the scientific world. With the research program of proof mining, logic and proof theory become a crucial link between mathematics and computer science with interdisciplinary consequences for practical applications. Above all, it opens new avenues in the philosophical debate on truth, effectivity, and constructivity.

In mathematics, the claims for truth and constructivity clash in existential theorems. Let us consider an existential theorem with the logical form $A \equiv \exists x\, B(x)$: A weaker requirement is to construct a list of terms $t_1, \ldots, t_n$ which are witnesses to A, such that $B(t_1) \vee \cdots \vee B(t_n)$ holds. More general, we are often confronted with a Π_2^0-statement, in order to extract an effective procedure [52, p. 14]:

> If $A \equiv \forall x \, \exists y \, B(x,y)$, then one can ask for an algorithm p such that $\forall x \, B(x, p(x))$ holds or — weaker — for a bounding function b such that $\forall x \, \exists y \leq b(x) \, B(x,y)$.

In general, the logical form of an existential theorem gives now hints how to extract effective procedures for computing solutions or, at least, bounds. Since Euclid, mathematical proofs were realized by logical conclusions and constructive procedures. In Greek geometry, figures are constructed by compass and ruler which were rooted in everyday technologies of practical engineering. With the development of new curves in modern times, constructive tools were extended beyond compass and ruler. Descartes, for example, suggested new kinematic instruments to construct algebraic curves which he defined by algebraic equations. Later on, constructions were supported by machines and nowadays we use computers for applications of algorithms. Thus, since the very beginning, proofs are referred to tools of construction [67]. In the previous sections, hierarchies of computational degrees were suggested to distinguish different degrees of constructivity.

Proofs are logical procedures to verify the truth of statements. Thus, it depends on the accepted tools of construction whether a proof of a statement is accepted. A proof can be defined as constructive procedure referring to the logical operations of a statement in the Brouwer-Heyting-Kolmogorov (BHK)-interpretation [39, 55]:

(1) A proof of $A \wedge B$ is given by a proof of A and a proof of B.

(2) A proof of $A \vee B$ is given by either a proof of A or a proof of B.

(3) A proof of $A \rightarrow B$ is a construction to transform any proof of A into a proof of B.

(4) A contradiction $\perp$ has no proof. A proof of $\neg A :\leftrightarrow A \rightarrow \perp$ is (according to (3)) a construction which transforms any assumed (hypothetical) proof of A into a proof of a contradiction.

(5) A proof of $\forall x \, A(x)$ is a construction which transforms a proof of the statement that a is an element of range M of the variable x into a proof of $A(a)$.

(6) A proof of $\exists x \, A(x)$ is given by a construction of an element a from M, and a proof of $A(a)$.

The BHK-interpretation explains the meaning of the logical constants $\perp, \wedge, \vee, \rightarrow, \forall, \exists$ in terms of proof constructions. The disadvantage of the

BHK-interpretation is the unexplained notion of construction resp. constructive proof. Gödel wanted that constructive proofs of existential theorems provide explicit realizers. Therefore, he replaced the notion of constructive proof by the more definite concept of computable functionals of finite type. In Section 5, the notion of a computable functional of finite type was mathematically defined as an ideal in an information system. Gödel's proof interpretation is largely independent of a precise definition of computable functionals: One only needs certain basic functionals as computable (e.g., primitive recursion in finite types) and their closure under composition.

In 1945, Kleene proposed a variant to the BHK-Interpretation [51]. He understood a construction as a machine procedure which can be coded by a machine number. Like in the BHK-interpretation, an existential statement $\exists x\, A(x)$ is considered to be an incomplete information. The statement $\exists x\, A(x)$ can only be made, if x can be constructed or "realized" by a Turing machine or, in general, by an information system. Kleene's historical concept of realizability was restricted to (partial) recursive functions and their machine numbers. It interprets arithmetic sentences in the mathematical structure $\mathcal{N} = (\mathbb{N}, 0, S, +, \cdot)$ of natural numbers. The elements (numbers) of $\mathbb{N}$ are denoted by $n, m, \ldots$, the corresponding numerals in formal language by $\bar{n}, \bar{m}, \ldots$. The statement "n realizes A" is defined by induction on the formal complexity of formula A. $(n)_i$ $(i = 1, 2)$ stands for the i-th component n_i of code number n which is computable.

(i) n realizes $(t = s)$ iff $t = s$ holds in $\mathcal{N}$.

(ii) n realizes $A \wedge B$ iff $(n)_1$ realizes A and $(n)_2$ realizes B.

(iii) n realizes $A \vee B$ iff $(n)_1 = 0$ and $(n)_2$ realizes A, or $(n)_1 \neq 0$ and $(n)_2$ realizes B.

(iv) n realizes $A \to B$ iff for each m realizing A, $[n](m)$ is defined and realizes B.

(v) n realizes $\exists x\, A(x)$ iff $(n)_2$ realizes $A\left(x/\overline{(n)_1}\right)$.

(vi) n realizes $\forall x\, A(x)$ iff for all m, $[n](m)$ is defined and realizes $A(\bar{m})$.

Instead of a semantical definition of realizability, we could reformulate realizability as a syntactical mapping assigning a formula $x\, r\, A$ ("x realizes A") to each formula of the intuitionistic Heyting arithmetic. Kreisel introduced modified realizability (mr) in the formal language $\mathcal{L}(\mathrm{E} - \mathrm{HA}^\omega)$ of (extensional) Heyting arithmetic (with ω-rule) $\mathrm{E} - \mathrm{HA}^\omega$ (with notation $\underline{y}\underline{x} := y_1\underline{x}, \ldots, y_n\underline{x}$, where $\underline{y} = y_1, \ldots, y_n$ and $\underline{x} = x_1, \ldots, x_n$ are tuples of functionals of certain types and $y_i \underline{x} := y_i x_1 \dots x_n)$ [58]. For each formula A of $\mathcal{L}(\mathrm{E} - \mathrm{HA}^\omega)$, formula $\underline{x}\,mr\,A$ ("x modified realizes A") is defined in $\mathcal{L}(\mathrm{E} - \mathrm{HA}^\omega)$ [52, Chapter 5]:

(i) $\underline{x}\,mr\,A$ iff A with empty tuple $\underline{x}$ and A prime formula.

(ii) $\underline{x}, \underline{y}\,mr\,(A \wedge B)$ iff $\underline{x}\,mr\,A \wedge \underline{y}\,mr\,B$.

(iii) $z^0, \underline{x}, \underline{y}\,mr\,(A \vee B)$ iff $(z =_0 0 \to \underline{x}\,mr\,A) \wedge (z \neq_0 0 \to \underline{y}\,mr\,B)$.

(iv) $\underline{y}\,mr\,(A \to B)$ iff $\forall \underline{x}\,(\underline{x}\,mr\,A \to \underline{y}\underline{x}\,mr\,B)$.

(v) $\underline{x}\,mr\,(\forall y^\rho A(y))$ iff $\forall y^\rho(\underline{x}y\,mr\,A(y))$.

(vi) $z^\rho, \underline{x}\,mr\,(\exists y^\rho A(y))$ iff $\underline{x}\,mr\,A(z)$.

In general, an interpretation of a formal system is sound, if any derivable formula of the formal system is true with respect to the interpretation. Actually, an interpretation is a mapping from a formal language into another one. Thus, the interpreted formula is "true" if it is derivable in the formal system the formula was mapped into. The soundness theorem of modified realization in $\mathcal{L}(\mathrm{E} - \mathrm{HA}^\omega)$ is proven by induction on the length of the derivation of formulas [102], [52, pp. 99–100].

The soundness and characterization theorems allow us to extract an effective procedure (program) from a given proof. Extraction of programs from given proofs can be automated by software. An example of automatic program extraction is MINLOG which was developed by Helmut Schwichtenberg and his research group in Munich. MINLOG is an interactive proof system which is equipped with tools to extract functional programs directly from proof terms [90]. The system is supported by automatic proof search and normalization by evaluation as an efficient term rewriting device. It uses minimal rather than classical or intuitionistic logic. Therefore, MINLOG can be extended to applications in intuitionistic as well as classical logic. There are also other interactive systems (e.g., Coq and Isabelle). In this book, a course is dedicated to interactive proof systems.

What is the practical use of soundness proofs and program extraction? In business, economy, and science, customers want software which solves a problem. Thus, they require a proof that it works. Suppliers answer with a proof of the existence of a solution to the problem. The proof has been automatically extracted from the formal specification of the problem by a proof mining software (e.g., MINLOG). But, the question arises whether the extraction mechanism of the proof is itself, in general, correct. The soundness theorem guarantees that every formal proof can be realized by a normalized extracted term.

The previous functional interpretation and several variations have different disadvantages which are overcome by Gödel's functional ('Dialectica') interpretation. For example, in contrast to the modified realizability interpretation, Gödel's functional interpretation also satisfies the Markov principle which is fundamental for applications in ordinary computational mathematics.

In constructive recursive mathematics, α is an algorithmic or recursive function and A_0 is a recursively decidable predicate. In this case, Markov's principle means: If there is an algorithm for testing A_0, and if we know by an indirect proof that the non-existence of an $\underline{x}$ must be refuted (i.e., we cannot avoid encountering $\underline{x}$ with $A_0(\underline{x})$), then, actually, we can find $\underline{x}$ with $A_0(\underline{x})$. The algorithm α is the procedure for finding $\underline{x}$. For an intuitionistic mathematician, the assumption $\neg\neg\exists\underline{x}\,\alpha(\underline{x}) = 0$ still provides no constructive procedure of finding a natural number as solution. But in recursive mathematics, the acceptance of a proof $\neg\neg\exists\underline{x}\,\alpha(\underline{x}) = 0$ together with an algorithm α is sufficient for finding $\underline{x}$. The Markov Principle in all finite types is defined by $M^\omega := \bigcup_{\rho\in T}\{M^{\underline{\rho}}\}$, where $M^{\underline{\rho}} : \neg\neg\exists\underline{x}^{\underline{\rho}}A_0(\underline{x}^{\underline{\rho}}) \to \exists\underline{x}^{\underline{\rho}}A_0(\underline{x}^{\underline{\rho}})$ with A_0 arbitrary quantifier-free formula of $\mathrm{WE-HA}^\omega$ and $\underline{x}$ of arbitrary types $\underline{\rho}$. The formula $A_0(\underline{x}^{\underline{\rho}})$ may contain further free variables in addition to $\underline{x}$. Functional interpretation satisfies the Markov principle M^ω.

Gödel's historical motivation of his functional ('Dialectica') interpretation was a relative consistency proof for Heyting's intuitionistic arithmetic [33]. Gödel's interpretation contains a formula translation and a proof translation. The formula translation describes how each statement A of Heyting arithmetic is mapped to a quantifier-free formula $A_D(\underline{x}, \underline{y})$ (with variables $\underline{x}, \underline{y}$ not free in A). Intuitively, A is interpreted by an existential statement $\exists\underline{x}\,\forall\underline{y}\,A_D(\underline{x}, \underline{y})$. The proof translation shows how a proof of A can be transformed into closed terms $\underline{t}$ and a proof of $A_D(\underline{t}, \underline{y})$.

The soundness theorem does only hold in a weakened version of extensional intuitionistic arithmetic $\mathrm{WE-HA}^\omega$ with functional interpretation of the Markov principle. A proof can be found in [52, pp. 130–135]. But the Markov principle is not accepted in (extensional) intuitionistic arithmetic. Thus, this soundnesss theorem does not hold in $\mathrm{E-HA}^\omega$.

Gödel's functional "Dialectica" interpretation can be used to extract functional programs from non-constructive proofs. For example, we consider Euclid's classical existence proof of the greatest common divisor with a quantifier-free kernel which does not contain an algorithm. The greatest common divisor (gcd) of two or more integers, when at least one of them is not zero, is the largest positive integer that divides the numbers without a remainder (Rem). In arithmetic, the remainder is the integer left over after dividing one integer by another to produce an integer quotient. In the case of a greatest common divisor, the remainder is zero. (For example, the gcd of 4 and 16 is 4 with remainder 0.) Thus, the gcd of the natural numbers a_1 and a_2 is a linear combination of the two numbers. Gödel's Dialectica (D) interpretation can be used to extract a formal term from the classical proof representing a computer program [75]. Extraction of programs from given proofs can be automated by software. An example of automatic program extraction is MINLOG which was already mentioned previously. For applying D-interpretation, we must analyze the informal existence proof in all details, in order to apply formal program extraction [89, p. 389 f.].

In [52] and [62], several functional interpretations with their different advantages and disadvantages are discussed. Before considering further applications of proof mining in mathematics, we remind the reader of the representation of real numbers by Cauchy sequences of rational numbers which is fundamental for classical, intuitionistic, constructive, and computational analysis [6, 47, 61, 89, 100]. The Archimedian ordered field of real numbers is an example of a complete metric space [2, Chapter 1] [54, Chapter 4].

In the formal representation of complete separable metric spaces, Kohlenbach could derive general metatheorems on the extractability of effective uniform bounds from proofs in analysis [52, Chapters 15, 17]. In this context, 'uniform' means that the bounds are independent of parameters in (compact) metric spaces. Metatheorems of proof mining open new avenues to applications of proof theory in mathematics. From a logical point of view, the metatheorems on abstract spaces are so-called semi-formal rules: The validity of the extracted bound follows from the provability of the premise of the rule in the sense of [52, p. 280].

The separation of mathematics and logic has been an unfortunate development during the last decades. Mathematicians are often not interested in logic and proof theory, because these fields seem to have no consequences for their research fields in, e.g., numerical or functional analysis. Vice versa, logicians and proof theorists are operating on an abstract and formal level which seems to be unrealizable in advanced mathematical research. Actually, metatheorems are practical links between logic and proof theory with mathematics.

In [52, p. 279], Kohlenbach discusses the following example of numerical analysis: Many problems demand the construction of solutions $x \in K$ for equations

$$A(x) :\equiv (F(x) = 0)$$

with a compact metric space K and $F : K \to \mathbb{R}$ continuous function. The solution involves two steps:

1) Construct approximate solutions $x_n \in K$ satisfying $A_n(x_n) :\equiv (|F(x_n)| < 2^{-n})$.
2) Conclude that either $(x_n)_{n \in \mathbb{N}}$ itself or some subsequence converges to a solution of $A(x)$, using the compactness of K and the continuity of F.

Metatheorems of proof theory can deliver general tools of solutions if we distinguish the following cases:

a) If F has exactly one root $\hat{x} \in K$, proof-theoretic analysis of a given proof of uniqueness of $\hat{x}$ can be applied to extract an effective rate of convergence under quite general circumstances.
b) If the solution $\hat{x}$ is not necessarily unique, one often cannot effectively obtain a solution but weaker tasks like obtaining effective rates of asymptotic regularity might be possible.

In approximation theory, we are interested in the extraction of rates of convergence towards a unique solution. A metatheorem delivers a general tool of proof mining to extract a bound from a given proof for complete separable spaces in the compact case. This metatheorem proves the extractability of effective uniform bounds which only depend on representatives of elements in complete separable metric spaces K. The total boundedness and the completeness of K are necessary for this result. In a next step of generalization, bounds can be obtained only under the assumption of abstract metric spaces. The bounds are independent even from noncompact but only metrically bounded spaces. Separability of the spaces need not be assumed. The proofs are given in [52, Chapter 17].

8. Constructive Reverse Mathematics

Since Euclid (mid-4th Century BC–mid-3rd Century BC), mathematicians used axioms to deduce theorems. But the "forward" procedure from axioms to theorems is not always obvious. How can we find appropriate axioms for a proof starting with a given theorem in a "backward" (reverse) procedure?

Pappos of Alexandria (290–350 AD) understood the "forward" procedure as "synthesis" (Greek for the Latin word "constructio" or English translation "construction"), because, in Euclid's geometry, logical deductions from axioms to theorems are connected with geometric constructions of corresponding figures (Pappos 1876–1878 II, 634 ff.). The reverse search procedure of axioms for a given theorem was called "analysis" which decomposes a theorem in its necessary and sufficient conditions and the corresponding geometric figures in their elementary building blocks (e.g., circle and straight line) [42], [65, p. 120 ff.].

In modern times, reverse mathematics is a research program to determine the minimal axiomatic system required to prove theorems [29, 30]. In general, it is not possible to start from a theorem τ to prove a whole axiomatic subsystem T_1. A weak base theory T_2 is required to supplement τ:

If $T_2 + \tau$ can prove T_1, this proof is called a reversal. If T_1 proves τ and $T_2 + \tau$ is a reversal, then T_1 and τ are said to be equivalent over T_2.

Reverse mathematics enables us to determine the proof-theoretic strength and complexity of theorems by classifying them with respect to equivalent theorems and systems. Many theorems of classical mathematics can be classified by subsystems of second-order arithmetic $\mathcal{Z}_2$ with variables of natural numbers and variables of sets of natural numbers [94, 95].

Arithmetical formulas can be classified according to the arithmetical hierarchy Σ_n^0, Π_n^0, and Δ_n^0. We can distinguish Σ_n^0, Π_n^0, and Δ_n^0-schemes of induction and comprehension. For example, in the induction axiom, let X be the set that exists due to the comprehension axiom with formula $\varphi(n)$. Then, the Σ_1^0-induction scheme is the restriction of the induction axiom to Σ_1^0-formulas $\varphi(n)$. A set X is the comprehension of a Σ_1^0-formula iff it is recursively enumerable. The Δ_1^0-comprehension scheme is the restriction of the comprehension axiom to Δ_1^0-formulas $\varphi(n)$. A set X is the comprehension of a Δ_1^0-formula iff it is recursively computable. Thus, Δ_1^0-comprehension means recursive comprehension (RC) [96].

We can distinguish subsystems of $\mathcal{Z}_2$ with different comprehension schemes. The arithmetical and analytical hierarchies yield classifications of axiomatic subsystems of $\mathcal{Z}_2$ with increasing proof-theoretic power and corresponding structures of $\mathcal{Z}_2$-models. We start with the base $\mathcal{Z}_2$-subsystem

RCA_0 consisting of the axioms of Peano arithmetic, the Σ_1^0-induction scheme, and the Δ_1^0-comprehension scheme.

In a next step of distinguished $\mathcal{Z}_2$-subsystems, we consider the Weak König's Lemma. König's Lemma states that every infinite, finitely branching tree has an infinite path. Weak König's Lemma restricts König's to binary trees. The $\mathcal{Z}_2$-subsystem WKL_0 consists of RCA_0 extended by weak König's Lemma. In the sense of reverse mathematics, WKL_0 is equivalent (over RCA_0) to, e.g.,

- Σ_1^0-separation
- separable Hahn/Banach theorem
- Brouwer/Schauder fixed point theorems.

Σ_1^0-separation means that given two Σ_1^0-formulas of number variable n which are exclusive, there exists a set containing all numbers n satisfying one formula and none satisfying the other. Actually, weak König's lemma proves Σ_1^0-separation and Σ_1^0-separation implies WKL_0 over RCA_0 which is the reversal.

In a next step of distinguished $\mathcal{Z}_2$-subsystems, we consider a comprehension scheme for all arithmetical formulas with no set quantifiers (not just Δ_1^0-formulas). The $\mathcal{Z}_2$-subsystem ACA_0 consists of RCA_0 extended by the arithmetical comprehension scheme. In the sense of reverse mathematics, ACA_0 is equivalent (over RCA_0) to, e.g.,

- existence of the strong algebraic closure of a countable field
- König's lemma for subtrees of $\mathbb{N}^{\mathbb{N}}$
- least upper bound principle for sequences of real numbers.

Reverse mathematics can be extended to the analytical hierarchy Σ_n^1, Π_n^1, and Δ_n^1. The Δ_1^1-comprehension axiom reduces the comprehension of sets to Δ_1^1-formulas, i.e., hyperarithmetic theories. From a proof-theoretic point of view, hyperarithmetic theories are distinguished by predicative proofs (compare Section 4). Arithmetic transfinite recursion means that the arithmetic comprehension scheme can be iterated transfinitely. We can also consider extended systems with Π_1^1- and Π_2^1-comprehension axioms which are no longer predicative. Thus, along the complexity degrees of arithmetic and analytic hierarchies, we can distinguish main systems of reverse mathematics like RCA_0 with Δ_1^0-comprehension, WKL_0 with weak König's Lemma, ACA_0 with arithmetic comprehension, ATR_0 with arithmetic comprehension iterated transfinitely, and Π_1^1-comprehension axiom:

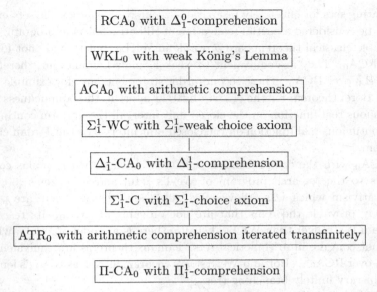

The five most commonly used $\mathcal{Z}_2$-subsystems in reverse mathematics correspond to philosophical programs in foundations of mathematics with increasing proof-theoretic power, starting with RCA_0. RCA_0 corresponds to the research program of computable mathematics. It consists of arithmetic axioms with induction principle reduced to Σ_1^0-formulas and the comprehension principle for Δ_1^0-formulas. Every set of natural numbers which can be proven to exist in RCA_0 is computable. Any theorem on non-computable sets is not provable in RCA_0. Therefore, RCA_0 is a constructive system, but not in the sense of intuitionism, because it accepts classical logic with the principle of excluded middle. It is close to Bishop's constructivism [6], but with classical logic.

Many classical theorems of arithmetic and analysis can be proven in RCA_0. For example, the set of rational numbers satisfy the axioms of an ordered field. If the real numbers are defined by Cauchy sequences of rational numbers, they can be proven to satisfy the axioms of an Archimedian ordered field. Karl T.W. Weierstrass (1815–1897) defined real numbers by nested sequences of closed intervals. Any nested sequence of closed intervals whose length tends to zero can be proven to have a single point in its intersection [20, Chapter 2]. In RCA_0, we can also prove the existence and uniqueness of the real closure of a countable ordered field.

WKL_0 is a stronger research program than RCA_0, because it implies that non-computable sets exist. For example, in WKL_0, we can prove that separating sets for effectively inseparable recursively enumerable sets exist. It can be considered a partial realization of Hilbert's finitistic program. Remarkable classical theorems are provable in WKL_0, but they do not follow from RCA_0. They are equivalent to weak König's Lemma and, therefore, to WKL_0 over RCA_0. These second-order statements are, for example, the Heine-Borel theorem for the closed unit real interval, the boundedness of a continuous real function on the closed unit interval, the uniform continuity of a continuous real function on the closed unit interval or the Jordan curve theorem.

ACA_0 with the comprehension scheme of arithmetical formulas corresponds to the research program of Weyl's [110] and Lorenzen's [60, 61] predicativism which was followed by [26]. Nevertheless, there are predicatively provable theorems that are not provable in ACA_0. Its research program is stronger than WKL_0, because there is a model of WKL_0 which does not contain all arithmetical sets. Famous theorems are equivalent to ACA_0 over RCA_0 — Bolzano-Weierstrass theorem as well as König's lemma for arbitrary finitely branching trees.

ATR_0 proves the consistency of ACA_0. Therefore, by Gödel's theorem, it is stronger than ACA_0. It has the proof-theoretic ordinal Γ_0 which is the supremum of ordinals of predicative systems. Δ_1^1-CA_0 yields systems of hyperarithmetical analysis with Δ_1^1-predicativism.

Π_1^1-CA_0 which consists of RCA_0 with the comprehension axiom for Π_1^1-formulas is stronger than arithmetical transfinite recursion. Therefore, it is impredicative. The Π_1^1-CA_0 comprehension principle is equivalent to theorems of set theory with impredicative proofs. The equivalence proves that there are impredicative proofs which cannot be reduced to predicative ones. Therefore, they are essentially impredicative.

Classical reverse mathematics uses classical logic and classification of proof-theoretic strength with RCA_0 (Δ_1^0-recursive comprehension) as weakest subsystem. The research program of reverse classical mathematics tries to clarify which set existence axioms are necessary to prove the theorems of ordinary mathematics.

In [44] and [45], Hajime Ishihara characterized constructive reverse mathematics by a constructive hierarchy which starts with intuitionistic logic and Bishop's constructive mathematics BISH as weakest subsystem, followed by intuitionistic mathematics INT of Brouwer, Heyting et al., constructive recursive mathematics RUSS ("Russian") of Markov et al., and classical mathematics CLASS of Hilbert et al.:

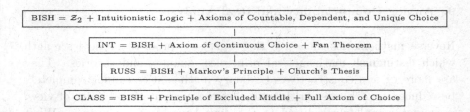

RUSS refers to "Russia" because of the Russian mathematician Markov and his school of constructive mathematics in the sense of Turing-computability. Therefore, RUSS is Bishop's constructive mathematics (BISH) with Markov's principle and Church's thesis. Markov's principle MP is an instance of the double negation elimination $\neg\neg A \to A$.

Intuitionistic mathematics INT strongly depends on the acceptance of the Fan theorem (FAN) (cf. Section 6) which has immense consequences for constructive analysis [5]:

- The binary fan is the collection of all (finite) sequences of 0s and 1s (including empty sequence).
- A bar of a binary fan is a set of cut-offs, such that for each infinite path (sequence) α through the fan there exists a natural number n such that $\overline{\alpha}(n)$ is in the bar.
- A bar is uniform if there exists a natural number k such that for each infinite path α through the fan, there is some $i \leq k$ such that $\overline{\alpha}(i)$ is in the bar.
- A bar is detachable iff, for all finite binary sequences u, u is element of the bar or u is not element of the bar.

Brouwer's fan theorem for detachable bars is stated as FAN_Δ which is a contrapositive form of weak König's lemma WKL. FAN_Δ is weaker than WKL, accepted in INT, but false in RUSS [100, p. 220].

From a philosophical point of view, constructive reverse mathematics strictly follows Ockham's principle of parsimony: It classifies mathematical principles according to their most efficient costs avoiding unnecessary abstractions. In the hierarchy of BISH, INT, RUSS, and CLASS, we can determine how far mathematical principles are away from constructive and computational proofs. In this sense, constructive reverse mathematics bridges logic, mathematics, and computer science.

9. Analog Computability and Real Analysis

Reverse mathematics measures the degrees of proof-theoretical strength which distinguish mathematical principles, axioms, and theories. They are more or less constructive. Ordinary mathematics cannot completely be reduced to 2nd order arithmetic. From a computational point of view, it cannot completely reduced to the digital world of computers. What about mathematical thinking beyond Turing computability? What about computability based on, e.g., the fields $\mathbb{R}$ and $\mathbb{C}$ of real and complex numbers [20]? The mathematical theory of complex dynamical systems is based on continuous differential equations over $\mathbb{R}$ and $\mathbb{C}$. Complex dynamical systems and their nonlinear differential equations are important tools of interdisciplinary modeling in science and technology [64]. Historically, many algorithms in mathematics (e.g., Newton, Euler, Gauss) were defined on $\mathbb{R}$ and $\mathbb{C}$. But the logical theory of decidability and computability is defined over digital numbers $\mathbb{Z}_2$.

The conflict of a discrete and continuous model of the world is deeply rooted in the beginning of modern science. Ancient philosophers of atomism (Democritus et al.) believed in a world of interacting discrete particles as indivisible building blocks (atoms). They were attacked by philosophers of the continuum (Aristotle et al.) which cannot be explained by a chain of coupled pearls. Aristotle described change in nature by continuous dynamics. In the beginning of modern science, physicists assumed a mechanistic world of interacting atoms. But, mathematically, atoms were considered mass points determined by continuous differential equations. Leibniz as inventor of the differential calculus tried to bridge the discrete and continuous world by his philosophical concept of monads which were mathematically represented by infinitesimally small, but nevertheless non-zero quantities called differentials. In modern mathematics, the assumption of their existence leads to non-standard analysis. For an overview on the development of standard and non-standard infinitesimal thinking compare [66].

Scientific computing in physics and chemistry was mainly based on continuous functions. In practical cases, the solutions of their (real or complex) equations could only be approximated by algorithms of numerical analysis (e.g., Newton's method). Therefore, the procedures of numerical analysis depend strongly on the continuous concept of real numbers. But until today, numerical analysis is still a highly efficient collection of successful procedures of problem solving. But, we also need logical-mathematical foundations of "real" (analog) computing. The discrete theory of computability is well-founded by the discrete concept of Turing machines (Church's thesis).

Besides algorithmic procedures, there is also a long standing tradition of decidability problems in analysis and algebra. Since antiquity, decidability of geometric constructions like the squaring of the circle, trisection of angles, or Deli's problem were discussed. But, it was the transcendental real number π and Galois' proof of the unsolvability by radicals of polynomial equations of degree 5 and more which decided these problems [67]. Thus, the foundations of algorithms with real numbers must be clarified.

John von Neumann, computer-pioneer and brilliant mathematician, felt the gap between the digital world of logic and computers and the "real" world of ordinary mathematics, when he proclaimed in the Hixon Symposium Lecture 1948:

"There exists today a very elaborate system of formal logic, and specifically, of logic as applied to mathematics. This is a discipline with many good sides, but also serious weakness. The reason for this is that it deals with rigid, all-or-none concepts, and has very little contact with the continuous concept of the real or of the complex number, that is, with mathematical analysis. Yet analysis is the technically most successful and best-elaborated part of mathematics..." [76]

Roger Penrose argued exactly on this line, when he considered chaos theory in his book "The Emperor's New Mind" [79, p. 124]:

"Now we witnessed ... a certain extraordinarily complicated-looking set, namely the Mandelbrot set. Although the rules which provide its definition are surprisingly simple, the set itself exhibits an endless variety of highly elaborate structures. Could this be an example of non-recursive set, truly exhibited before our mortal eyes?"

Penrose rejects recursive analysis, because in this case, a real (or complex) number must be input to a Turing machine bit by bit: Problems arise when one wants to decide if two numbers are equal. It is also problematic to reduce a real (resp. complex) problem to its rational "skeleton", because, e.g., the curve $x^3 + y^3 = 1$ has no rational points with both x and y positive.

Thus, we need a generalized concept of machines over a ring R (e.g., ring $\mathbb{Z}$ or fields $\mathbb{R}$ and $\mathbb{C}$), to unify the theory of complex dynamical systems with computational complexity in numerical analysis and mathematical physics. A canonical model of computation over the reals was introduced by Leonore Blum, Mike Shub, and Steve Smale [8].

Newton's method is a typical search algorithm of numerical analysis and scientific computation [35]. It is an iterative method to approximate the roots of nonlinear equations. Thus, it is classical example to bridge the gap between digital algorithms of logic and computability theory and real algorithms of ordinary mathematics. Given an initial approximation a to a root of the polynomial equation $f(z) = 0$, Newton's method replaces a by the exact solution a' of the best linear approximation to f which is given by the tangent to the graph of f at point $(a, f(a))$ in Fig. 4. With a' the approximation is iterated to generate a'' etc.

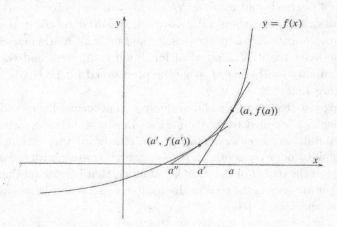

Fig. 4.   Newton's method

The algorithm is defined by Newton's endomorphism $N_f : \mathbb{C} \to \mathbb{C}$ with

$$N_f(z) = z - \frac{f(z)}{f'(z)}.$$

Newton's method is not generally convergent. The Newton machine is represented by a finite directed graph in Fig. 5 [7, p. 38] with four types of nodes — input, computation, branch, and output — each with associated functions and conditions on incoming and outgoing edges. For an input z_0 the machine generates the orbit z_0, $z_1 = N_f(z_0)$, $z_2 = N_f(z_1), \dots, z_{k+1} = N_f(z_k) = N_f^{k+1}(z_0), \dots$ The stopping rule is "stop if $|f(z_k)| < \epsilon$ and output z_k." If $N_f(z_k)$ is undefined at some stage, there is no output.

The machine will not in general halt on all inputs. The set $\Omega \subset \mathbb{C}$ is called the halting set of the machine iff it contains exactly all inputs for which the machine halts with an output. The input-output map φ is defined on Ω with $\varphi : \Omega \to \mathbb{C}$.

Newton's machine (Fig. 6 [7, p. 40]) is actually a machine over $\mathbb{R}$ with $\mathbb{C} = \mathbb{R}^2$ as input, output and state space. Newton's endomorphism

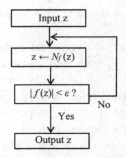

Fig. 5.   Search algorithm for Newton's method

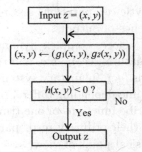

Fig. 6.   Computation node of Newton's machine

(computation node) is given by a rational function (quotient of two polynomials) $g = (g_1, g_2) : \mathbb{R}^2 \to \mathbb{R}^2$ with $g_1(x, y) = \operatorname{Re} N_f(x + iy)$ and $g_2(x, y) = \operatorname{Im} N_f(x + iy)$.

These examples of $\mathbb{R}$- and $\mathbb{C}$-based machines can mathematically be generalized for an ordered commutative ring R. Examples are the integers $R = \mathbb{Z}$ and the real numbers $R = \mathbb{R}$. The direct sum of R with itself n times is denoted by R^n. These examples of graphic systems can be generalized in a finite-dimensional machine over a ring [7]. In an infinite dimensional machine, the input space and output space are equal to infinite countable direct sums R^∞ with infinite sequences as elements. The four nodes are the same as before in the finite dimensional case. A fifth type of nodes is supplemented, in order to handle finite, but unbounded sequences. A classical Turing machine is a machine over $\mathbb{Z}_2$. Thus a graphic machine over $\mathbb{R}$ might be considered a "real" Turing machine. With graphic machines, the concepts of computability, decidability, and enumerability can be generalized over a ring. We are now able to solve decidability problems of, e.g., Newton's method.

The mathematical theory of dynamical systems is based on real and complex numbers. Therefore, with real and ·complex computing, general problems of decidability and enumerability can be solved for dynamical systems. From a mathematical point of view, graphic machines on a ring R are convenient to illustrate algebraic properties of real computing. But, they are somewhat abstract with respect to the architectures of computers. Actually, a graphic machine on a ring R is a kind of register machine with unlimited storage which processes elements of a ring with exact arithmetic. In computer science, it is more convenient to illustrate computing processes by generalized register machines which are equivalent to graphic machines on a ring R [13, 43]. Analytic computations are infinite computations generating a convergent sequence of outputs.

As already mentioned before, complex dynamical systems are fundamental modeling tools in natural, economic, and social sciences. Changing states of dynamical systems (e.g., planetary systems, cellular systems, populations) are modeled by time-depending differential equations. In general, differential equations describe functions of one (time-depending) resp. several (spatial) variables by their ordinary resp. partial derivations of 1st or higher order. From a physical point of view, systems of differential equations with finitely many degrees of freedom (e.g., finitely many points of mass in Newtonian mechanics of planets) correspond to ordinary differential equations. Systems with infinitely many degrees of freedom (e.g., fluids, gases, or electromagnetic fields) correspond to partial differential equations.

Real computable functions are important for determining solutions of differential equations. Unique solutions of differential equations are computable under certain conditions. In scientific modeling, solutions of differential equations are used to explain or to predict effects of dynamical systems under certain constraints. Examples are initial value problems and stability problems of dynamical systems [32]. In initial value problems, we search for mathematical models of time-depending processes with given initial states of dynamical systems. Mathematically, an initial value problem is an ordinary differential equation together with a specified value, called the initial condition, of the unknown function at a given point in the domain of the solution.

In initial value problems, the question arises whether the dynamical system converges to a fixed value in the long run and becomes stable. In general, this question is not decidable with graphic machines or register machines over $\mathbb{R}$ which accept real numbers as whole entities. Another aspect of stability asks whether dynamical systems remain stable after local perturbations of their initial states. In this case, the solution of an initial value problem must mathematically converge to a fixed point. If small perturbations are damped in the long run, the system is called asymptotically stable [18].

Many processes in science and technology depend essentially on time. Therefore, modelling and predicting time-depending processes are important targets of the dynamical systems approach [64]. In examples like the planetary system, processes do not depend on the initial point of time, but on the initial state of the dynamical system. Mathematically, these processes are modeled by autonomous differential equations which do not depend on time. The results on undecidability and stability can be transmitted to autonomous differential equations.

Classical results on decidability and undecidability in logic have deep impact on real computing. Examples are Alfred Tarski's result that the first-order theory of reals is decidable, but the integers are not according to Gödel. In technology and engineering sciences, real computing relates to analog systems with continuous parameters. In the life sciences, organisms are equipped with analog sensors. Analog models are unlikely to solve NP-hard problems, but they are more realistic for analog components of human cognition, brains, and bio-inspired technical networks.

According to Church's thesis, every discrete effective procedure can be simulated by a Turing machine. But, processes in the world are successfully modeled by continuous (analog) dynamical systems (e.g., differential equations) with different degrees of real computational complexity. Scientific computing uses real and complex models which need foundations of real computing. In the engineering sciences, sensor technology needs analog models with rigorous computational foundations. Last but not least, in philosophy, the question arises whether human brains and human thinking can be restricted to the digital paradigm of Turing computability. With respect to artificial intelligence, degrees of intelligence depend on degrees of computational complexity. Human-like AI-systems will need digital and analog abilities [63]. Therefore, the extension of Church's thesis to real (analog) computation is well motivated with respect to the complexity degrees of the world [62].

10. Perspectives of Proof and Computation for Mathematics, Computer Science and Philosophy

Classification of constructive and computational problem solving is deeply rooted in fundamental questions of logic and philosophy. In the Middle Ages, William of Ockham (Ockham 1287 – Munich 1347), Franciscan friar and scholastic philosopher, was an early forerunner of this research program [77]. He strongly argued that only individuals exist, rather than supra-individual universals, essences, or forms. He denied the real existence of metaphysical universals and demanded a reduction of ontological assumptions. With respect to constructive positions, he argued for efficient reasoning with a principle of parsimony in explanation and theory building. In modern times, this principle became popular as Ockham's razor. In the interpretation of Bertrand Russell, Ockham's razor only allows an explanation in terms of the fewest possible causes, factors, or variables [85].

Historically, Ockham argued that universals are generated by abstraction from individuals by the human mind. Thus, for him, they do not exist outside the human mind, but they have a mental existence as internal representation. In epistemology, Ockham is in the middle between two extreme positions [97]. On the one side, Platonic philosophers advocated an extramental ("real") existence of ideas. On the other side, radical nominalists like the medieval philosopher Roscelin stated that universals even have no existence in the human mind, but they are only names and signs as used in sentences. Obviously, these early epistemic positions foreshadow modern positions in the philosophy of mathematics. There are still believers in a Platonic world of abstract mathematical sets and structures, but there are also radical formalists reducing mathematics to a game of symbols. Following Ockham, mathematics is a product of the human mind. Historically, he was sometimes called a conceptualist or "terminist" contrary to a "nominalist" (from Latin "nomen" for "name") [106].

Coming back to mathematics, theorems and proofs should be reduced to constructive principles as far as possible. But ordinary mathematics cannot be totally reduced to constructive principles. In constructive reverse mathematics, we learnt that there are degrees of constructivity and computability. This insight is also supported by constructive proof mining and program extraction. Therefore, in mathematics, Ockham's principle of parsimony demands to reduce the number of abstract entities (e.g., sets and functionals of any type), axioms, and principles as far as possible. But, like in economy, a proof has its "price" of necessary and sufficient principles which may be more or less constructive. There is no free lunch — even in mathematics.

The crucial question of how far abstractions are necessary arises. Without any doubt, in ordinary mathematics, real and complex numbers, continuous and analog concepts are extremely useful to find robust and reliable problem solving. From a constructive point of view, it is only a question of "price" which we are ready to "pay" with the assumption of axioms and principles. But, Ockham's principle of parsimony is only a useful methodological demand of efficient reasoning. It cannot exclude Platonism in principle.

With respect to modern quantum physics, physical reality seems to be discrete. Quantum processes could be considered quantum algorithms — the universe as a quantum computer. Nevertheless, quantum field theories use continuous functionals and spaces, real and complex numbers, to compute and predict events with great precision. Are analog and continuous concepts only useful inventions of the human mind, in order to solve problems in a discrete world with approximations? In the sense of constructive reverse mathematics, the question arises how far these abstractions of the human mind are away from constructive concepts.

Mathematics is not only used to model, but also to change the world. Since Antiquity, mathematics opens new avenues of technologies. In engineering sciences, real and complex numbers are used to compute sure and reliable technical machines. Digital technologies and the analog world of life, humans and society should converge to master the increasing complexity of our civilization. We are living in a world with exponentially growing power of algorithms! Therefore, we end with a plea for more foundational and interdisciplinary research to get controllable and reliable tools. Since its very beginning, mathematical science is deeply rooted with the foundations of human existence.

References

[1] Aigner, M.; Ziegler, G.M. (2001): *Proofs from The Book*. Berlin 2nd edition.
[2] Beeson, M. (1985): *Foundations of Constructive Mathematics*. Springer: Berlin.
[3] Berger, J. (2005): The fan theorem and uniform continuity. In: Cooper, S.B.; Löwe, B.; Torenvliet, L. (Eds.): *New Computational Paradigms*. Springer: Berlin, 18–22.
[4] Berger, U.; Eberl, M.; Schwichtenberg, H. (2003): Term rewriting for normalization by evaluation. In: *Information and Computation* 183, 19–42.

[5] Berger, J.; Ishihara, H. (2005): Brouwer's fan theorem and unique existence
     in constructive analysis. In: *MLQ Math. Logic Q.* 51, 360–364.

[6] Bishop, E. (1967): *Foundations of Constructive Analysis.* McGraw-Hill:
     New York.

[7] Blum, L.; Cucker, F.; Shub, M.; Smale, S. (1998): *Complexity and Real
     Computation.* Springer: New York.

[8] Blum, L.; Shub, M.; Smale, S. (1989): On a theory of computation and
     complexity over the real numbers: NP-completeness, recursive functions
     and universal Machines. In: *Bulletin of the American Mathematical Society*
     21 1, 1–46.

[9] Brouwer, L.E.J. (1981): *Brouwer's Cambridge Lectures on Intuitionism*
     (ed. D. van Dalen). Cambridge University Press: Cambridge.

[10] Brouwer, L.E.J. (1975): *Collected Works I* (ed. A. Heyting). North-
     Holland: Amsterdam.

[11] Brouwer, L.E.J. (1927): Über Definitionsbereiche von Funktionen. In:
     Mathematische Annalen 97, 60–75.

[12] Brouwer, L.E.J. (1907): *Over de Grondslagen der Wiskunde.* [On the foun-
     dations of Mathematics]. Maas & Van Suchtelen: Amsterdam.

[13] Chadzelek, T.; Hotz, G. (1997): *Analytic machines.* Technical Report
     12/97, Sonderforschungsbereich 124 (VLSI-Entwurfsmethoden und Paral-
     lelität), Universität des Saarlandes: Saarbrücken.

[14] Chaitin, G.J. (1998): *The Limits of Mathematics.* Springer: Singapore.

[15] Church, A.; Kleene, S.C. (1936): Formal definitions in the theory of ordinal
     numbers. In: *Fundamenta Mathematicae* 28, 11–21.

[16] Davis, M. (1958): *Computability & Unsolvability.* McGraw-Hill: New York.

[17] Davis, M.; Putnam, H.; Robinson, J. (1961): The decision problem for
     exponential Diophantine equations. In: *Ann. of Math.* II 74, 425–436.

[18] Deuflhard, P.; Bornemann, F. (1994): *Integration gewöhnlicher Differen-
     tialgleichungen II (Numerische Mathematik).* De Gruyter: Berlin.

[19] Dummett, M. (1977): *Elements of Intuitionism.* Clarendon Press: Oxford.

[20] Ebbinghaus, H.-D.; Hermes, H.; Hirzebruch, F.; Koecher, M.; Mainzer, K.;
     Neukirch, J.; Prestel, A.; Remmert, R. (1991): *Numbers.* Springer: Berlin
     (German 1983 3rd edition).

[21] Enderton, H.; Luckham, D. (1964): Hierarchies over recursive well-
     orderings. In: *Journal of Symbolic Logic* 29, 183–190.

[22] Euclid (1782): *The Elements of Euclid (with Dissertations).* J. Williamson
     (translator and commentator). Clarendon Press: Oxford.

[23] Feferman, S. (2006): Turing's Thesis. In: *Notices of the American Mathe-
     matical Society* 53 10, 1200–1206.

[24] Feferman, S. (2002): Predicativity. Lecture at a meeting of the American
     Philosophical Association March 28, 2002.

[25] Feferman, S. (1996): Kreisel's "unwinding" Program. In: P. Odifreddi
     (Ed.): *Kreiseliana. About and Around Georg Kreisel.* Review of Modern
     Logic, 247–273.

[26] Feferman, S. (1968a): Systems of predicative analysis II: Representations
     of ordinals. In: *Journal of Symbolic Logic* 53, 193–213.

[27] Feferman, S. (1968b): Lectures on Proof Theory. In: *Proceedings of the Summer School in logic (Leeds)*. Springer: Berlin, 1–107.

[28] Feferman, S. (1964): Systems of predicative analysis. In: *Journal of Symbolic Logic* 29, 1–30.

[29] Friedman, H. (1975): Some systems of second order arithmetic and their use. In: *Proceedings of the International Congress of Mathematicians* (Vancouver, B.C., 1974), 1. Canad. Math. Congress. Montreal, 235–242.

[30] Friedman, H.; Simpson, S.G. (2000): Issues and Problems in Reverse Mathematics. In: *Computability Theory and Its Applications. Contemporary Mathematics* 257, 127–144.

[31] Gerhardy, P.; Kohlenbach, U. (2008): General logical metatheorems for functional analysis. In: *Trans. Amer. Math. Soc.* 360, 2615–2660.

[32] Glendinning, P. (1994): *Stability, Instability, and Chaos. An Introduction to the Theory of Nonlinear Differential Equations.* Cambridge University Press: Cambridge.

[33] Gödel, K. (1958): Über eine bisher noch nicht benützte Erweiterung des finiten Standpunktes. In: *Dialectica* 12, 280–287.

[34] Gödel, K. (1931): Über formal unentscheidbare Sätze der Principia Mathematica und verwandter Systeme I. In: *Monatshefte für Mathematik und Physik* 38 1, 173–198.

[35] Goldstine, H. (1977): *A History of Numerical Analysis from the 16th through the 19th Century.* Springer: New York.

[36] Goodstein, R.L. (1964): *Recursive Number Theory.* North-Holland: Amsterdam.

[37] Herken, R. (Hrsg.) (1995): *The Universal Turing Machine. A Half-Century Survey.* Springer: Wien.

[38] Hermes, H. (1969): *Enumerability, Decidability, Computability.* Springer: Berlin.

[39] Heyting, A. (1934): *Mathematische Grundlagenforschung. Intuitionismus. Beweistheorie.* Springer: Berlin repr. 1974.

[40] Hilbert, D. (1900): Mathematische Probleme. In: *Nachrichten der Königlichen Gesellschaft der Wissenschaften zu Göttingen, mathematisch-physikalische Klasse.* Heft 3, 253–297.

[41] Hinman, P.G. (1978): *Recursion-Theoretic Hierarchies.* Springer: Berlin.

[42] Hintikka, J.; Remes, U. (1974): *The Method of Analysis Its Geometrical Origin and Its General Significance.* North-Holland: Dordrecht.

[43] Hotz, G.; Schieffer, B.; Vierke, G. (1995): *Analytical Machines.* Technical Report TR95-025. Electronic Colloquium on Computational Complexity. Universität des Saarlands: Saarbrücken.

[44] Ishihara, H. (2006): Reverse mathematics in Bishop's constructive mathematics. In: *Philosophia Scientiae* 6, 43–59.

[45] Ishihara, H. (2005): *Constructive Reverse Mathematics. Compactness Properties.* Oxford University Press. Oxford Logic Guides 48, 245–267.

[46] Kant, I. (1787): Werke. Akademie Textausgabe. De Gruyter: Berlin 1968. Band 3: *Kritik der reinen Vernunft.* (Nachdruck der 2. Auflage 1787).

[47] Kleene, S.C. (1974): *Introduction to Metamathematics.* North Holland: Amsterdam 7th edition.

[48] Kleene, S.C. (1959): Quantification of number-theoretic functions. In: *Compositio Math.* 14, 23–40.

[49] Kleene, S.C. (1958): Arithmetical predicates and function quantifiers. In: *Transactions of the American Mathem. Society*, 312–340.

[50] Kleene, S.C. (1955a): Hierarchies of number theoretic predicates. In: *Bull. of the American Math. Soc.* 61, 193–213.

[51] Kleene, S.C. (1945): On the interpretation of intuitionistic number theory. In: *Journal of Symbolic Logic* 10, 109–124.

[52] Kohlenbach, U. (2008): *Applied Proof Theory: Proof Interpretations and their Use in Mathematics.* Springer: Berlin.

[53] Kohlenbach, U. (2005): Some logical metatheorems with application to functional analysis. In: *Trans. Amer. Math. Soc.* 357 (1), 89–128.

[54] Kohlenbach, U. (1993): Effective moduli of uniqueness from ineffective uniqueness proofs. An unwinding of de La Valle Poussin's proof for Chebycheff approximation. In: *Ann. Pure Appl.Log.*, 27–94.

[55] Kolmogorov, A.N. (1932): Zur Deutung der intuitionistischen Logik. In: *Math. Z.* 35, 58–65.

[56] Kreisel, G. (1967): Informal rigour and completeness proofs. In: Lakatos, I. (ed.): *Problems in the Philosophy of Mathematics. Proceedings of the International Colloquium in the Philosophy of Science 1965.* North-Holland: Amsterdam, 138–186.

[57] Kreisel, G. (1960): La prédicativité. In: *Bull. de la Soc. Math. de France* 88, 371–391.

[58] Kreisel, G. (1959): An interpretation of analysis by means of constructive functionals of finite types. In: Heyting, A. (ed.): *Constructivity in Mathematics.* North-Holland. Amsterdam, 101–128.

[59] Leibniz, G.W. (1998): *Philosophical Texts.* Edited and translated by R.S. Woolhouse and R. Francks. Oxford University Press: Oxford.

[60] Lorenzen, P. (1955): *Einführung in die operative Logik und Mathematik.* Springer: Berlin.

[61] Lorenzen, P. (1965): *Differential und Integral. Eine konstruktive Einführung in die klassische Analysis.* Akademische Verlagsgesellschaft. Frankfurt.

[62] Mainzer, K. (2018): *The Digital and the Real World. Computational Foundations of Mathematics, Science, Technology, and Philosophy.* World Scientific: Singapore.

[63] Mainzer, K. (2014): *Die Berechnung der Welt. Von der Weltformel zu Big Data.* C.H. Beck: München.

[64] Mainzer, K. (2007): *Thinking in Complexity. The Computational Dynamics of Matter, Mind, and Mankind.* Springer: Berlin 5th edition.

[65] Mainzer, K. (1994): *Computer – Neue Flügel des Geistes?* De Gruyter: Berlin, New York.

[66] Mainzer, K. (1981): *Grundlagen und Geschichte der exakten Wissenschaften.* Universitätsverlag: Konstanz.

[67] Mainzer, K. (1980): *Geschichte der Geometrie.* B.I. Wissenschaftsverlag: Mannheim.

[68] Mainzer, K. (1977): Is the intuitionistic bar-induction a constructive principle? In: *Notre Dame Journal of Formal Logic* 18 (4), 583–588.

[69] Mainzer, K. (1973): *Mathematischer Konstruktivismus*, Ph.D. U. Münster.

[70] Mainzer, K. (1972): Mathematischer Konstruktivismus im Lichte Kantischer Philosophie, in: *Philosophia Mathematica* 9 (no. 1) 1972, 3–26.

[71] Mainzer, K. (1970): Der Konstruktionsbegriff in der Mathematik. In: *Philosophia Naturalis* 12, 367–412.

[72] Matijasevich, Y.V. (1970): Enumerable sets are diophantic. In: *Soviet Math. Dokl.* 11, 345–357.

[73] Matijasevich, Y.V. (1993): *Hilbert's Tenth Problem.* The MIT Press: Cambridge MA.

[74] Minsky, M.L. (1961): Recursive unsolvability of Post's problem of tag and other topics in the theory of Turing machines. In: *Annals of Math.* 74, 437–454.

[75] Moschovakis, Y. (1997): The logic of functional recursion. In: *Logic and Scientific Methods. 10th Intern. Congr. Logic, Methodology and Philosophy of Science.* Kluwer Academic Publishers: Dordrecht, 179–208.

[76] Neumann, J. von (1948): The general and logical theory of automata. In: Jeffress, L.A. (Ed.). *Cerebral Mechanisms in Behavior The Hixon Symposium.* California Institute of Technology: Pasadena, 1–31.

[77] Ockham, W. of (1967–1988): *Opera philosophica et theologica.* Ed. G. Gl et al. 17 vols. Franciscan Institute St. Bonaventure: New York.

[78] Pappi Alexandrini *collectionis quae supersunt* 3 vols. (ed. F.O. Hultsch). Berlin 1876–1878.

[79] Penrose, R. (1991): *The Emperor's New Mind.* Penguin: London.

[80] Pinasco, J.P. (2009): New Proofs of Euclid's and Euler's theorems. In: *American Mathematical Monthly* 116 (2), 172–173.

[81] Plotkin, G.D. (1977): LCF considered as a programming language. In: *Theoretical Computer Science* 5, 223–255.

[82] Pohlers, W. (1989): *Proof Theory.* Springer: Berlin.

[83] Post, E.L. (1936): Finite combinatory processes Formulation I. In: *Symbolic Logic* I, 103–105.

[84] Rogers, H. (1967): *Theory of Recursive Functions and Effective Computability.* McGraw-Hill.

[85] Russell, B. (2000): *History of Western Philosophy.* Allen & Unwin: London, 462–463.

[86] Scholz, H. (1961): *Mathesis Universalis. Abhandlungen zur Philosophie als strenger Wissenschaft*, hrsg. von H. Hermes, F. Kambartel, J. Ritter. Benno Schwabe & Co: Basel.

[87] Schütte, K. (1977): *Proof Theory.* Springer: Berlin.

[88] Schütte, K. (1965): Predicative well-orderings. In: *Formal Systems and Recursive Functions.* North-Holland: Amsterdam, 280–303.

[89] Schwichtenberg, H.; Wainer, S.S. (2012): *Proofs and Computations.* Cambridge.

[90] Schwichtenberg, H. (2006): Minlog. In: F. Wiedijk (ed.): *The Seventeen Provers of the World. Lecture Notes in Artificial Intelligence* vol. 3600. Springer: Berlin, 151–157.

[91] Scott, D. (1982): Domains for denotational semantics. In: Nielsen, E.; Schmidt, E.M. (eds.): *Automata, Languages, and Programming*. Lecture Notes in Computer Science 140. Springer: Berlin, 577–613.

[92] Sheperdson, J.C.; Sturgis, H.E. (1963): Computability of recursive functions. In: *J. Assoc. Comp. Mach.* 10, 217–255.

[93] Shoenfield, J.R. (1967): *Mathematical Logic*. Reading (Mass.).

[94] Simpson, S.G. (2005): *Reverse Mathematics*. Lecture Notes in Logic 21. The Association of Symbolic Logic 2005.

[95] Simpson, S.G. (1999): *Subsystems of Second Order Arithmetic*. Perspectives in Mathematical Logic. Springer: Berlin.

[96] Soare, R.I. (2016): *Turing-Computability. Theory and Applications*. Springer. New York.

[97] Spade, P.V. (1999): *The Cambridge Companion to Ockham*. Cambridge University Press: Cambridge.

[98] Spector, C. (1955): Recursive well-orderings. In: *Journal of Symbolic Logic* 20, 151–163.

[99] Sundholm, G. (2014): Constructive recursive functions, Church's thesis, and Brouwer's theory of the creative subject: Afterthoughts on a Parisian Joint session. In: Dubucs, J.; Bourdeau, M. (Eds.): *Constructivity and Computability in Historical and Philosophical Perspective*. Springer: Heidelberg, 1–35.

[100] Troelstra, A.S.; van Dalen, D. (1988): *Constructivism in Mathematics*. An Introduction. 2 vols. North-Holland: Amsterdam.

[101] Troelstra, A.S. (1974): Note on the fan theorem. In: *Journal of Symbolic Logic* 39, 584–596.

[102] Troelstra, A.S. (ed.) (1973): *Metamathematical Investigation of Intuitionistic Arithmetic and Analysis*. Springer: Berlin.

[103] Troelstra, A.S. (1968): The theory of choice sequences. In: van Rootselaar, B.; Staal, J.F. (Eds.). *Logic, Methodology, and Philosophy of Science* 3. North-Holland: Amsterdam, 201–223.

[104] Turing, A.M. (1936–1937): On computable numbers, with an application to the Entscheidungsproblem. In: *Proc. London Math. Soc.* Ser. 2 (42), 230–265.

[105] Turing, A.M. (1939): Systems of logic based on ordinals. In: *Proc. London Math. Soc.* 2, 161–228.

[106] Turner, W. (1913): William of Ockham. In: Herbermann, C.: *Catholic Encyclopedia*. Robert Appleton Company: New York.

[107] Veblen, O. (1908): Continuous increasing functions of finite and transfinite ordinals. In: *Transactions of the American Mathematical Society* 9 (3), 280–292.

[108] Weber, H. (1893): Leopold Kronecker. In: *Jahresbericht der Deutschen Mathematiker-Vereinigung* 2, 19.

[109] Weyl, H. (1921): Über die neue Grundlagenkrise der Mathematik. In: *Mathematische Zeitschrift* 10 1921, 39–79.

[110] Weyl, H. (1918): *Das Kontinuum. Kritische Untersuchungen über die Grundlagen der Analysis*. De Gruyter: Leipzig.

Chapter 2

Constructive Convex Programming

Josef Berger and Gregor Svindland

Mathematisches Institut
Ludwig-Maximilians-Universität München
Theresienstraße 39, 80333 München, Germany
jberger@math.lmu.de
svindla@math.lmu.de

Working within Bishop-style constructive mathematics, we show that positive-valued, uniformly continuous, convex functions defined on convex and compact subsets of $\mathbb{R}^n$ have positive infimum. This gives rise to a separation theorem for convex sets. Based on these results, we show that the fundamental theorem of asset pricing is constructively equivalent to Markov's principle. The philosophical background behind all this is a constructively valid convex version of Brouwer's fan theorem. The emerging comprehensive yet concise overall picture of assets, infima of functions, separation of convex sets, and the fan theorem indicates that mathematics in convex environments has some innate constructive nature.

Contents

1. Introduction

When mathematics is employed to solve real world problems — for instance in a decision-making process — the derived results should be directly applicable. But there is an issue: mathematics as we normally practice it is based on the law of excluded middle (LEM), which says that either a statement is true or its negation is true. This rule allows for proving the existence of an object by merely showing that the assumption of non-existence of that object is false. In view of applicability, deriving the existence of objects with such indirect proof methods has the disadvantage that the proof does not tell us how to find the object, as it only rejects non-existence. This is where constructive mathematics, which is crudely characterized by not using LEM — or in other words which is based on intuitionistic rather than classical logic — becomes important. A constructive existence proof of an object always comes with an algorithm to compute it. Of course, proving results constructively is often more challenging than proving the same results with LEM as an admitted proof tool. For a mathematical theorem the question arises whether there is a constructive proof of that result, and if not, how far away it is from being constructive. Constructive reverse mathematics (CRM) addresses this question.

CRM as we apply it in this survey classifies theorems by logical axioms, that is fragments of LEM.[1] Given a theorem, the idea is to find such an axiom which is sufficient and necessary to prove it constructively. This answers the question how far away the theorem is from being constructive. The information obtained is also important from an applied point of view, because if a theorem is not constructive, we would like to discover a constructive version of it, typically by requiring some additional initial information. The constructive version of the fundamental theorem of asset pricing which we derive at the end of Section 4 is an example for this.

Whereas a fair amount of mathematics has been investigated from a constructive point of view, this is not yet the case for the field of financial mathematics. One reason for this is that constructive mathematics is considered a foundational or even philosophical issue and thus far away from highly applied disciplines like finance. However, there are strong arguments for carefully looking into the proofs of applied theorems, since constructive proofs are close to algorithms for constructing the desired objects.

As an initial case study we chose to investigate the fundamental theorem of asset pricing, which says that in absence of arbitrage trading strategies there exists a martingale measure. This theorem is the backbone of mathematical finance. Since it is normally proved by means of highly non-constructive separation results we first assumed that it is far away from being

---

[1] Another view of CRM is to take function existence axioms and induction axioms into account as well. This is more refined. In a first approach to calibrate theorems of finance, we chose the 'simple' variant.

constructive. However, it turns out to be equivalent to Markov's principle, which is the double-negation-elimination for purely existential formulas,

$$\neg\neg\exists n\, A(n) \to \exists n\, A(n), \tag{1}$$

where $A(n)$ is quantifier-free for each $n \in \mathbb{N}$. This principle amounts to unbounded search and is considered acceptable from a computational point of view. This equivalence was shown in [4] and is the content of Section 4. The proofs therein are quite easy. However, this is possible only since the required mathematical background was outsourced to Section 3.

The equivalence between the fundamental theorem of asset pricing and Markovs principle is based on the observation that positive-valued, uniformly continuous, convex functions defined on convex and compact subsets of $\mathbb{R}^n$ have positive infimum. This as well as a crucial consequence — a separation theorem for convex sets — was published in [3] and is worked out in Section 3.

In Section 5, which is based on [5], all theses results are traced back to a novel constructively valid version of Brouwer's fan theorem. We introduce co-convexity as a property of subsets B of $\{0,1\}^*$, the set of finite binary sequences, and prove that co-convex bars are uniform. Moreover, we establish a canonical correspondence between detachable subsets B of $\{0,1\}^*$ and uniformly continuous functions f defined on the unit interval such that B is a bar if and only if the corresponding function f is positive-valued, B is a uniform bar if and only if f has positive infimum, and B is co-convex if and only if f satisfies a weak convexity condition.

2. Bishop-style constructive mathematics

Constructive mathematics in the tradition of Errett Bishop [6, 7] is characterised by not using the law of excluded middle as a proof tool. As a major consequence, properties of the real number line $\mathbb{R}$ like the *limited principle of omniscience*

LPO $\forall x, y \in \mathbb{R}\, (x < y \lor x > y \lor x = y)$,

the *lesser limited principle of omniscience*

LLPO $\forall x, y \in \mathbb{R}\, (x \leq y \lor x \geq y)$,

and *Markov's principle* (the following formulation of Markov's principle in terms of real numbers is equivalent to the formulation (1))

MP $\forall x \in \mathbb{R} \left(\neg \left(x = 0 \right) \Rightarrow |x| > 0 \right)$

are no longer provable propositions but rather considered additional axioms. Many properties of the reals still hold constructively.

Lemma 1. *For all real numbers x, y, z,*

a) $x = 0 \Leftrightarrow |x| = 0$

b) $x \geq y \Leftrightarrow \neg \left(x < y \right)$

c) $x = y \Leftrightarrow \neg\neg \left(x = y \right)$

d) $x < y \Rightarrow x < z \lor z < y$

e) $|x| \cdot |y| > 0 \Leftrightarrow |x| > 0 \land |y| > 0$

f) $|x| > 0 \Leftrightarrow x > 0 \lor x < 0$.

If $\mathbb{R}$ is replaced with $\mathbb{Q}$ in the statement of LPO, then the resulting proposition can be proved constructively. Fix an inhabited subset S of $\mathbb{R}$ (*inhabited* means that there exists s with $s \in S$) and $x \in \mathbb{R}$. The real number x is a *lower bound* of S if

$$\forall s \in S \left(x \leq s \right)$$

and the *infimum* of S if it is a lower bound of S and

$$\forall \varepsilon > 0 \, \exists s \in S \left(s < x + \varepsilon \right).$$

In this case we write $x = \inf S$. The notions *upper bound, supremum,* and $x = \sup S$ are defined analogously. We cannot assume that every inhabited set with a lower bound has an infimum. However, under some additional conditions, this is the case. See [9, Corollary 2.1.19] for a proof of the following criterion.

Lemma 2. *Let S be an inhabited set of real numbers which has a lower bound. Assume further that for all $p, q \in \mathbb{Q}$ with $p < q$ either p is a lower bound of S or else there exists $s \in S$ with $s < q$. Then S has an infimum.*

Set $\mathbb{N} = \{0, 1, 2, \ldots\}$, $\mathbb{N}^+ = \{1, 2, \ldots\}$, and $\mathbb{R}^+ = \{x \in \mathbb{R} \mid x > 0\}$. For $X \subseteq \mathbb{R}$, a function $h : X \to \mathbb{R}$ is *weakly increasing* if

$$\forall s, t \in X \left(s < t \; \Rightarrow \; f(s) \leq f(t) \right)$$

and *strictly increasing* if

$$\forall s, t \in X \left(s < t \; \Rightarrow \; f(s) < f(t) \right).$$

The properties *weakly decreasing* and *strictly decreasing* are defined analogously.

Lemma 3. *For every weakly increasing function $h : \mathbb{N} \to \{0, 1\}$ with $h(0) = 0$ the set*

$$S = \left\{ 3^{-k} \mid h(k) = 0 \right\}$$

has an infimum. If $\inf S > 0$, there exists k such that

(1) $h(k) = 0$
(2) $h(k + 1) = 1$
(3) $\inf S = 3^{-k}$.

Moreover,

$$h(n) = 1 \quad \Leftrightarrow \quad \inf S \geq 3^{-n+1} \quad \Leftrightarrow \quad \inf S > 3^{-n}$$

for all n.

Proof. We apply Lemma 2. Note that $1 \in S$ and that 0 is a lower bound of S. Fix $p, q \in \mathbb{Q}$ with $p < q$. If $p \leq 0$, p is a lower bound of S. Now assume that $0 < p$. Then there exists k with $3^{-k} < p$. If $h(k) = 0$, there exist $s \in S$ (choose $s = 3^{-k}$) with $s < q$. If $h(k) = 1$, we can compute the minimum s_0 of S. If $p < s_0$, p is a lower bound of S; if $s_0 < q$, there exists $s \in S$ (choose $s = s_0$) with $s < q$.

If $\inf S > 0$, there exists l such that $3^{-l} < \inf S$. Therefore, $h(l) = 1$. Let k be the largest number such that $h(k) = 0$.

Assume that $h(n) = 1$. Let l be the largest natural number with $h(l) = 0$. Then $l \leq n - 1$ and thus $\inf S = 3^{-l} \geq 3^{-n+1}$.

Assume that $\inf S > 3^{-n}$. Then there exists k with (1), (2), and (3). We obtain $k < n$ and therefore $h(n) = 1$. $\qquad \square$

Let X be a metric space with metric d. Fix $\varepsilon > 0$ and sets $D \subseteq C \subseteq X$. D is an *ε-approximation* of C if for every $c \in C$ there exists $d \in D$ with $d(c, d) < \varepsilon$. The set C is

- *totally bounded* if for every $\varepsilon > 0$ there exist elements $x_1, \ldots, x_m$ of C such that $\{x_1, \ldots, x_m\}$ is an ε-approximation of C
- *complete* if every Cauchy sequence in C has a limit in C
- *closed in X* if every sequence in C which converges in X also converges in C

- *compact* if it is totally bounded and complete
- *located* if it is inhabited and if for every $x \in X$ the distance

$$d(x, C) = \inf \{d(x, c) \mid c \in C\}$$

exists.

If X is complete, a subset C of X is closed in X if and only if it is complete. Note further that totally bounded sets are inhabited. Totally boundedness is another crucial criterion for the existence of suprema and infima, see [9, Proposition 2.2.5].

Lemma 4. *If $C \subseteq \mathbb{R}$ is totally bounded, then $\inf C$ and $\sup C$ exist.*

Lemma 5. *Suppose that C is a located subset of X. Then the function*

$$f : X \to \mathbb{R}, \, x \mapsto d(x, C)$$

is uniformly continuous.

Proof. Fix $x, y \in X$. For every $c \in C$ we have

$$d(x, c) \leq d(x, y) + d(y, c).$$

We obtain

$$d(x, C) \leq d(x, y) + d(y, c),$$

and therefore

$$d(x, C) \leq d(x, y) + d(y, C).$$

This implies

$$f(x) - f(y) \leq d(x, y).$$

$\square$

We learn from [9, Proposition 2.2.6] that totally boundedness is preserved by uniformly continuous functions.

Lemma 6. *If $C \subseteq X$ is totally bounded and $f : C \to Y$ is uniformly continuous, where Y is also a metric space. Then*

$$\{f(x) \mid x \in C\}$$

is totally bounded.

The following combination of Lemma 4 and Lemma 6 is used frequently.

Lemma 7. *If $C \subseteq X$ is totally bounded and $f : C \to \mathbb{R}$ is uniformly continuous, then the infimum of f,*

$$\inf f = \inf \{f(x) \mid x \in C\}$$

and the supremum of f,

$$\sup f = \sup \{f(x) \mid x \in C\}$$

exist.

We refer to [7, Chapter 4, Proposition 4.4] for a proof of the following result.

Lemma 8. *A totally bounded subset C of a metric space X is located.*

A subset C of a linear space Z is *convex* if $\lambda x + (1 - \lambda)y \in C$ for all $x, y \in C$ and $\lambda \in [0, 1]$. The following lemma is of great importance in Section 3.

Lemma 9. *Let Y be an inhabited convex subset of a Hilbert space H and $x \in H$ such that $d = d(x, Y)$ exists. Then there exists a unique a in the closure $\overline{Y}$ of Y such that $\|a - x\| = d$. Furthermore, for all $c \in Y$ we have*

$$\langle a - x, c - a \rangle \geq 0$$

and therefore

$$\langle a - x, c - x \rangle \geq d^2.$$

Proof. Fix a sequence (c_l) in Y such that $\|c_l - x\| \to d$. Since

$$\|c_m - c_l\|^2 = \|(c_m - x) - (c_l - x)\|^2$$

$$= 2\|c_m - x\|^2 + 2\|c_l - x\|^2 - 4\underbrace{\left\|\frac{c_m + c_l}{2} - x\right\|^2}_{\geq 4d^2}$$

$$\leq 2\left(\|c_m - x\|^2 - d^2\right) + 2\left(\|c_l - x\|^2 - d^2\right),$$

(c_l) is a Cauchy sequence and therefore converges to an $a \in \overline{Y}$. Since $\|c_l - x\| \to \|a - x\|$, we obtain $\|a - x\| = d$. Now fix $b \in \overline{Y}$ with $\|b - x\| = d$. Then

$$\|a - b\|^2 = \|(a - x) - (b - x)\|^2$$

$$= 2\|a - x\|^2 + 2\|b - x\|^2 - 4\underbrace{\left\|\frac{a + b}{2} - x\right\|^2}_{\geq 4d^2} \leq 0,$$

thus $a = b$.

Fix $c \in Y$ and $\lambda \in (0,1)$. Since

$$\|a - x\|^2 \leq \|(1 - \lambda)a + \lambda c - x\|^2 = \|(a - x) + \lambda(c - a)\|^2$$

$$= \|a - x\|^2 + \lambda^2 \|c - a\|^2 + 2\lambda\langle a - x, c - a\rangle,$$

we obtain

$$0 \leq \lambda \|c - a\|^2 + 2\langle a - x, c - a\rangle.$$

Since λ can be arbitrarily small, we can conclude that

$$\langle a - x, c - a\rangle \geq 0.$$

This also implies that

$$\langle a - x, c - x\rangle = \langle a - x, c - a\rangle + \langle a - x, a - x\rangle \geq d^2.$$

$\square$

For $x, y \in \mathbb{R}^n$, we define the scalar product $\langle x, y\rangle = \sum_{i=1}^{n} x_i \cdot y_i$, the norm $\|x\| = \sqrt{\langle x, x\rangle}$, and the metric $d(x, y) = \|y - x\|$.

Vectors $x_1, \ldots, x_n \in \mathbb{R}^n$ are *linearly independent* if for all $\lambda \in \mathbb{R}^n$ the implication

$$\sum_{i=1}^{n} |\lambda_i| > 0 \Rightarrow \|\sum_{i=1}^{n} \lambda_i x_i\| > 0$$

is valid. Such vectors span located subsets [9, Lemma 4.1.2].

Lemma 10. *If $x_1, \ldots, x_m \in \mathbb{R}^n$ are linearly independent, then the set*

$$\left\{ \sum_{i=1}^{m} \xi_i x_i \mid \xi \in \mathbb{R}^m \right\}$$

is closed in $\mathbb{R}^n$ and located.

3. Convexity and constructive infima

Fix $n \in \mathbb{N}^+$. In this section, the variable i denotes elements of $\{1, \ldots, n\}$. A function $f : C \to \mathbb{R}$, where C is a convex subset of $\mathbb{R}^n$, is called *quasi-convex* if

$$f(\lambda x + (1 - \lambda)y) \leq \max(f(x), f(y))$$

for all $\lambda \in [0,1]$ and $x, y \in C$. Hence, in particular, any convex function $f : C \to \mathbb{R}$ — a function is *convex* if

$$f(\lambda x + (1 - \lambda)y) \leq \lambda f(x) + (1 - \lambda)f(y)$$

for all $\lambda \in [0,1]$ and $x, y \in C$ — is quasi-convex.

Theorem 1. *If $C \subseteq \mathbb{R}^n$ is compact and convex and*

$$f : C \to \mathbb{R}^+$$

is quasi-convex and uniformly continuous, then $\inf f > 0$.

In order to prove Theorem 1, we start with some technical lemmas. For a subset C of $\mathbb{R}^n$ and $t \in \mathbb{R}$ we define

$$C_i^t = \{x \in C \mid x_i = t\}.$$

Lemma 11. *Fix a convex subset C of $\mathbb{R}^n$ and $t \in \mathbb{R}$. Suppose further that there are $y, z \in C$ with $y_i < t < z_i$. Then there exists $\lambda \in \,]0,1[$ such that*

$$\lambda y + (1 - \lambda)z \in C_i^t.$$

Proof. Set $\lambda = \frac{z_i - t}{z_i - y_i}$. $\square$

We call C_i^t *admissible* if there exist $y, z \in C$ with $y_i < t < z_i$.

Lemma 12. *Let $n > 1$. Fix a subset C of $\mathbb{R}^n$ and suppose that C_i^t is convex and compact. Then there exists a convex compact subset $\hat{C}$ of $\mathbb{R}^{n-1}$ and a uniformly continuous bijection*

$$g : \hat{C} \to C_i^t$$

which is affine in the sense that

$$g(\lambda x + (1 - \lambda)y) = \lambda g(x) + (1 - \lambda)g(y)$$

for all $\lambda \in [0,1]$ and $x, y \in \hat{C}$.

Proof. We can assume that $i = 1$. Set

$$\hat{C} = \{(x_2, \ldots, x_n) \in \mathbb{R}^{n-1} \mid (t, x_2, \ldots, x_n) \in C_1^t\}$$

and

$$g(x_2, \ldots, x_n) = (t, x_2, \ldots, x_n).$$

$\square$

The next lemma is crucial for the proof of Theorem 1, and of interest on its own.

Lemma 13. *If $C \subseteq \mathbb{R}^n$ is convex and compact and C_i^t is admissible, then C_i^t is convex and compact.*

Proof. Let $C \subseteq \mathbb{R}^n$ be convex and compact and let C_i^t be admissible. Without loss of generality, we may assume that $t = 0$ and $i = 1$. There exist $y, z \in C$ with $y_1 < 0 < z_1$. Define

$$\mathcal{M} = C_1^0, \; \mathcal{L} = \{x \in C \mid x_1 \leq 0\}, \; \mathcal{R} = \{x \in C \mid x_1 \geq 0\}.$$

We show that the sets $\mathcal{L}$, $\mathcal{R}$ and $\mathcal{M}$ are convex and compact. It is clear that these sets are convex and complete. By applying Lemma 6 repeatedly, we show that they are totally bounded as well.
We start with the case of $\mathcal{R}$. Set

$$\kappa : \mathbb{R} \to \mathbb{R}, \; s \mapsto \max(-s, 0)$$

and

$$f : \mathbb{R}^n \to \mathbb{R}^n, \; x \mapsto \frac{z_1}{z_1 + \kappa(x_1)} x + \frac{\kappa(x_1)}{z_1 + \kappa(x_1)} z$$

and note that

- f is uniformly continuous

- f maps C onto $\mathcal{R}$.

In order to prove the latter, we proceed step by step and show that

(1) $f(C) \subseteq C$
(2) $f(C) \subseteq \mathcal{R}$
(3) $f(C) = \mathcal{R}$.

The property (1) follows from the convexity of C. In order to show (2), fix $x \in C$. We show that the assumption that the first component of $f(x)$ is negative is contradictory. So assume that

$$\frac{z_1}{z_1 + \kappa(x_1)} x_1 + \frac{\kappa(x_1)}{z_1 + \kappa(x_1)} z_1 < 0.$$

Then $x_1 < 0$ and therefore $\kappa(x_1) = -x_1$. We obtain

$$z_1 \cdot x_1 - x_1 \cdot z_1 < 0,$$

a contradiction. The property (3) follows from the fact that f leaves the elements of $\mathcal{R}$ unchanged. So we have shown that $\mathcal{R}$ is totally bounded. Analogously, we can show that $\mathcal{L}$ is totally bounded. Next, we show that

$$\mathcal{M} = f(\mathcal{L}),$$

which implies that $\mathcal{M}$ is totally bounded as well. To this end, fix $x \in \mathcal{L}$. Then $\kappa(x_1) = -x_1$ and therefore

$$\frac{z_1}{z_1 + \kappa(x_1)} x_1 + \frac{\kappa(x_1)}{z_1 + \kappa(x_1)} z_1 = 0,$$

which implies that $f(x) \in \mathcal{M}$. □

The following Lemma 14 basically already proves Theorem 1.

Lemma 14. *Fix a convex compact subset C of $\mathbb{R}^n$ and suppose that*

$$f : C \to \mathbb{R}^+$$

is quasi-convex and uniformly continuous. Assume further that

$$\inf \left\{ f(x) \mid x \in C_i^t \right\} > 0$$

for every admissible C_i^t. Then $\inf f > 0$.

Proof. Note that $\inf f$ exists by Lemma 7. We define a sequence (x^m) in C and a binary sequence (λ^m) such that

- $\lambda^{m+1} = 0 \Rightarrow \lambda^m = 0$ and $f(x^{m+1}) < \min\left(2^{-(m+1)}, f(x^m)\right)$
- $\lambda^{m+1} = 1 \Rightarrow \inf f > 0$ and $x^{m+1} = x^m$

for every m. Note that under these conditions the sequence $(f(x^m))$ is weakly decreasing.

Let x^0 be an arbitrary element of C and set $\lambda^0 = 0$. Assume that x^m and λ^m have already been defined.

case 1 If $\lambda^m = 1$, set $x^{m+1} = x^m$ and $\lambda^{m+1} = 1$.

case 2 If $\lambda^m = 0$ and $0 < \inf f$, set $x^{m+1} = x^m$ and $\lambda^{m+1} = 1$.

case 3 If $\lambda^m = 0$ and

$$\inf f < \min\left(2^{-(m+1)}, f(x^m)\right),$$

choose x^{m+1} in C with

$$f(x^{m+1}) < \min\left(2^{-(m+1)}, f(x^m)\right)$$

and set $\lambda^{m+1} = 0$.

We show that the sequence (x^m) converges.

It is sufficient to show that for each component i the sequence $(x_i^m)_{m \in \mathbb{N}}$ is a Cauchy sequence. We consider the case $i = 1$. Fix $\varepsilon > 0$. Let D be the image of C under the projection onto the first component, i.e.

$$D = \{x_1 \mid x \in C\}.$$

Note that D is a totally bounded interval. Denote its infimum by a and its supremum by b.

<u>case 1</u> If $b - a < \varepsilon$, then $\left| x_1^k - x_1^l \right| \leq \varepsilon$ for all k, l.

<u>case 2</u> If $b - a > 0$, there exists a finite $\frac{\varepsilon}{2}$-approximation F of $]a, b[$. Note that for every t with $a < t < b$ the set $\mathcal{C}_1^t$ is admissible. Hence, we can choose an l_0 such that

$$f(x) > 2^{-l_0}$$

for all $t \in F$ and all $x \in C_1^t$. Fix $k, l \geq l_0$. We show that $\left| x_1^k - x_1^l \right| \leq \varepsilon$.

<u>case 2.1</u> If $\lambda^{l_0} = 1$, then $x^k = x^l$.

<u>case 2.2</u> If $\lambda^{l_0} = 0$, then $f(x^k) < 2^{-l_0}$ and $f(x^l) < 2^{-l_0}$. Suppose that $x_1^k - x_1^l > \varepsilon$. Then there exists $t \in F$ with $x_1^k < t < x_1^l$. According to Lemma 11 there is $\mu \in]0, 1[$ such that $\mu x^k + (1 - \mu) x^l \in C_1^t$, and by quasi-convexity of f we obtain

$$f(\mu x^k + (1 - \mu) x^l) \leq \max \left(f(x^k), f(x^l) \right) < 2^{-l_0}$$

which contradicts the construction of l_0. Therefore, $x_1^k - x_1^l \leq \varepsilon$, and similarly also $x_1^l - x_1^k \leq \varepsilon$.

Let $x \in C$ be the limit of the sequence (x^m). There exists l such that

$$f(x) > 2^{-l}$$

and a k such that

$$d(x, y) < 2^{-k} \ \Rightarrow \ |f(x) - f(y)| < 2^{-(l+1)}$$

for all $y \in C$. Finally, pick $N > l$ such that

$$d(x, x^N) < 2^{-k}.$$

Then $f(x^N) \geq 2^{-N}$, therefore $\lambda_N = 1$, which implies that $\inf f > 0$. $\square$

Proof of Theorem 1. We use induction over the dimension n.

If $n = 1$, then every admissible set C_1^t equals $\{t\}$, so inf $f > 0$ follows from Lemma 14.

Now fix $n > 1$ and assume the assertion of Theorem 1 holds for $n - 1$. Furthermore, let C be a convex compact subset of $\mathbb{R}^n$, and suppose that

$$f : C \to \mathbb{R}^+$$

is convex and uniformly continuous. Fix an admissible subset C_i^t of C. By Lemma 13, C_i^t is convex and compact. Using Lemma 12 construct the convex compact set $\hat{C} \subseteq \mathbb{R}^{n-1}$ and the uniformly continuous affine bijection

$$g : \hat{C} \to C_i^t.$$

Then $F : \hat{C} \to \mathbb{R}^+$ given by $F = f \circ g$ is quasi-convex and uniformly continuous. The induction hypothesis now implies that

$$\inf \left\{ f(x) \mid x \in C_t^i \right\} = \inf \left\{ F(x) \mid x \in \hat{C} \right\} > 0 .$$

Thus, inf $f > 0$ follows from Lemma 14. $\qquad\square$

Theorem 2. *Let C and Y be subsets of $\mathbb{R}^n$ and suppose that*

(1) C is convex and compact
(2) Y is convex, complete, and located
(3) $d(c, y) > 0$ for all $c \in C$ and $y \in Y$.

Then there exist $p \in \mathbb{R}^n$ and reals α, β such that

$$\langle p, c \rangle < \alpha < \beta < \langle p, y \rangle$$

for all $c \in C$ and $y \in Y$. In particular, the sets C and Y are strictly separated by the hyperplane

$$H = \{ x \in \mathbb{R}^n \mid \langle p, x \rangle = \gamma \} ,$$

with $\gamma = \frac{1}{2}(\alpha + \beta)$.

Proof. By Lemma 5, the function

$$f : C \to \mathbb{R}, \, c \mapsto d(c, Y)$$

is uniformly continuous. Since Y is closed, Lemma 9 implies that for every $c \in C$ there is a unique $y \in Y$ with

$$f(c) = d(c, y).$$

Therefore, f is positive-valued and also convex, as we can see as follows. Fix $c_1, c_2 \in C$ and $\lambda \in [0, 1]$. There are $y_0, y_1, y_2 \in Y$ such that

$$f(c_1) = d(c_1, y_1), \quad f(c_2) = d(c_2, y_2),$$

and

$$f(\lambda c_1 + (1 - \lambda)c_2) = d(\lambda c_1 + (1 - \lambda)c_2, y_0).$$

We obtain

$$\begin{aligned} f(\lambda c_1 + (1 - \lambda)c_2) &= d(\lambda c_1 + (1 - \lambda)c_2, y_0) \\ &\leq d(\lambda c_1 + (1 - \lambda)c_2, \lambda y_1 + (1 - \lambda)y_2) \\ &\leq \lambda d(c_1, y_1) + (1 - \lambda)d(c_2, y_2) \\ &= \lambda f(c_1) + (1 - \lambda)f(c_2). \end{aligned}$$

By Theorem 1, $\inf f > 0$. The set

$$Z = \{y - c \mid x \in C, y \in Y\}$$

is inhabited and convex. Since we have

$$\inf\{\, \|y - c\| \mid x \in C, y \in Y\} = \inf f,$$

we can conclude that $\delta = d(0, Z)$ exists and is positive. By Lemma 9, there exists $p \in \mathbb{R}^n$ such that

$$\langle p, y \rangle \geq \delta^2 + \langle p, c \rangle$$

for all $y \in Y$ and $c \in C$. By Lemma 7, $\eta = \sup\{\langle p, c \rangle \mid c \in C\}$ exists. Setting

$$\alpha = \frac{\delta^2}{3} + \eta \text{ and } \beta = \frac{\delta^2}{2} + \eta,$$

we obtain

$$\langle p, c \rangle < \alpha < \beta < \langle p, y \rangle$$

for all $c \in C$ and $y \in Y$. $\qquad \square$

4. The fundamental theorem of asset pricing

Fix $m, n \in \mathbb{N}^+$. Set $I_n = \{1, \ldots, n\}$. In this section, the variable i always stands for an element of I_m and the variable j always stands for an element of I_n. The linear space of real matrices A with m rows and n columns is denoted by $\mathbb{R}^{m \times n}$. For such an A, we denote its entry at row i and column j by a_{ij}. For $l \in \mathbb{N}^+$ and $B \in \mathbb{R}^{n \times l}$, we denote the matrix product of A

and B (which is an element of $\mathbb{R}^{m \times l}$) by $A \cdot B$. Let C be the convex hull of the unit vectors of $\mathbb{R}^n$.

Our market consists of m assets. Their value at time 0 (present) is known. Their value at time 1 (future) is unknown. There are n possible developments and we know the prices in each of the n cases. Define a matrix $A \in \mathbb{R}^{m \times n}$ as follows: the value of the entry a_{ij} is the price development (price at time 1 minus price at time 0) of asset i in case j. Set

$$P = \left\{ p \in \mathbb{R}^n \mid \sum_{j=1}^{m} p_j = 1 \text{ and } 0 < p_j \text{ for all } j \right\}.$$

A vector $p \in P$ is a *martingale measure* if $A \cdot p = 0$. Under a martingale measure the average profit is zero, that is today's price of the assets is reasonable in the sense of being the expected value of the assets tomorrow. For $x \in \mathbb{R}^n$ we define

$$x > 0 \quad :\Leftrightarrow \quad \forall j \, (x_j \geq 0) \wedge \exists j \, (x_j > 0).$$

A vector $\xi \in \mathbb{R}^m$ is an *arbitrage trading strategy* if $\xi \cdot A > 0$.

Note that every $\xi \in \mathbb{R}^m$ corresponds to a trading strategy, where ξ_i denotes the number of shares of asset i that the trader buys. Hence, the payoff at time 1 over all possible future scenarios is $\xi \cdot A$. Thus arbitrage strategies are trading strategies which correspond to riskless gains, since they never produce any losses, and even a strict gain for at least one possible future scenario.

The *fundamental theorem of asset pricing* says that the absence of an arbitrage trading strategy is equivalent to the existence of a martingale measure.

FTAP Fix an $\mathbb{R}^{m \times n}$-matrix A with linearly independent rows. Then

$$\neg \exists \xi \in \mathbb{R}^m \, (\xi \cdot A > 0) \quad \Leftrightarrow \quad \exists p \in P \, (A \cdot p = 0).$$

Note that "$\Leftarrow$" is clear: assume there exist both p in P with $A \cdot p = 0$ and ξ in $\mathbb{R}^m$ with $\xi \cdot A > 0$. Then from $A \cdot p = 0$ we can conclude that $\xi \cdot A \cdot p = 0$ and from $\xi \cdot A > 0$ and $p \in P$ we can conclude that $\xi \cdot A \cdot p > 0$. This is a contradiction.

Theorem 3.

$$\text{FTAP} \Leftrightarrow \text{MP}$$

Proof. Fix an $\mathbb{R}^{m \times n}$-matrix A such that

$$\neg \exists \xi \in \mathbb{R}^m \, (\xi \cdot A > 0).$$

Let Y be the linear subspace of $\mathbb{R}^n$ which is generated by the rows of A. This set is convex. Since we have assumed that the rows of A are linearly independent, we can conclude from Lemma 10 that Y is convex, closed, and located. The set C is convex and compact. By $\neq$, we obtain

$$\forall c \in C, y \in Y \, (d(c, y) > 0) \, .$$

By Theorem 2, there exists a vector $p \in \mathbb{R}^n$ and reals α, β such that

$$\forall y \in Y, c \in C \, (\langle p, c \rangle > \alpha > \beta > \langle p, y \rangle) \, .$$

This implies that $A \cdot p = 0$ and that all components of p are positive. We can assume further that $p_1 + \ldots + p_n = 1$.

In order to prove the converse direction, fix a real number a with $\neg(a = 0)$. Apply FTAP to the matrix

$$A = (|a|, -1) \, .$$

Note that A has linearly independent rows. The no-arbitrage condition is satisfied: assume that there exists $\xi \in \mathbb{R}$ with

$$(\xi \cdot |a|, -\xi) > 0. \tag{2}$$

We obtain that $\xi \cdot |a| \geq 0$ and $\xi \leq 0$, which implies that $\xi = 0$, a contradiction to (2). Now FTAP yields the existence of a $p \in P$ with

$$p_1 \cdot |a| = p_2 \, .$$

This implies that $|a| > 0$. $\square$

Note that we obtain the following constructively valid version of FTAP, see [4, Corollary 3].

Theorem 4. *Fix an $\mathbb{R}^{m \times n}$-matrix A with linearly independent rows. Then*

$$\forall c \in C, \, \xi \in \mathbb{R}^m \, (d(c, \xi \cdot A) > 0 \; \Rightarrow \; \exists p \in P \, (A \cdot p = 0)) \, .$$

This theorem says that we can construct a martingale measure if we exclude the existence of arbitrage strategies in a stricter way.

5. Brouwer's fan theorem and convexity

We write $\{0,1\}^*$ for the set of all finite binary sequences u, v, w. Let $\emptyset$ be the empty sequence and let $\{0,1\}^{\mathbb{N}}$ be the set of all infinite binary sequences α, β, γ. For every u let $|u|$ be the *length* of u, that is $|\emptyset| = 0$ and for $u = (u_0, \ldots, u_{n-1})$ we have $|u| = n$. For $v = (v_0, \ldots, v_{m-1})$, the *concatenation* $u * v$ of u and v is defined by

$$u * v = (u_0, \ldots, u_{n-1}, v_0, \ldots, v_{m-1}).$$

The *restriction* $\overline{\alpha}n$ of α to n bits is given by

$$\overline{\alpha}n = (\alpha_0, \ldots, \alpha_{n-1}).$$

Thus $|\overline{\alpha}n| = n$ and $\overline{\alpha}0 = \emptyset$. For u with $n \leq |u|$, the restriction $\overline{u}n$ is defined analogously. A subset B of $\{0,1\}^*$ is *closed under extension* if $u * v \in B$ for all $u \in B$ and for all v. A sequence α *hits* B if there exists n such that $\overline{\alpha}n \in B$. B is a *bar* if every α hits B. B is a *uniform bar* if there exists N such that for every α there exists $n \leq N$ such that $\overline{\alpha}n \in B$. Often one requires B to be *detachable*, that is for every u the statement $u \in B$ is decidable. Now we are ready to introduce Brouwer's *fan theorem for detachable bars*.

FAN Every detachable bar is a uniform bar.

In Bishop's constructive mathematics, FAN is neither provable nor falsifiable, see [8, Section 3 of Chapter 5]. In their seminal paper [11], Julian and Richman established a correspondence between FAN and functions on $[0,1]$ as follows.

Proposition 1. *For every detachable subset B of $\{0,1\}^*$ there exists a uniformly continuous function $f : [0,1] \to [0, \infty[$ such that*

(1) B is a bar $\Leftrightarrow$ f is positive-valued
(2) B is a uniform bar $\Leftrightarrow$ f has positive infimum.

Conversely, for every uniformly continuous function $f : [0,1] \to [0, \infty[$ there exists a detachable subset B of $\{0,1\}^*$ such that (1) and (2) hold.

Consequently, FAN is equivalent to the statement that every uniformly continuous, positive-valued function defined on the unit interval has positive infimum. Now, in view of Theorem 1, the question arises whether there is a constructively valid 'convex' version of the fan theorem. To this end, we define

$$u < v \; :\Leftrightarrow \; |u| = |v| \wedge \exists k < |u| \, (\overline{u}k = \overline{v}k \wedge u_k = 0 \wedge v_k = 1)$$

and

$$u \le v \; :\Leftrightarrow \; u = v \vee u < v.$$

A subset B of $\{0,1\}^*$ is *co-convex* if for every α which hits B there exists n such that either

$$\{v \mid v \le \overline{\alpha}n\} \subseteq B \quad \text{or} \quad \{v \mid \overline{\alpha}n \le v\} \subseteq B.$$

Note that, for detachable B, co-convexity follows from the convexity of the complement of B, where $C \subseteq \{0,1\}^*$ is *convex* if for all u, v, w we have

$$u \le v \le w \wedge u, w \in C \Rightarrow v \in C.$$

Define the *upper closure* B' of B by

$$B' = \{u \mid \exists k \le |u| \, (\overline{u}k \in B)\}.$$

Note that B is a (detachable) bar if and only if B' is a (detachable) bar and B is a uniform bar if and only B' is a uniform bar. Therefore, we may assume that bars are closed under extension.

Theorem 5. *Every co-convex bar is a uniform bar.*

Proof. Fix a co-convex bar B. Since the upper closure of B is also co-convex, we can assume that B is closed under extension. Define

$$C = \{u \mid \exists n \, \forall w \in \{0,1\}^n \, (u * w \in B)\}.$$

Note that $B \subseteq C$ and that C is closed under extension as well. Moreover, B is a uniform bar if and only if there exists n such that $\{0,1\}^n \subseteq C$. First, we show that

$$\forall u \, \exists i \in \{0,1\} \, (u * i \in C). \tag{3}$$

Fix u. For

$$\beta = u * 1 * 0 * 0 * 0 * \ldots$$

there exist an l such that either

$$\{v \mid v \le \overline{\beta}l\} \subseteq B,$$

or

$$\{v \mid \overline{\beta}l \le v\} \subseteq B.$$

Since B is closed under extension, we can assume that $l > |u| + 1$. Let $m = l - |u| - 1$. If $\{v \mid v \le \overline{\beta}l\} \subseteq B$, we can conclude that

$$u * 0 * w \in B$$

for every w of length m, which implies that $u * 0 \in C$. If $\{v \mid \overline{\beta}l \le v\} \subseteq B$, we obtain

$$u * 1 * w \in B$$

for every w of length m, which implies that $u * 1 \in C$. This concludes the proof of (3).

By countable choice, there exists a function $F : \{0, 1\}^* \to \{0, 1\}$ such that

$$\forall u \ (u * F(u) \in C).$$

Define α by

$$\alpha_n = 1 - F(\overline{\alpha}n).$$

Next, we show by induction on n that

$$\forall n \forall u \in \{0, 1\}^n \ (u \ne \overline{\alpha}n \Rightarrow u \in C). \tag{4}$$

If $n = 0$, the statement clearly holds, since in this case the statement $u \ne \overline{\alpha}n$ is false. Now fix some n such that (4) holds. Moreover, fix $w \in \{0, 1\}^{n+1}$ such that $w \ne \overline{\alpha}(n + 1)$.

<u>case 1.</u> $\overline{w}n \ne \overline{\alpha}n$. Then $\overline{w}n \in C$ and therefore $w \in C$.

<u>case 2.</u> $w = \overline{\alpha}n * (1 - \alpha_n) = \overline{\alpha}n * F(\overline{\alpha}n)$. This implies $w \in C$. So we have established (4).

There exists n such that $\overline{\alpha}n \in B$. Applying (4) to this n, we can conclude that every u of length n is an element of C, thus B is a uniform bar. $\square$

Remark 1. Note that we do not need to require that the co-convex bar in Theorem 5 be detachable.

In order to include convexity in the list of Proposition 1, we introduce a notion of weakly convex functions. Let S be a subset of $\mathbb{R}$. A function $f : S \to \mathbb{R}$ is *weakly convex* if for all $t \in S$ with $f(t) > 0$ there exists $\varepsilon > 0$ such that either

$$\forall s \in S \ (s \le t \ \Rightarrow \ f(s) \ge \varepsilon)$$

or

$$\forall s \in S \ (t \le s \ \Rightarrow \ f(s) \ge \varepsilon).$$

Remark 2. Fix a dense subset D of $[0, 1]$. A uniformly continuous function $f : [0, 1] \to \mathbb{R}$ is weakly convex if and only its restriction to D is weakly convex.

The following generalisation of Proposition 1 links Theorem 1 with Theorem 5.

Theorem 6. *For every detachable subset B of $\{0,1\}^*$ which is closed under extension there exists a uniformly continuous function $f : [0,1] \to \mathbb{R}$ such that*

(1) B is a bar $\Leftrightarrow$ f is positive-valued
(2) B is a uniform bar $\Leftrightarrow$ $\inf f > 0$
(3) B is co-convex $\Leftrightarrow$ f is weakly convex.

Conversely, for every uniformly continuous function $f : [0,1] \to \mathbb{R}$ there exists a detachable subset B of $\{0,1\}^*$ which is closed under extension such that (1), (2), and (3) hold.

We split the proof of Theorem 6 into two parts.

Part I: Construction of a function f for given B

Fix a detachable subset B of $\{0,1\}^*$ which is closed under extension. We can assume that $\emptyset \notin B$. (Otherwise, let f be the constant function $t \mapsto 1$.) First, we define a function $g : [0,1] \to \mathbb{R}$ which satisfies the properties (1) and (2) of Theorem 6. Then, we introduce a refined version f of g which satisfies all properties of Theorem 6. Define metrics

$$d_1(s,t) = |s-t|, \quad d_2((x_1,x_2),(y_1,y_2)) = |x_1-y_1| + |x_2-y_2|$$

on $\mathbb{R}$ and $\mathbb{R}^2$, respectively. The mapping

$$(\alpha, \beta) \mapsto \inf\left\{2^{-k} \mid \overline{\alpha}k = \overline{\beta}k\right\}$$

is a compact metric on $\{0,1\}^{\mathbb{N}}$. See [8, Section 1 of Chapter 5] for an introduction to basic properties of this metric space. Define a uniformly continuous function $\kappa : \{0,1\}^{\mathbb{N}} \to [0,1]$ by

$$\kappa(\alpha) = 2 \cdot \sum_{k=0}^{\infty} \alpha_k \cdot 3^{-(k+1)}.$$

The next lemma immediately follows from the definition of κ.

Lemma 15. *For all α, β, and n, we have*

- $\overline{\alpha}n = \overline{\beta}n \implies |\kappa(\alpha) - \kappa(\beta)| \leq 3^{-n}$
- $\overline{\alpha}n = \overline{\beta}n \land \alpha_n < \beta_n \implies \kappa(\alpha) + 3^{-(n+1)} \leq \kappa(\beta)$
- $\overline{\alpha}n \neq \overline{\beta}n \implies |\kappa(\alpha) - \kappa(\beta)| \geq 3^{-n}$
- $\overline{\alpha}n < \overline{\beta}n \implies \kappa(\alpha) < \kappa(\beta).$

Now define

$$\eta_B : \{0,1\}^{\mathbb{N}} \to [0,1], \; \alpha \mapsto \inf \left\{ 3^{-k} \mid \overline{\alpha} k \notin B \right\}.$$

The following lemma is an immediate consequence of Lemma 3.

Lemma 16. *The function η_B is well-defined — the infimum in the definition of η_B always exists — and uniformly continuous. If $\eta_B(\alpha) > 0$, there exists k such that*

> *(1)* $\overline{\alpha} k \notin B$
> *(2)* $\overline{\alpha}(k+1) \in B$
> *(3)* $\eta_B(\alpha) = 3^{-k}$.

Moreover,

$$\overline{\alpha} n \in B \; \Leftrightarrow \; \eta_B(\alpha) \geq 3^{-n+1} \; \Leftrightarrow \; \eta_B(\alpha) > 3^{-n}$$

for all α and n.

Set

$$C = \left\{ \kappa(\alpha) \mid \alpha \in \{0,1\}^{\mathbb{N}} \right\}$$

and

$$K = \left\{ (\kappa(\alpha), \eta_B(\alpha)) \mid \alpha \in \{0,1\}^{\mathbb{N}} \right\}.$$

Lemma 17. *The sets C and K are compact.*

Proof. Both sets are uniformly continuous images of the compact set $\{0,1\}^{\mathbb{N}}$ and therefore totally bounded, by Lemma 6. Suppose that $\kappa(\alpha^n)$ converges to t and $\eta_B(\alpha^n)$ converges to s. By Lemma 15, the sequence (α^n) is Cauchy, therefore it converges to a limit α. Then $\kappa(\alpha^n)$ converges to $\kappa(\alpha)$ and $\eta_B(\alpha^n)$ converges to $\eta_B(\alpha)$. Therefore $t = \kappa(\alpha)$ and $s = \eta_B(\alpha)$. Thus we have shown that both C and K are complete. $\qquad\square$

In the following, we will use Bishop's lemma, see [7, Ch. 4, Lemma 3.8].

Lemma 18. *Let A be a complete, located subset of a metric space X, and x a point of X. Then there exists a point a in A such that $d(x,a) > 0$ entails $d(x, A) > 0$.*

Define

$$g : [0,1] \to [0, \infty[, \; t \mapsto d_2((t,0), K).$$

Proposition 2.

(1) B is a bar $\Leftrightarrow$ g is positive-valued
(2) B is a uniform bar $\Leftrightarrow$ $\inf g > 0$.

Proof. Assume that B is a bar. Fix $t \in [0,1]$. In view of Bishop's lemma and the compactness of K, it is sufficient to show that

$$d_2((t,0),(\kappa(\alpha),\eta_B(\alpha))) > 0$$

for each α. This follows from $\eta_B(\alpha) > 0$.

Now assume that g is positive-valued. Fix α. Since

$$d_2((\kappa(\alpha),0),K) = g(\kappa(\alpha)) > 0,$$

we can conclude that

$$d_2((\kappa(\alpha),0),(\kappa(\alpha),\eta_B(\alpha))) > 0.$$

Thus $\eta_B(\alpha)$ is positive which implies that α hits B.

The second equivalence follows from Lemma 16 and the fact that $\inf g = \inf \eta_B$. $\qquad\qquad\qquad\qquad\qquad\qquad\qquad\qquad\qquad\qquad\qquad\qquad\quad\square$

Set

$$-C = \{t \in [0,1] \mid d_1(t,C) > 0\}.$$

We would like to include the statement

- B is co-convex $\Leftrightarrow$ g is weakly convex

into Proposition 2. Note, however, that g is positive on $-C$. Thus we introduce a new function f by

$$f : [0,1] \to \mathbb{R}, \ t \mapsto g(t) - d_1(t,C).$$

The next lemma lists up a few properties of f and g.

Lemma 19. *For all α, n, and t we have*

- $g(\kappa(\alpha)) = f(\kappa(\alpha)) \leq \eta_B(\alpha)$
- $f(\kappa(\alpha)) > 3^{-n} \ \Rightarrow \ \overline{\alpha}n \in B$
- $\overline{\alpha}n \in B \ \Rightarrow \ f(\kappa(\alpha)) \geq 3^{-n}$
- $d_1(t,C) \leq g(t)$.

Next, we clarify how f behaves on $-C$.

Lemma 20. *The set $-C$ is dense in $[0,1]$. For every $t \in -C$ there exist unique elements a, a' of C such that*

(1) $t \in \,]a, a'[\, \subseteq -C$.

(2) $d_1(t, C) = \min\left(d_1(t, a), d_1(t, a')\right)$.

Moreover, setting $\gamma = \kappa^{-1}(a)$ *and* $\gamma' = \kappa^{-1}(a')$, *we obtain*

(3) $\forall n \left(\overline{\gamma}n \in B \land \overline{\gamma'}n \in B \Rightarrow f(t) \geq 3^{-n}\right)$

(4) if $d_1(t, a) < d_1(t, a')$, *then*

$$\gamma \text{ hits } B \iff f(t) > 0 \iff \inf\left\{f(s) \mid a \leq s \leq t\right\} > 0$$

(5) if $d_1(t, a') < d_1(t, a)$, *then*

$$\gamma' \text{ hits } B \iff f(t) > 0 \iff \inf\left\{f(s) \mid t \leq s \leq a'\right\} > 0.$$

Proof. Fix $t \in [0, 1]$ and $\delta > 0$. If $d_1(t, C) > 0$, then $t \in -C$. Now assume that there exists α such that $d_1(t, \kappa(\alpha)) < \delta/2$. There exists u such that $d_1(\kappa(\alpha), t_u) < \delta/2$ where

$$t_u = \tfrac{1}{2} \cdot \kappa(u * 0 * 1 * 1 * 1 * \ldots) + \tfrac{1}{2} \cdot \kappa(u * 1 * 0 * 0 * 0 * \ldots).$$

Note that $t_u \in -C$ and that $d_1(t, t_u) < \delta$. So $-C$ is dense in $[0, 1]$.

Fix $t \in -C$. Since for any α it is decidable whether $\kappa(\alpha) > t$ or $\kappa(\alpha) < t$, the sets $C_{<t} = \{s \in C \mid s < t\}$ and $C_{>t} = \{s \in C \mid s > t\}$ are compact. Let a be the maximum of $C_{<t}$ and let a' be the minimum of $C_{>t}$. Clearly, a and a' fulfil (1) and (2).

In order to show (3), assume that $\overline{\gamma}n \in B$ and $\overline{\gamma'}n \in B$. Fix α. We show that

$$d_2((t, 0), (\kappa(\alpha), \eta_B(\alpha))) - d_1(t, C) \geq 3^{-n}. \tag{5}$$

First, assume that $\kappa(\alpha) < t$. Then we have

$$d_2((t, 0), (\kappa(\alpha), \eta_B(\alpha))) - d_1(t, C) \geq \kappa(\gamma) - \kappa(\alpha) + \eta_B(\alpha).$$

If $\overline{\alpha}n = \overline{\gamma}n$, then $\overline{\alpha}n \in B$ and we can conclude that $\eta_B(\alpha) \geq 3^{-n+1}$, by Lemma 16. On the other hand, Lemma 15 implies that $\kappa(\gamma) - \kappa(\alpha) \leq 3^{-n}$. This proves (5). If $\overline{\alpha}n \neq \overline{\gamma}n$, then $\kappa(\gamma) - \kappa(\alpha) \geq 3^{-n}$, by Lemma 15. This also proves (5). The case $t < \kappa(\alpha)$ can be treated similarly.

In order to show (4), set $\iota = d_1(t, a') - d_1(t, a)$ and suppose that $\overline{\gamma}n \in B$. Set $\varepsilon = \min(\iota, 3^{-n})$. Fix s with $a \leq s \leq t$. We show that $f(s) \geq \varepsilon$. Note that $d_1(s, C) = s - a$. Fix α. We show that

$$d_2((s, 0), (\kappa(\alpha), \eta_B(\alpha))) - (s - a) \geq \varepsilon.$$

If $a' \leq \kappa(\alpha)$, we obtain

$$d_2((s,0),(\kappa(\alpha),\eta_B(\alpha))) - (s-a)$$

$$\geq \kappa(\alpha) - s - (s-a) \geq \iota \geq \varepsilon.$$

If $\kappa(\alpha) \leq a$, we obtain

$$d_2((s,0),(\kappa(\alpha),\eta_B(\alpha))) - (s-a) = s - \kappa(\alpha) + \eta_B(\alpha) - (s-a)$$

$$= \eta_B(\alpha) + a - \kappa(\alpha) \geq 3^{-n} \geq \varepsilon,$$

where $\eta_B(\alpha) + a - \kappa(\alpha) \geq 3^{-n}$ is derived by looking at the cases $\overline{\alpha}n = \overline{\gamma}n$ and $\overline{\alpha}n \neq \overline{\gamma}n$ separately.

Now assume that $f(t) > 0$. We show that γ hits B. If $f(t) > 0$, then $g(t) > t - a$. On the other hand, we have

$$g(t) \leq d_2((t,0),(a,\eta_B(\gamma))) = t - a + \eta_B(\gamma),$$

so $\eta_B(\gamma) > 0$. By Lemma 16, this implies that γ hits B.

The statement (5) is proved analogously to (4). □

The next lemma is very easy to prove, we just formulate it to be able to refer to it.

Lemma 21. *For real numbers $x < y < z$ and $\delta > 0$ there exists a real number y' such that*

- $x < y' < z$
- $d_1(y,y') < \delta$
- $d_1(x,y') < d_1(y',z)$ *or* $d_1(x,y') > d_1(y',z)$.

For a function F defined on $\{0,1\}^{\mathbb{N}}$, set

$$F(u) = F(u * 0 * 0 * 0 * \ldots). \tag{6}$$

Now we can show that f has all the desired properties.

Proposition 3.

- *(1) B is a bar $\Leftrightarrow$ f is positive-valued*
- *(2) B is a uniform bar $\Leftrightarrow$ $\inf f > 0$*
- *(3) B is co-convex $\Leftrightarrow$ f is weakly convex.*

Proof. (1) "$\Rightarrow$". Suppose that B is a bar and fix t. By Proposition 2, we obtain $g(t) > 0$. If $d_1(t, C) < g(t)$, then $f(t) > 0$, by the definition of f. If $0 < d_1(t, C)$, we can apply Lemma 20 to conclude that $f(t) > 0$.

(1) "$\Leftarrow$". If f is positive-valued, then g is positive-valued as well and Proposition 2 implies that B is a bar.

(2) "$\Rightarrow$". If B is a uniform bar, Proposition 2 yields

$$\varepsilon := \inf g > 0.$$

Moreover, there exists n such that $\{0, 1\}^n \subseteq B$. Fix $\delta > 0$ such that

$$|s - t| < \delta \quad \Rightarrow \quad |f(s) - f(t)| < \varepsilon/2$$

for all s and t. Fix t. If $d_1(t, C) < \delta$, we can conclude that

$$f(t) \geq \varepsilon/2$$

by the choice of ε and δ. If $d_1(t, C) > 0$, Lemma 20 and $\{0, 1\}^n \subseteq B$ imply that

$$f(t) \geq 3^{-n}.$$

So we have shown that $\inf f \geq \min\left(\varepsilon/2, 3^{-n}\right)$.

(2) "$\Leftarrow$". If $\inf f > 0$, then $\inf g > 0$, and Proposition 2 implies that B is a uniform bar.

(3) "$\Rightarrow$". In view of Remark 2 and Lemma 20, it is sufficient to show that the restriction of f to $-C$ is weakly convex. Fix $t \in -C$ and assume that $f(t) > 0$. Choose a, a', γ, and γ' according to Lemma 20. In view of Lemma 21 and the uniform continuity of f, we may assume without loss of generality that either

$$d_1(a, t) < d_1(t, a') \quad \text{or} \quad d_1(a, t) > d_1(t, a').$$

Consider the first case. The second case can be treated analogously. By Lemma 20, we obtain

$$\iota = \inf \{f(s) \mid a \leq s \leq t\} > 0.$$

In particular, $f(\kappa(\gamma)) > 0$, so γ hits B. There exists n such that either

$$\{v \mid v \leq \overline{\gamma}n\} \subseteq B \tag{7}$$

or

$$\{v \mid \overline{\gamma}n \leq v\} \subseteq B. \tag{8}$$

Set $\varepsilon = \min\left(\iota, 3^{-n}\right)$. In case (7), we show that

$$\forall s \in -C\left(s \leq t \;\Rightarrow\; f(s) \geq \varepsilon\right),$$

as follows. Assume that there exists $s \in -C$ with $s \leq t$ such that $f(s) < \varepsilon$. Then, by the definition of ι, we obtain that $s < a$. Applying Lemma 20 again, we can choose α and α' such that

$$s \in \,]\kappa(\alpha), \kappa(\alpha')[\, \subseteq -C.$$

Then $\overline{\alpha}n \leq \overline{\alpha'}n \leq \overline{\gamma}n$. Thus both $\overline{\alpha}n$ and $\overline{\alpha'}n$ are in B. This implies $f(s) \geq 3^{-n}$, which is a contradiction. In case (8), a similar argument yields

$$\forall s \in -C\left(t \leq s \;\Rightarrow\; f(s) \geq \varepsilon\right).$$

(3) "$\Leftarrow$". Assume that f is weakly convex. Fix α and suppose that α hits B. Then Lemma 19 implies that $f(\kappa(\alpha)) > 0$. There exists n with $\overline{\alpha}n \in B$ such that

$$\forall s \left(s \leq \kappa(\alpha) \;\Rightarrow\; f(s) > 3^{-n}\right)$$

or

$$\forall s \left(\kappa(\alpha) \leq s \;\Rightarrow\; f(s) > 3^{-n}\right).$$

Assume the first case. Fix v with $v \leq \overline{\alpha}n$. Then $\kappa(v) \leq \kappa(\alpha)$. If $v \notin B$, then, by Lemma 16 and Lemma 19,

$$f(\kappa(v)) = g(\kappa(v)) \leq \eta_B(v) \leq 3^{-n}.$$

This contradiction shows that

$$\{v \mid v \leq \overline{\alpha}n\} \subseteq B.$$

Now, consider the second case. Fix v with $\overline{\alpha}n < v$. Then $\kappa(\alpha) \leq \kappa(v)$. If $v \notin B$, then $f(\kappa(v)) \leq 3^{-n}$. This contradiction shows that

$$\{v \mid \overline{\alpha}n \leq v\} \subseteq B.$$

$\square$

Part II: Construction of a set B for given f

Set

$$\kappa' : \{0,1\}^{\mathbb{N}} \to [0,1], \; \alpha \mapsto \sum_{k=0}^{\infty} \alpha_k \cdot 2^{-(k+1)}.$$

One cannot prove that κ' is surjective, since this would imply LLPO. Note, however, that every rational $q \in [0,1]$ is in the range of κ'. Moreover, we make use of the following lemma, see [1, Lemma 1].

Lemma 22. *Let S be a subset of $[0,1]$ such that*

$$\forall \alpha \, \exists \varepsilon > 0 \, \forall t \in [0,1] \, (|t - \kappa'(\alpha)| < \varepsilon \Rightarrow t \in S) \, .$$

Then $S = [0,1]$.

The next lemma is a typical application of Lemma 22.

Lemma 23. *Fix a uniformly continuous function $f : [0,1] \to \mathbb{R}$ and define*

$$F : \{0,1\}^{\mathbb{N}} \to \mathbb{R}, \, \alpha \mapsto f(\kappa'(\alpha)).$$

Then

(1) f is positive-valued $\Leftrightarrow$ F is positive-valued
(2) $\inf f > 0$ $\Leftrightarrow$ $\inf F > 0$.

Proof. In (1), the direction "$\Rightarrow$" is clear. For "$\Leftarrow$", apply Lemma 22 to the set

$$S = \{t \in [0,1] \mid f(t) > 0\} \, .$$

The equivalence (2) follows from the density of the image of κ' in $[0,1]$ and the uniform continuity of f. $\qquad\qquad\square$

In the following proposition, we use a similar construction as in [2].

Proposition 4. *For every uniformly continuous function*

$$f : [0,1] \to \mathbb{R}$$

there exists a detachable subset B of $\{0,1\}^*$ which is closed under extension such that

(i) B is a bar $\Leftrightarrow$ f is positive-valued
(ii) B is a uniform bar $\Leftrightarrow$ $\inf f > 0$
(iii) B is co-convex $\Leftrightarrow$ f is weakly convex.

Proof. Since the function

$$F : \{0,1\}^{\mathbb{N}} \to \mathbb{R}, \, \alpha \mapsto f(\kappa'(\alpha))$$

is uniformly continuous, there exists a strictly increasing function $M : \mathbb{N} \to \mathbb{N}$ such that

$$|F(\alpha) - F(\overline{\alpha}(M(n)))| < 2^{-n}$$

for all α and n, recalling the convention given in (6). Since M is strictly increasing, for every k the statement

$$\exists n\, (k = M(n))$$

is decidable. Therefore, for every u we can choose $\lambda_u \in \{0, 1\}$ such that

$$l_u = 0 \quad \Rightarrow \quad \forall n\, (|u| \neq M(n)) \ \lor\ \exists n\, (|u| = M(n) \ \land\ F(u) < 2^{-n+2})$$
$$l_u = 1 \quad \Rightarrow \quad \exists n\, (|u| = M(n) \ \land\ F(u) > 2^{-n+1}).$$

The set

$$B = \{u \in \{0,1\}^* \mid \exists l \leq |u|\, (\lambda_{\overline{u}l} = 1)\}$$

is detachable and closed under extension. Note that

$$F(\alpha) \geq 2^{-n+3} \quad \Rightarrow \quad \overline{\alpha}(M(n)) \in B \tag{9}$$

and

$$\overline{\alpha}(M(n)) \in B \quad \Rightarrow \quad F(\alpha) \geq 2^{-n} \tag{10}$$

for all α and n. In view of Lemma 23, (9) and (10) yield (1) and (2).

In order to show (3), fix a co-convex set B. Moreover, fix $t \in [0, 1]$ and assume that $f(t) > 0$. By Remark 2, we may assume that t is a rational number, which implies that there exists α such that $\kappa'(\alpha) = t$. Now $F(\alpha) > 0$ implies that α hits B. Therefore, there exists n such that either

$$\{v \mid v \leq \overline{\alpha}n\} \subseteq B$$

or

$$\{v \mid \overline{\alpha}n \leq v\} \subseteq B.$$

In the first case, we show that

$$\inf\{f(s) \mid s \in [0, t]\} \geq \min\left(2^{-n}, F(\alpha)\right). \tag{11}$$

Assume that there exists $s \leq t$ such that $f(s) < 2^{-n}$ and $f(s) < F(\alpha)$. The latter implies that $s < t$. Choose a β with the property that $\kappa'(\beta)$ is close enough to s such that

$$\kappa'(\beta) < \kappa'(\alpha) \tag{12}$$

and

$$F(\beta) = f(\kappa'(\beta)) < 2^{-n}. \tag{13}$$

Now (10) and (13) imply that $\overline{\beta}n \notin B$. On the other hand, (12) implies that $\overline{\beta}n \leq \overline{\alpha}n$ and therefore $\overline{\beta}n \in B$. This is a contradiction, so we have shown (11).

In the case

$$\{v \mid \overline{\alpha}n \le v\} \subseteq B$$

we can similarly show that

$$\inf\{f(s) \mid s \in [t,1]\} \ge \min\left(2^{-n}, F(\alpha)\right).$$

Now assume that f is weakly convex. Fix an α which hits B. Then there exists n with $\overline{\alpha}(M(n)) \in B$ and (10) implies that $f(\kappa'(\alpha)) > 0$. We choose n large enough such that either

$$\inf\{f(t) \mid t \in [0, \kappa'(\alpha)]\} \ge 2^{-n+3}$$

or

$$\inf\{f(t) \mid t \in [\kappa'(\alpha), 1]\} \ge 2^{-n+3}.$$

By (9), we obtain

$$\{v \mid v \le \overline{\alpha}(M(n))\} \subseteq B$$

in the first case and

$$\{v \mid \overline{\alpha}(M(n)) \le v\} \subseteq B.$$

in the second. Therefore, B is co-convex. □

Thus the proof of Theorem 6 is completed. We conclude this section with a discussion about weakly convex functions.

Remark 3. Uniformly continuous, (quasi-)convex functions $f : [0,1] \to \mathbb{R}$ are weakly convex. To this end, we recall that f is *convex* if we have

$$f(\lambda s + (1-\lambda)t) \le \lambda f(s) + (1-\lambda)f(t)$$

and *quasiconvex* if we have

$$f(\lambda s + (1-\lambda)t) \le \max\left(f(s), f(t)\right)$$

for all $s, t \in [0,1]$ and all $\lambda \in [0,1]$. Clearly, convexity implies quasi-convexity. Now assume that f is quasi-convex. Fix $t \in [0,1]$ and assume that $f(t) > 0$. Set $\varepsilon = f(t)/2$. The assumption that both

$$\inf\{f(s) \mid s \in [0,t]\} < f(t) \quad \text{and} \quad \inf\{f(s) \mid s \in [t,1]\} < f(t)$$

is absurd, because in that case by uniform continuity there exists $s < t < s'$ such that $f(s) < f(t)$ and $f(s') < f(t)$. Compute $\lambda \in (0,1)$ such that $t = \lambda s + (1 - \lambda)s'$, and note that quasi-convexity of f implies $f(t) \le \max(f(s), f(s')) < f(t)$ which is absurd. Hence, either $\inf\{f(s) \mid s \in [0,t]\} > \varepsilon$ or $\inf\{f(s) \mid s \in [t,1]\} > \varepsilon$.

Pointwise continuous functions on $[0,1]$ which are weakly decreasing on $[0,s]$ and weakly increasing on $[s,1]$ for some s are weakly convex. See [10] for a detailed discussion of various notions of convexity.

If f is weakly convex, then the set $\{t \mid f(t) \leq 0\}$ is convex. With classical logic, the reverse implication holds as well, if f is continuous. This illustrates that weak convexity is indeed a convexity property.

Acknowledgements

We thank the Excellence Initiative of the LMU Munich, the Japan Advanced Institute of Science and Technology, and the European Commission Research Executive Agency for supporting the research.

References

[1] Josef Berger and Douglas Bridges, *A fan-theoretic equivalent of the antithesis of Specker's theorem.* Indag. Mathem., N.S., 18(2) (2007) 195–202.

[2] Josef Berger and Hajime Ishihara, *Brouwer's fan theorem and unique existence in constructive analysis.* Math. Log. Quart. 51, No. 4 (2005) 360–36.

[3] Josef Berger and Gregor Svindland, *Convexity and constructive infima.* Arch. Math. Logic 55 (2016) 873–881.

[4] Josef Berger and Gregor Svindland, *A separating hyperplane theorem, the fundamental theorem of asset pricing, and Markov's principle.* Annals of Pure and Applied Logic 167 (2016) 1161–1170.

[5] Josef Berger and Gregor Svindland, *Brouwer's fan theorem and convexity.* Preprint.

[6] Errett Bishop, *Foundations of Constructive Analysis.* McGraw-Hill, New York (1967) xiii + 370 pp.

[7] Errett Bishop and Douglas Bridges, *Constructive Analysis.* Springer-Verlag (1985) 477 pp.

[8] Douglas Bridges and Fred Richman, *Varieties of Constructive Mathematics.* London Math. Soc. Lecture Notes 97, Cambridge Univ. Press (1987) 160 pp.

[9] Douglas S. Bridges and Luminiţa Simona Vîţă, *Techniques of Constructive Analysis.* Universitext, Springer-Verlag New York (2006) 215 pp.

[10] Lars Hörmander, *Notions of Convexity.* Birkhäuser (2007) 416 pp.

[11] William H. Julian and Fred Richman, *A uniformly continuous function on* [0, 1] *that is everywhere different from its infimum.* Pacific Journal of Mathematics 111, No. 2 (1984) 333–340.

Chapter 3

Exploring Predicativity

Laura Crosilla

Department of Philosophy
ERI Building
University of Birmingham
Edgbaston, Birmingham, B15 2TT, UK
Laura.Crosilla@gmail.com

Prominent constructive theories of sets as Martin-Löf type theory and Aczel and Myhill constructive set theory, feature a distinctive form of constructivity: predicativity. This may be phrased as a constructibility requirement for sets, which ought to be finitely specifiable in terms of some uncontroversial initial "objects" and simple operations over them. Predicativity emerged at the beginning of the 20th century as a fundamental component of an influential analysis of the paradoxes by Poincaré and Russell. According to this analysis the paradoxes are caused by a vicious circularity in definitions; adherence to predicativity was therefore proposed as a systematic method for preventing such problematic circularity. In the following, I sketch the origins of predicativity, review the fundamental contributions by Russell and Weyl and look at modern incarnations of this notion.

Contents

1. Introduction

Since recent years the word "constructive" is typically employed as synonym of "using intuitionistic logic". Indeed, influential constructive mathematical theories, as Martin-Löf type theory and Aczel and Myhill constructive set theory employ intuitionistic logic.[1] However, there is a more fundamental sense of constructivity that these theories also aim at capturing, which is deeply rooted in the mathematical tradition, and is commonly expressed by stating that they are *predicative*. According to one way of spelling out the notion of constructivity, this relates to a *finitary process of "construction" or specification* of a mathematical entity. For example, the Oxford English Dictionary so defines the word "constructive" (in the mathematical case): "*Relating to, based on, or denoting mathematical proofs which show how an entity may in principle be constructed or arrived at in a finite number of steps.*" There is a sense in which an intuitionistic approach to mathematics satisfies this notion of constructivity, as it assigns a fundamental role in mathematics to the availability of explicit or constructive proofs of mathematical statements. For instance, an intuitionistic proof of an existential statement is usually read as embodying an algorithm which provides (at least in principle) a witness to the statement in question and a proof that the witness does satisfy the relevant condition expressed by the statement. In addition, an intuitionistic proof of a disjunction ought to offer the mean for deciding which of the disjuncts holds true.[2] The requirement that an intuitionistic proof of an existential statement also ought to include a witness is a paradigmatic example of the constructivity of the resulting proof, as the latter shows how to "construct", or specify, a witness. Similarly, the requirement of availability of a decision procedure for a disjunction ensures that a constructive proof is fully explicit.

The issue that predicativity rises is distinct from, though related to, that of the availability of an intuitionistic proof, and may be seen as directly pertaining to the question of how we specify domains of quantification.[3] For this reason, debates on predicativity have been traditionally perceived as

---

[1] See, for example, [1, 13, 40, 41, 43]. For surveys see [5, 14].

[2] This understanding of the notion of intuitionistic proof is manifested, for example, by the so-called Brouwer-Heyting-Kolmogorov (BHK) interpretation of intuitionistic logic. See e.g. [18, 70]. The ideas underlying the BHK interpretation are made more precise in Martin-Löf type theory and in models of intuitionistic theories, as realizability models [33].

[3] The interrelation between predicativity and intuitionistic logic is complex, and for this reason an analysis of this issue is postponed to subsequent work.

disputes pertaining the concept of set.[4] The thought is that an analysis of constructivity, which relates to "how an entity may in principle be constructed or arrived at in a finite number of steps", ought to include also a clarification of the concept of set, in as much as sets are domains of quantification. It is then clear that the case of universal quantification on infinite domains is particularly problematic, as according to this constructive perspective we require that a domain of quantification be *finitarily* specified.

If sets are to be understood constructively, then two crucial issues need to be examined: which methods of "construction" of sets are considered admissible, and which initial entities can be taken as legitimate starting points of the construction process. As I clarify below, the predicative literature witnesses a number of distinct answers to these questions.

The principal aim of this article is to survey and discuss predicativity.[5] In the following, I outline the origins of predicativity, and review an informal characterization of this notion. I then briefly sketch the main traits of two fundamental predicativist proposals by Russell and Weyl, respectively [60, 73]. Finally, I delineate the principal steps of a logical analysis of predicativity that began in the 1950's, and conclude with a comparison between two forms of predicativity: predicativity given the natural numbers and strict predicativity.

2. The emergence of predicativity

The notion of predicativity has its origins within a remarkable exchange between Poincaré and Russell at the beginning of the last century.[1] The wider context of Poincaré and Russell's debate on predicativity are well-known reflections by prominent mathematicians of the time on the new concepts and methods of proof which had made their way in mathematics from the 19th century.[2] Predicativity specifically was forged as part of Poincaré and Russell's attempts to analyse and counter the paradoxes that afflicted the foundations of mathematics.[3] Russell's influential analysis of the paradoxes imputes them to the illegitimate assumption that *any propositional function gives rise to a class*, the class of all the objects which

---

[4]See the final section for a brief discussion of this point.
[5]This note is part of a wider project of clarification and assessment of predicativity. See also [15–17].
[1]See [50–54, 58–60].
[2]See e.g. [20, 69, 72].
[3]See below for an emblematic example of paradox: Russell's paradox.

satisfy it.[4] Russell therefore introduced the term **predicative** to denote those *propositional functions* which define a class and distinguish them from the non-predicative or **impredicative** ones, which do not define a class [58]. Poincaré and Russell's debate witnesses the difficulties involved in spelling out the notion of impredicativity, and clarifying its perceived problematic status. Russell, in particular, turned to devise formal ways of capturing the distinction between predicative and impredicative propositional functions. A clarification of the notion of predicativity became a fundamental component of a thorough analysis of the foundations of mathematics, and adherence to predicativity the main instrument for blocking the paradoxes. Russell's efforts culminated in his ramified type theory, as further discussed in Section 3.1.

2.1. *Circularity and Russell's VCP*

A crucial feature of Russell's type theory is its implementation of the so-called Vicious Circle Principle (VCP), introduced to ban vicious-circular definitions (see below). The VCP was prompted by the fundamental observation by Richard, further developed by Poincaré, that paradoxes typically manifest a form of *vicious circularity, or self-reference* [50, 52, 57]. An example may help clarify this point.

Richard's paradox. This paradox arises form the definition of the least non-definable real number, r, by reference to the class of all definable real numbers. More precisely, let us consider all the real numbers which are definable in English by a finite number of words and let D be their collection. D is countable. We can then list all the elements of D, and mimic Cantor's diagonal proof of the non-denumerability of the real numbers to produce a new real number, r, which is different from each element of D. However, one can easily express in English a rendering of the "algorithm" that allows for the definition of r, so that r turns out to be a definable real number after all, and a contradiction arises.

According to the present analysis, Richard's paradox is engendered by a form of circularity: we define r by reference to the *whole* D, and therefore, so it is claimed, by reference to r itself. In fact, Russell introduced the

---

[4]In the present context for simplicity we may identify a propositional function with an open formula, i.e. a formula with a free variable (see e.g. [25]). Note, however, that the interpretation of the notions of proposition and propositional function in Russell is complex. See e.g. [37]. Note also that the term "class" is here used as in the original literature, to refer to a collection of elements. Therefore it should be carefully distinguished from the notion of "proper class" that is often used in contemporary set theory.

VCP to prevent the formation of collections as D, and claimed that these are *ill-formed*. He gave a number of variants of the VCP. For example: "no totality can contain members defined in terms of itself" [60, p. 237].

Another version is to be found in [61, p. 198]:

> ... whatever in any way concerns *all* or *any* or *some* of a class must not be itself one of the members of a class.

The latter rendering of the VCP clearly highlights the fundamental link between impredicativity and quantification.[5]

2.2. *Characterising predicativity*

Having introduced the VCP and explained its motivation with an example, I can now review an informal characterisation of predicativity which figures in the early foundational debates. Predicativity is typically presented negatively: we specify what is impredicativity and term predicativity the negation of impredicativity. We say that *a definition is **impredicative** if it defines an entity by reference to a class to which the entity itself belongs.* In particular, a definition is impredicative if it defines an entity by *quantifying over* a class which includes the entity to be defined. Given this notion of impredicative definition, one can further specify a notion of *impredicative entity*: this is an entity which can *only* be defined by an impredicative definition. We can hence talk of impredicative classes, propositions, properties. A definition or entity is **predicative** *if it is not impredicative.*[6]

2.2.1. *Examples*

In the previous section, I have reviewed Richard's paradox, which arises from an impredicative definition, as we define a new element r of D by reference to the whole D. Further examples may help clarify the notion of impredicativity. In presenting these examples I closely follow [60, 76], but

---

[5]This is perhaps the best know expression of the VCP. However, Russell gave other renderings, some of which, like the first one above, do not directly involve quantification over, but reference to a class that includes the definiendum. The plurality of formulations of the VCP induces difficulties for an exegesis of Russell's thought, and, indeed, for a thorough clarification of the notion of predicativity, as already noted by Gödel [29].

[6]This negative characterisation of impredicativity is clearly unsatisfactory. As discussed below, subsequent technical work, starting from Russell's type theory, aimed at offering more informative characterisations of predicativity. Note also that the late writings by Poincaré [53, 54] feature another, apparently distinct, characterisation of predicativity which does not appeal to circularity. A predicative set is "invariant under extension": the addition of new elements does not "disorder" the set itself. See [8, 9, 16, 17, 35].

also [10, 21], as their analysis clearly highlights a number of reasons that are typically adduced against impredicativity.

1. *Russell's paradox.* The first example of impredicative definition I consider is that of **Russell's "set"**, R. This can be so defined in modern terminology:

$$R = \{x \mid x \notin x\}.$$

Here R arises from an application of the **Unrestricted Comprehension schema**: given any formula φ in the language of set theory, we form the set of all the x's that satisfy φ, that is, $\{x \mid \varphi(x)\}$. In R's definition, in particular, one takes φ to be $x \notin x$.

In his analysis of this paradox Russell observes that R is defined impredicatively as it *refers* to the class of all classes [60, p. 225]. He also states that if we tried to block the paradox by deciding that no class is a member of itself, then R would become the class of all classes. But then the question arises whether R is a member of itself, and we have to conclude that R is not a member of itself, that is, that R is not a class. Russell therefore draws the conclusion that there is no class of all classes, since if we supposed there is, then this very assumption would give rise to a new class lying outside the presumed class of all classes.

2. *Burali-Forti paradox.* This is so described by Russell: we can show that every well-ordered class has an ordinal number, and that the ordinal of the class of ordinals up to and including any given ordinal exceeds the given ordinal by one. But "on certain very natural assumptions" the class of all ordinals itself is well-ordered and has an ordinal number, say Ω. However, the class of all ordinals including Ω turns out to have ordinal number $\Omega + 1$, contradicting the assumption that Ω is the ordinal of the class of all ordinals. Russell's assessment of this paradox is that it shows that "all ordinals" is an "illegitimate notion; for if not, all ordinals in order of magnitude form a well-ordered series which must have an ordinal number greater than all ordinals." [60, p. 225]

3. *The logicist definition of natural number.* The logicist definition of natural number may be so expressed: n is a natural number if it satisfies all properties which hold of 0 and which are closed under the successor operation. In modern terminology:

$$N(n) := \forall F[F(0) \wedge \forall x(F(x) \rightarrow F(Suc(x))) \rightarrow F(n)].$$

The predicate N expressing the property "to be a natural number" is here defined by reference to all predicates, F, expressing properties of the natural numbers. A circularity arises as the predicate N itself is within the range of the first quantifier: the definition is impredicative, as N is defined by reference to all predicates expressing properties of the natural numbers and thus, so it is contended, to itself.[7]

In the following, I draw on Carnap's clear analysis of this example in [10]. Carnap rightly stresses the importance of this example, which shows that impredicativity affects not only paradoxical cases, but one of the most fundamental concepts in mathematics, that of natural number.[8] Carnap remarks that the above definition induces difficulties which are best seen if we consider a *specific* natural number, say 3, and check whether it satisfies this definition. In order to do so, we need to check if *each* property of the natural numbers holds of 3, that is, if: $\forall F[F(0) \wedge \forall x(F(x) \to F(Suc(x))) \to F(3)]$. However, the property "to be a natural number", which is expressed by the predicate N, is one of the properties of the natural numbers. That is, to find out whether $N(3)$ holds, we need to be able to clarify whether the following holds:

$$N(0) \wedge \forall x(N(x) \to N(Suc(x))) \to N(3).$$

Hence it would seem that we need first to ascertain whether the property of being a natural number holds of 3, in order to assess whether it holds of 3. Carnap concludes that this definition of natural number is therefore "circular and useless" [10, p. 48].[9]

4. *Napoleon's qualities.* Another example of impredicative definition which does not involve a paradox is given by the sentence: **Napoleon had all the qualities that make a great general** [76, p. 59]. We might wish to compare the expression above with the following: *Napoleon was Corsican*, or *Napoleon was brave*. These are utterly unproblematic, as the properties expressed by "being Corsican" and "being brave" do not refer to other properties. However, the property expressed by "having all the

---

[7]Poincaré is well-known for having noted this difficulty with this definition of natural number, and for suggesting that the principle of mathematical induction and the natural numbers can not be reduced to something more primitive.

[8]A worry for Carnap is that the impredicativity of the logicist definition of natural number seems to compromise the logicist programme.

[9]Carnap [10] hints at the possibility of an alternative reading of universal quantification, which would vindicate the usefulness of impredicative definitions as the one above. The idea is to consider a reading of universal quantification which avoids the presupposition that a universal quantification also entails reference to each individual element of a domain of quantification. Carnap's proposal is briefly discussed by Gödel [29]. See also [28] for a contemporary perspective into this issue inspired by Carnap's discussion. [16] also analyses, under the light of [19], the prospects of eliminating these difficulties by reading universal quantification intuitionistically.

qualities that make a great general" would seem to be itself a quality of a great general, and therefore refer to itself.[10]

5. *Least Upper Bound principle.* Finally an example from analysis: the **Least Upper Bound principle** (LUB). This states that:

> Every bounded, non–empty subset $\mathcal{M}$ of the real numbers has a least upper bound.

To explain the impredicativity of the (LUB) I follow [21, p. 2]. In a standard way, we can codify rational numbers by suitable natural numbers and identify real numbers with certain sets of rational numbers, via Dedekind sections or Cauchy sequences. If we identify real numbers with the upper parts of Dedekind sections, we see that the least upper bound of a bounded, non-empty set $\mathcal{M}$ of real numbers is given by $S = \cap X[X \in \mathcal{M}]$. The difficulty with this definition can be seen as follows. The set $\mathcal{M}$ will be typically given by a condition, $\mathcal{C}(X)$, such that

$$\forall X[X \in \mathcal{M} \iff \mathcal{C}(X)].$$

Now S above is such that

$$\forall x[x \in S \iff \forall X(\mathcal{C}(X) \to x \in X)].$$

Feferman writes:

> However, to answer the question "What are the members of S?" we would, in general, first have to know what sets X satisfy $\mathcal{C}(X)$, and in particular whether or not $\mathcal{C}(S)$ holds; this would, in turn, in general depend on knowing what members S has. [21, p. 2]

As further discussed below, if impredicativity is seen as problematic, this particular example is critical, as it goes at the very heart of mathematics, affecting a core discipline as analysis.

To conclude this section, I outline Russell's verdict on examples as 1 and 2 above. As already anticipated in [58], Russell claims that a solution to the paradoxes lies in countering the assumption that *any* propositional function gives rise to a set. More precisely, Russell observes that a typical common feature of the paradoxes is that they involve a form of "self-reference of reflexiveness". "In each contradiction something is said about *all* cases of some kind, and from what is said a new case seems to be generated, which both is and is not of the same kind as the cases of which *all* were

---

[10]This example is often used by Russell to explain the workings of "ramification" in ramified type theory, as discussed below.

concerned in what was said." [60, p. 224] Russell concludes that viciously circular definitions do not give rise to a class, on pain of contradiction, and therefore are illegitimate:

> Thus all our contradictions have in common the assumption of a totality such that, if it were legitimate, it would at once be enlarged by new members defined in terms of itself. This leads us to the rule: "Whatever involves *all* of a collection must not be one of the collection"; or, conversely; "If, provided a certain collection had a total, it would have members only definable in terms of that total, then the said collection has no total."

3. Shedding light on predicativity: Russell's ramified type theory and Weyl's "Das Kontinuum"

The above characterisation of impredicativity and the examples suffice to convey a general idea of this notion. However, the above characterisation is insufficiently precise to fully clarify what counts as predicative and what does not. In fact, as already noted by Zermelo, mathematical notions are typically defined in a number of alternative but equivalent ways [77]. It is therefore to be expected that some mathematical notions which are prima facie impredicative may, under closer scrutiny, turn out to be predicative after all. A detailed assessment of this issue requires therefore a more sophisticated approach which makes use of a precise logical machinery.

Two main lines of research arise from this observation. First of all, from a perspective that finds faults with impredicativity, one would like to have some general criteria which systematically guarantee that when developing mathematics we do not introduce impredicative notions or entities. One way to achieve this is to develop a suitable foundational theory (e.g. a set theory) which complies with predicativity, so that working within it fully guarantees adherence to predicativity.

Secondly, as the examples of the logicist definition of natural number and the LUB principle clarify, impredicativity is to be found in everyday mathematics. Therefore, from both a perspective that favours and one that objects to predicativity, it becomes crucially important to assess what is the impact of complying with predicativity. A clarification of the latter point turns out to be more complex than the above informal characterisation of predicativity may suggest, especially because of the possibility of developing a portion of mathematics in a number of alternative ways. In the following, I briefly review two particularly significant steps in the development of these two lines of research: Russell's type theory and Weyl's predicative analysis. In Section 4, I discuss more recent developments that directly tackle the second issue.

3.1. *Russell's type theory*

The need to replace the purely negative characterisation above with a positive one was fully acknowledged by Russell [60], who claimed that "our positive doctrines [...] must make it plain that 'all propositions' and 'all properties' are meaningless phrases." [60, p. 226] Russell's "positive doctrine" finds full expression in his ramified type theory [60, 76]. This is a careful and complex formulation of a concept of set which introduces a number of restrictions compared with the naive concept of set given by unrestricted comprehension. Retrospectively, Russell [60] introduces *simultaneously* two kinds of regimentation: a type restriction and an order restriction for propositional functions.

Type restrictions have the effect of producing a hierarchy of "levels" or "types". As a consequence, membership is now a relation between, on the one side, members of a type and, on the other, the type itself. A type is "[...] the range of significance of a propositional function, i.e. [...] the collection of arguments for which the said function has values." [60, 236] The underlying idea is that we start from a type of individuals, and then consider types which are ranges of significance of propositional functions defined on the individuals, and so on. This apparently suffices to block set-theoretic paradoxes as, for example, Russell's paradox: expressions as $x \in x$ or $x \notin x$ are simply ill-formed.[1]

In addition to type restrictions, Russell also introduced further constraints on propositional functions, whose effect is to block impredicativity more generally. Example 4 above highlights the difference between expressions as "being Corsican" and "being brave" on the one hand, and expressions as "having all the qualities that make a great general" on the other hand. The first are utterly unproblematic, while the latter refers to "all qualities" of a great general, including the property it refers to, and therefore is problematic from a predicative perspective. The strategy underpinning ramification may be concisely summarised as follows: to avoid defining a property in terms of an expression which refers to "all properties", we subdivide the propositional functions (referring to properties) in

---

[1] Set-theoretic paradoxes are paradoxes, as Russell's and Burali-Forti's, that relate to the concept of set. See also page 96.

different "orders". First order propositional functions are those, as "being Corsican" and "being brave" which do not refer to other propositional functions. At the second order we have propositional functions which quantify over all propositional functions of the first order, and so on.[2] This is accounted for by introducing a hierarchical structure, ramification, at the level of propositional functions; quantification is then constrained to range over propositional functions of lower orders. Consequently, cases of impredicativity as that involved in Example 4 and in the definition of the natural numbers discussed above can now be eliminated. However, as quickly realised by Russell [60], the device of ramification causes difficulties as soon as we attempt to prove statements by induction on the natural numbers. For example, we can not refer to all the properties of the natural numbers, but only to those of some given order. These difficulties propagate to the case of the real numbers, as witnessed by the impredicativity of the LUB. As a consequence, many "fundamental theorems not only could not be proved but could not even be expressed." [10, p. 46] In fact, "many of the most important definitions and theorems of real number theory are lost" [10, p. 49]. Russell [60] felt compelled to introduce the axiom of reducibility, "by means of which the different orders of a type could be reduced in certain respects to the lowest order of the type." [10, p. 46] A frequent criticism of this axiom today, is that from an extensional point of view it has the effect of reintroducing impredicativity. Soon after its introduction, reducibility was met by stark criticism for its lack of satisfactory justification: "[t]he sole justification for this axiom was the fact that there seemed to be no other way out of this particular difficulty engendered by the ramified theory of types." [10, p. 46] More forcefully Weyl wrote:

> Russell, in order to extricate himself from the affair, causes reason to commit hara-kiri, by postulating the above assertion [the axiom of reducibility] in spite of its lack of support by any evidence. [75, p. 50]

3.2. Weyl's "Das Kontinuum"

A fresh attempt at developing mathematics from a predicative point of view was proposed by Weyl in his book "Das Kontinuum" [73]. Weyl may be seen as contributing in original ways to both the above mentioned lines of research: he put forth a concept of predicative set, that is, a systematic predicative foundation, but also explored the mathematical extent of

---

[2]On ramification see e.g. [31, 32, 36, 38, 42].

predicativity, developing (a portion of) analysis on the basis of this predicative concept of set.

Weyl was fully aware of the difficulties introduced by ramification for
the development of mathematics, and severely criticized the axiom of reducibility, as also witnessed by the quotation above; he thus refrained from
both. As to ramification, in Section 6 of "Das Kontinuum", Weyl does
consider this possibility, but concludes: "A 'hierarchical' version of analysis is artificial and useless. It loses sight of its proper object, i.e. number
[...]. Clearly, we must take the other path [...] *to abide the narrower iteration procedure.*" [73, p. 32] In the following, I present the main characters
of Weyl's approach and clarify what is the "narrower iteration procedure"
mentioned in the above quotation.[3]

In setting up his predicative analysis, Weyl's starting point are the natural numbers, which are assumed as given together with the principle of
mathematical induction.[4] Weyl refers back, approvingly, to Poincaré, who
had strongly criticised any attempts to justify the principle of induction as
viciously circular.[5] As for Poincaré, also for Weyl we have no option but
to take the natural numbers with the principle of induction at the start:
*"the idea of iteration, i.e., of the sequence of the natural numbers, is an
ultimate foundation of mathematical thought, which can not be further reduced"* [73, p. 48]. In fact, Weyl clearly highlights the fundamental role
that the principle of induction has for the natural numbers, as it is this
principle that allows us to characterize uniquely each natural number in
terms of its position in the number sequence.

While the natural numbers are assumed as "given", Weyl imposes restrictions, motivated by predicativity concerns, at the next level of idealiza-
tion beyond the natural numbers: the continuum. As the real numbers can
be represented by sets or sequences of rational numbers, and the rational
numbers, in turn, by natural numbers, the question underlying Weyl's predicative approach may be phrased as follows: which sets of natural numbers
can be justified predicatively? From an impredicative perspective, Weyl's
predicative analysis may be seen as introducing predicative restrictions on
the *powerset of the natural numbers*, which is an emblematic manifestation
of the concept of "arbitrary" set underpinning ZFC [3, 27, 72]. For Weyl,
an arbitrary set is "a 'gathering' brought together by infinitely many individual arbitrary acts of selection, assembled and then surveyed as a whole

---

[3]See also [22, 24, 46], and [17, 25] for overviews.
[4]I shall also write "induction" for "mathematical induction".
[5]See e.g. [52].

by consciousness" and, as such, it is "nonsensical" [73, p. 23]. Weyl proposes an alternative concept of set, one which can be portrayed as if it were produced by a step-by-step process from the safety of the natural numbers by application of well-understood logical operations. He is very clear that only by reforming the concept of set so to anchor it to the safe domain of the natural numbers, can we be confident that the edifice of mathematics stands on "pillars of enduring strength". For example, Weyl notes the impredicativity of the LUB and writes:

> But the more distinctly the logical fabric of analysis is brought to givenness and the more deeply and completely the glance of consciousness penetrates it, the clearer it becomes that, given the current approach to foundational matters, every cell (so to speak) of this mighty organism is permeated by the poison of contradiction and that a thorough revision is necessary to remedy the situation. [73, p. 32]

The process of "formation" of predicative subsets of the natural numbers suggested by Weyl may be concisely summarised as follows: we start from the natural numbers with full mathematical induction and use the ordinary logical operations to form judgements expressing properties of (and relations between) natural numbers. Crucially, *quantification is restricted to the domain of natural numbers.*[6] Sets are then *extensions* of properties (and relations) expressed by such judgements, modulo extensionality. More precisely, a set is the collection of all and only those objects which satisfy a property affirmed by one such judgement, and sets are identified if and only if they have the same elements. The restriction to quantification on the natural numbers witnesses Weyl's fundamental choice of following the "narrower iteration procedure". One could, in fact, imagine that once a set has been justified in the above manner, it would be legitimate to quantify over it, therefore using it to define new predicatively justified sets, that can again act as domains of quantification, and so on. As mentioned above, Weyl does consider this possibility but suggests that for the purpose of developing real mathematics, it is unnatural. He therefore explores how far can we go by restricting quantification to the natural numbers.

In modern terminology, in the language of second order arithmetic, Weyl introduced restrictions on how we form subsets of the natural numbers, that, in practice, justify only applications of the comprehension schema to

---

[6]Weyl in [73] takes a more general approach, by proposing a concept of set which is built on any "definite" basic category of object. The natural numbers are a paradigmatic example of basic category, and the fundamental one for analysis.

arithmetical formulas, that is, those formulas which do not quantify over sets (but may quantify over natural numbers). In this way one justifies sets of the form $\{x \mid \varphi(x)\}$ only if φ does not contain set quantifiers. This restriction prevents vicious-circular definitions of subsets of the natural numbers: the restriction to number quantifiers in the comprehension principle does not permit the definition of a new set by quantification over a collection of sets to which the definiendum belongs.

Weyl's fundamental realisation was that adopting this very restrictive concept of set does not impair the development of large parts of 19th century analysis. In fact, Solomon Feferman [22] has extended Weyl's work to include large portions of contemporary analysis, as further discussed below.

4. The re-emergence of predicativity

The interest in predicativity sharply declined soon after the publication of Weyl's book for a number of reasons, among which, for example, Weyl's brief conversion to Brouwer's intuitionism [74].[1] Peharps the most significant circumstance that determined predicativity's loss of appeal was the rapid accreditation of impredicative set theory as standard foundation.[2] In addition to historical and sociological reasons, the widespread neglect for predicativity is also due to two kinds of objections that were quickly leveled against it: one of a mathematical and one of a philosophical nature. The mathematical objection was that adherence to predicativity is unnecessary to avoid the paradoxes that afflict the foundations of mathematics. This followed a proposed distinction between set-theoretic and semantic paradoxes [48, 55]. Set-theoretic paradoxes are those which directly relate to the concept of set, and include e.g. Russell's and Burali-Forti paradoxes; semantic paradoxes involve linguistic or semantic notions, and include e.g. Richard's paradox and the Liar paradox. It was then suggested that only the first kind of paradoxes constitutes a serious threat to the foundations of mathematics [48, 55] (see also [10]). Furthermore, Chwistek and Ramsey [12, 55] observed that Russell's ramified type theory could be simplified by introducing type restrictions without also imposing ramification. The resulting formalism goes under the name of *simple type theory* and its formulation was subsequently simplified by Church [11], among

---

[1] Note that Weyl's classical approach in "Das Kontinuum" had lasting influence on Lorenzen [39].

[2] See also [25] for additional historical and sociological reasons for the decline of interest in predicativity.

others. Simple type theory does not eliminate all impredicativity, but seems sufficient to block all known set-theoretic paradoxes. This observation was then taken to undermine what is typically seen as the principal motivation for predicativity: to avoid the paradoxes. It certainly intimates more care in construing an argument for predicativism, the philosophical position according to which only predicative mathematics is justified.

As to the philosophical objection, a frequent interpretation of impredicativity is that it becomes a genuine difficulty only if the role of definitions is to produce or create their definiendum, rather than select it from a previously given domain of mathematical entities. In this context, one often mentions Ramsey's example: "the tallest man in this room" [55].[3] This is an impredicative definition, but its impredicativity is harmless. Its innocent nature is often explained by claiming that its purpose is *singling out* rather than *creating* a particular individual. Ramsey's objection to predicativity is so summarised, for critical purposes, by [10, p. 50]: "That we men are finite beings who cannot name individually each of infinitely many properties is an empirical fact that has nothing to do with logic." In other terms, it is often argued that it is only from a throughly constructive perspective that impredicativity is problematic; however, a realist attitude to mathematical entities grants the legitimacy of impredicativity. This interpretation of predicativity is very common, so much so that it might be termed the "received view" on predicativity. Its prevalence is probably also due to Gödel's well-known analysis of predicativity [29], that is often read along these lines. A discussion of these objections is beyond the aims of this note. As to the philosophical objection, I argue elsewhere that both the legitimacy of impredicativity from a realist perspective and its illegitimacy from a constructive perspective require further scrutiny.

4.1. *A new stage for predicativity*

Notwithstanding these objections, renewed interest in predicativity emerged from the 1950's, when fresh attempts were made to obtain a clearer demarcation of the boundary between predicative and impredicative mathematics, by making use of state-of-the-art logical machinery. The literature from the 1950's and 1960's witnesses the complexity of the task of clarifying the limit of predicativity, which saw the involvement of a number of prominent logicians, as Feferman, Gandy, Kleene, Kreisel, Lorenzen, Myhill, Schütte, Spector, and Wang. In the following, I sketch the most

---

[3]See also the discussion of this example in [10].

salient characters of this new phase for predicativity.[4] As argued in [17], a very striking aspect of the logical analysis of predicativity is the radical change in purpose, compared with the first discussions on predicativity. The mathematical logicians who approached predicativity in the second half of the last century typically aimed not at rectifying the foundations of mathematics, but at further clarifying the notion of predicativity and distinguishing it from impredicativity. The principal aim was to establish the limit of predicativity, clarifying how far predicativity goes. This was approached by two distinct, but related, strategies. A first objective was to devise formal instruments for capturing the notion of predicativity and establish their theoretical limit. A second purpose was to clarify which parts of contemporary mathematics can be developed without any appeal to impredicativity, by systematically analysing mathematical theorems from a logical perspective.

The efforts to establish the limit of predicativity culminated in a fundamental chapter in proof theory. Here Russell's original idea of ramification had a crucial role, as it lead to the definition of a transfinite progression of systems of ramified second order arithmetic indexed by ordinals. As in [73], the natural numbers were assumed as starting point, and constraints were introduced at the next level, restricting the powerset of the natural numbers. However, there was also a departure from Weyl's "narrower iteration" procedure, as the aim was now to assess *how far* can we go from a predicative perspective: the ascent to sets beyond the arithmetically definable ones was therefore permitted, as long as it could be predicatively justified. The principal difficulty was, however, devising suitable criteria for justifying predicatively the extension beyond Weyl's original approach. Here a notion of "predicative ordinal" played a pivotal role, as ascent along the progression of ramified systems was admitted only along predicative ordinals. The intuition underlying the notion of predicative ordinal is that this is an ordinal which can be recognized by exclusive appeal to notions that have already been secured. Roughly, one introduces a "boot-strapping" condition, requiring that we progress up along the hierarchy of ramified systems to a stage α only if α has already been recognized as predicative at a previous stage of the hierarchy, i.e. if α has been proved to be an ordinal at a previous stage of the hierarchy. Following a proposal by Kreisel [34], Feferman and Schütte (independently) determined the so-called **limit of predicativity** in terms of the first non-predicative ordinal, known as Γ_0 [21, 62, 63].[5] The

---

[4]See e.g. [17, 25] for surveys and further references.
[5]See e.g. [49, 64] for more details.

claim was that theories whose proof theoretic strength is below Γ_0 could be predicatively justified.

The second component of the logical analysis of predicativity was a detailed logical investigation of the underlying assumptions which are implicit in ordinary mathematics, with the purpose of elucidating the role of impredicativity in ordinary mathematics. The expression "ordinary mathematics" refers to mainstream mathematics, that is, those areas of mathematics which make no essential use of the concepts and methods of abstract set theory and, in particular, the theory of uncountable cardinal numbers. Weyl's pioneering work in "Das Kontinuum" constituted fundamental reference, especially for Feferman's investigations [22, 25]. Feferman [22] has carefully analysed Weyl's text and proposed a system, W (for Weyl), which can be used to codify the analysis in "Das Kontinuum". He has verified that not only Weyl's analysis, but large portions of contemporary analysis can be carried out on the basis of system W.[6] The significant point is that W is very weak proof-theoretically, as it is no stronger than Peano Arithmetic (and hence well within the Γ_0 limit). Another important source of insight on the mathematical extent of predicativity are the findings obtained within Friedman and Simpson's programme of Reverse Mathematics [68], which also produced important independence results.[7] This research overall confirms that if we confine our attention to ordinary mathematics, then impredicativity is largely unnecessary.[8] The situation bears similarity to the one we encounter in constructive mathematics. Brouwer and, subsequently, Bishop [4] realised that notwithstanding the fact that the principle of excluded middle is extensively used in ordinary mathematics, we can re-develop a large body of interesting and useful mathematics without any appeal to this logical principle. Here it is tempting to draw the moral that at least for a substantial portion of ordinary mathematics, the apparent necessity of certain features of ordinary mathematics, like the use of the principle of excluded middle or impredicativity, turns out to be a by-product of the context in which it is developed, and might also depend on the specific formulation of their statements. In particular, in the constructive case a careful choice of definitions allows for a constructive redevelopment of parts of mathematics that are non-constructive. In the case at hand, many instances of prima facie impredicativity become amenable to predicative

---

[6]See also [26].
[7]In this context, Weyl's predicative analysis can be recast within the system ACA_0.
[8]See [66] for details and [67] for examples of mathematical theorems that lie beyond predicativity.

treatment once we work within sufficiently weak systems.[9] In addition, like in constructive mathematics, we need to rely on *individual case studies* for our findings, so that any general conclusion can only be achieved on the basis of a thorough investigation of the mathematical practice.

5. Plurality

The logical analysis of predicativity aimed at determining the limits and the extent of a notion of predicativity *given the natural numbers*. Here one takes an approach to predicativity analogous to Weyl's, in that the natural numbers with full (i.e. unrestricted) induction are assumed at the start, and appropriate predicatively motivated constraints are imposed on the formation of subsets of the natural numbers. In the case of predicativity à la Weyl, an appeal to the natural numbers together with simple logical operations enables for the articulation of a predicative concept of set which offers a secure foundation for mathematics, as its certainty is grounded on the reliability of the natural numbers and those simple logical operations. In particular, by restricting quantification to the natural numbers, we do not appeal to dubious sets, definable only by vicious circles. The legitimate subsets of the natural numbers may be portrayed as if they were the result of a bottom-up procedure, or the application of an arithmetical rule which, starting from the natural numbers, prescribes step-by-step which elements belong to the given set (and when two such elements are equal). The comparison with the full powerset of the natural numbers is instructive, as in that case one collects together *all* the subsets of the natural numbers, irrespective of whether we can offer, even in principle, a finitary rule of formation. The arithmetical sets are therefore a particularly clear exemplification of a predicative and constructive notion of set grounded on the natural numbers (and the first order logical operations).

The assumption of the natural numbers with full induction as starting point for predicativity has, however, not gone unchallenged. Different forms of predicativity have been proposed in the relevant literature. For example, systems as Martin-Löf type theory and Aczel and Myhill constructive set theory embody a particularly generous notion of predicativity, and combine

---

[9]There has been extensive cross-fertilisation between reverse and constructive mathematics. Simpson, however, also emphasizes a difference with constructive mathematics, in that the aim in reverse mathematics is "to draw out the set existence assumptions which are implicit in the ordinary mathematical theorems *as they stand*". Bishop's goal, according to Simpson, is instead "to replace ordinary mathematical theorems by their "constructive" counterparts." [66, p. 137]

it with the rejection of the principle of excluded middle. The thought here is that compliance with intuitionistic logic makes predicatively legitimate constructions, as generalised inductive definitions, which are problematic from a classical predicativist perspective.[1] Another variant of predicativity instead preserves the adherence to classical logic, but restricts rather than extends the domain of predicatively justifiable mathematics. This variant of predicativity has been called *strict predicativity* [47] and arises from a criticism of the natural numbers from a predicativist perspective. An analysis of these variants of predicativity and their respective relations is beyond the remits of this note. Here I only briefly discuss strict predicativity with the aim of placing predicativity given the natural numbers under perspective.

Edward Nelson [44] has proposed a form of predicative arithmetic which imposes severe restrictions on the principle of induction. Nelson's principal motivation for his predicative arithmetic is the complaint that the natural numbers hide a form of circularity. Charles Parsons has also argued that impredicativity already makes its way at the level of the natural numbers, and that induction is the culprit [45, 47]. Nelson and Parsons' criticism of induction bears similarities with the objection to the impredicativity of the logicist definition of natural numbers which was discussed above, since it highlights a circularity in the definition of natural number. Nelson's complaint with induction is put forth in a dense paragraph at the beginning of [44], whose interpretation is complex. In the following, I propose a possible way of arguing for the impredicativity of induction which is inspired by [44] and [45].

The logicist definition of natural number is impredicative as it has a second order quantifier at the start. It is therefore problematic from a predicativist perspective. A more promising definition of natural number is its inductive specification: a natural number is either 0, or the successor of a natural number, and nothing else. Here the closure condition is given by the induction principle. Induction is expressed as follows in Peano Arithmetic:

$$[\varphi(0) \wedge \forall x(\varphi(x) \to \varphi(Suc(x)))] \to \forall x\varphi(x),$$

where φ is an *arbitrary* formula in the language of PA, and $Suc(x)$ is the successor of x. First of all, one claims that the principle of induction plays a fundamental role in clarifying what are the natural numbers. This point was already stressed by Poincaré and Weyl: the principle of induction is

---

[1]The proof-theoretic strength of theories of inductive definitions is well above the Kreisel-Feferman-Schütte limit of predicativity mentioned above [6]. See [16, 21, 45] for discussion.

a crucial component of the natural number structure. One could also say that induction is required to determine the extension of the natural number concept (i.e. what falls under that concept). Secondly, one observes that in the induction principle, the induction formula, φ, is arbitrary, so that it might be instantiated by a formula with unrestricted (number) quantifiers. Now a difficulty arises as follows: in order to make sense of an instance of induction with an unrestricted universal quantifier ranging over the natural numbers, we would seem to require a prior clarification of what falls under the natural number concept. That is, it would seem that we need first to have a specification of what belongs to the natural number concept before we can make sense of a statement of the form "for all natural numbers ...". However, if induction plays an essential role in determining the extension of the natural number concept, we end up with a vicious circle: we need induction to clarify what belongs to the natural numbers, but we need to already know what the natural numbers are, in order to make sense of crucial instances of induction.

Nelson's rejection of the induction principle on the grounds of circularity leads him to justify only weak subsystems of Peano Arithmetic. Interpretability in a fragment of primitive recursive arithmetic, Robinson's system Q, seems to be the main criteria Nelson adopts for assessing the predicativity of a formal system. Nelson predicative arithmetic therefore lays within the realm of bounded arithmetic [7].[2]

6. Conclusion

Strict predicativity is philosophically interesting as it suggests that from a thoroughly predicativist point of view already the theory of Peano Arithmetic, with its unrestricted induction, is problematic. In other terms, the above argument, if granted, would imply that if we were to appeal to predicativity as a way of avoiding all forms of vicious circularity, we would have to impose restrictions already on the principle of induction. This suggests that a predicativist given the natural numbers would have to offer suitable argumentation for the exemption of the natural numbers from predicativity constraints.[1]

The comparison between strict predicativity and predicativity given the natural numbers is particularly significant also from a perspective that does not attempt to argue for predicativism. From this "external" perspective,

---

[2]See [71] for examples of systems which would seem to conform with Nelson's views.
[1] [73] offers a possible strategy.

predicativity may be viewed as a notion that may be applied to a number of different contexts.[2] In general, predicativity imposes a requirement on domains of quantification: to be specifiable without vicious circularity. When spelling out the notion of predicative domain of quantification, we might choose, however, different initial assumptions. In the case of strict predicativity we only take a bare minimum. This could be expressed in terms of an initial element, say 0 and a successor operation. We then see how far we can go in building further mathematical constructions from these two initial "ingredients" and the usual first order logical operations. In the case of predicativity given the natural numbers, instead, we take as starting point not only 0 and successor, but also the principle of mathematical induction (unrestricted) and start building new sets from that. In all cases, once a certain initial "base" has been granted (or agreed on), the aim is to proceed without incurring in vicious circularity as far as we can, in predicatively justified ways. The crucial point is that with respect to a certain initial "base", sets are defined explicitly by application of a fixed set of simple operations. We could also say that sets which are so specified are predicative, relative to a certain "base".[3] When it is so formulated, predicativity may become a useful instrument in the philosophy of mathematics, as it may help clarify the assumptions underlying a number of philosophical positions.

I conclude with a remark. During the Autumn School, Professor Schwichtenberg organised a discussion on predicativity and raised a number of stimulating questions, which also go in the direction of broadening standard interpretations of predicativity. The notion of predicativity is customarily seen as pertaining primarily to the concept of set. In addition, the concept of set is often taken as prior to that of function, due to our familiarity with theories as ZFC, in which functions are codified as particular sets: graphs. However, it would be important to assess whether a comparison with other foundational contexts in which functions, possibly partial, are primitive, would help furthering our understanding of predicativity.[4] In fact, the early discussions on predicativity were typically framed within complex contexts, as the underlying systems displayed forms of intensionality or assumed a primitive concept of function. It would seem that a thorough discussion of predicativity would benefit from a deeper analysis of issues of identity and a comparison between different approaches to the notion of function, including partiality.

---

[2] See also [16, 23, 25].
[3] See also [15].
[4] See e.g. [65]. See also the notion of predicativity in Martin-Löf type theory.

Acknowledgements

This note arises from a course I gave at the Autumn school "Proof and Computation", 3 to 8 October 2016 in Fischbachau, Germany. I would like to thank the organisers of the school, Klaus Mainzer, Peter Schuster, and Helmut Schwichtenberg for their kind invitation and for stimulating discussions. The material in this article grew out of my PhD thesis [16]. I gratefully acknowledge funding by the School of Philosophy, Religion and History of Science of the University of Leeds, and the European Research Council under the European Union's Seventh Framework Programme (FP/2007-2013) / ERC Grant Agreement no. 312938, led by Robert Williams. Thanks to Robert Williams for many helpful discussions on predicativity.

References

[1] P. Aczel. The type theoretic interpretation of constructive set theory. In eds. A. MacIntyre, L. Pacholski, and J. Paris, *Logic Colloquium '77*, pp. 55–66, North-Holland, Amsterdam-New York (1978).

[2] P. Benacerraf and H. Putnam, *Philosophy of Mathematics: Selected Readings*. Cambridge University Press (1983).

[3] P. Bernays, Sur the platonisme dans les mathématiques, *L'Enseignement mathématique*. (34), 52–69 (1935). Translated in [2] with the title: On Platonism in Mathematics. (Page references are to the reprinting).

[4] E. Bishop, *Foundations of constructive analysis*. McGraw-Hill, New York (1967).

[5] D. S. Bridges and E. Palmgren. Constructive mathematics. In ed. E. N. Zalta, *The Stanford Encyclopedia of Philosophy* (2013), winter 2013 edn.

[6] W. Buchholz, S. Feferman, W. Pohlers, and W. Sieg, *Iterated inductive definitions and subsystems of analysis*. Springer, Berlin (1981).

[7] S. Buss, *Bounded Arithmetic*. Studies in Proof Theory Lecture Notes, Bibliopolis, Naples (1986).

[8] A. Cantini, Una teoria della predicatività secondo Poincaré, *Rivista di Filosofia*. **72**, 32–50 (1981).

[9] A. Cantini. Paradoxes, self-reference and truth in the 20th century. In ed. D. Gabbay, *The Handbook of the History of Logic*, pp. 5–875. Elsevier (2009).

[10] R. Carnap, Die Logizistische Grundlegung der Mathematik, *Erkenntnis*. **2** (1), 91–105 (1931). Translated in [2]. (Page references are to the reprinting).

[11] A. Church, A Formulation of the Simple Theory of Types, *Journal of Symbolic Logic*. **5** (1940).

[12] L. Chwistek, Über die Antinomien der Prinzipien der Mathematik, *Mathematische Zeitschrift*. **14**, 236–43 (1922).

[13] T. Coquand and G. Huet. The calculus of constructions. Technical Report RR-0530, INRIA (May, 1986).

[14] L. Crosilla. Set theory: Constructive and intuitionistic ZF. In ed. E. N. Zalta, *The Stanford Encyclopedia of Philosophy* (2014).

[15] L. Crosilla. Error and predicativity. In eds. A. Beckmann, V. Mitrana, and M. Soskova, *Evolving Computability*, vol. 9136, *Lecture Notes in Computer Science*, pp. 13–22. Springer International Publishing (2015).

[16] L. Crosilla. *Constructivity and Predicativity: Philosophical foundations*. PhD thesis, School of Philosophy, Religion and the History of Science, University of Leeds (2016).

[17] L. Crosilla. Predicativity and Feferman. In eds. G. Jäger and W. Sieg, *Feferman on Foundations: Logic, Mathematics, Philosophy*, Outstanding Contributions to Logic, Springer (2018).

[18] M. Dummett, *Elements of Intuitionism*. Oxford University Press, Oxford (1977).

[19] M. Dummett, *Frege: Philosophy of Mathematics*. Cambridge MA, Harvard University Press (1991).

[20] W. B. Ewald, *From Kant to Hilbert: A Source Book in the Foundations of Mathematics*. Oxford University Press (1996).

[21] S. Feferman, Systems of predicative analysis, *Journal of Symbolic Logic*. **29**, 1–30 (1964).

[22] S. Feferman. Weyl vindicated: Das Kontinuum seventy years later. In eds. C. Cellucci and G. Sambin, *Temi e prospettive della logica e della scienza contemporanee*, pp. 59–93 (1988).

[23] S. Feferman and G. Hellman, Predicative foundations of arithmetic, *Journal of Philosophical Logic*. **22**, 1–17 (1995).

[24] S. Feferman. The significance of Hermann Weyl's *Das Kontinuum*. In eds. V. Hendricks, S. A. Pedersen, and K. F. Jørgense, *Proof Theory*, Dordrecht, Kluwer (2000).

[25] S. Feferman. Predicativity. In ed. S. Shapiro, *Handbook of the Philosophy of Mathematics and Logic*, Oxford University Press, Oxford (2005).

[26] S. Feferman. Why a little bit goes a long way: predicative foundations of analysis. Unpublished notes dating from 1977–1981, with a new introduction. Retrieved from the address: https://math.stanford.edu/~feferman/papers.html (2013).

[27] J. Ferreirós, On arbitrary sets and ZFC, *The Bulletin of Symbolic Logic*. **17** (3), 361–393 (2011).

[28] T. Fruchart and G. Longo. Carnap's remarks on Impredicative Definitions and the Genericity Theorem. In eds. A. Cantini, E. Casari, and P. Minari, *Logic and Foundations of Mathematics: Selected Contributed Papers of the Tenth International Congress of Logic, Methodology and Philosophy of Science, Florence, August 1995*, Kluwer (1997).

[29] K. Gödel. Russell's mathematical logic. In ed. P. A. Schlipp, *The philosophy of Bertrand Russell*, pp. 123–153, Northwestern University, Evanston and Chicago (1944). Reprinted in [2]. (Page references are to the reprinting).

[30] J. van Heijenoort, *From Frege to Gödel*. Cambridge, Harvard University Press (1967).

[31] H. T. Hodes, Why ramify?, *Notre Dame J. Formal Logic.* **56** (2), 379–415 (2015). doi: 10.1215/00294527-2864352.

[32] F. Kamareddine, T. Laan, and R. Nederpelt, Types in logic and mathematics before 1940, *Bulletin of Symbolic Logic.* **8** (2), 185–245 (2002).

[33] S. C. Kleene, On the interpretation of intuitionistic number theory, *Journal of Symbolic Logic.* **4** (10), 109–124 (1945).

[34] G. Kreisel. Ordinal logics and the characterization of informal concepts of proof. In *Proceedings of the International Congress of Mathematicians (August 1958)*, pp. 289–299, Gauthier–Villars, Paris (1958).

[35] G. Kreisel, La prédicativité, *Bulletin de la Societé Mathématique de France.* **88**, 371–391 (1960).

[36] G. Link. Formal Discourse in Russell: From Metaphysics to Philosophical Logic. In ed. G. Link, *Formalism and Beyond: On the Nature of Mathematical Discourse*, pp. 119–182, De Gruyter (2014).

[37] B. Linsky, Propositional functions and universals in principia mathematica, *Australasian Journal of Philosophy.* **66** (4), 447–460 (1988).

[38] B. Linsky, *Russell's Metaphysical Logic.* Stanford: CSLI (1999).

[39] P. Lorenzen, *Einführung in die Operative Logik und Mathematik.* Berlin, Springer-Verlag (1955).

[40] P. Martin-Löf. An intuitionistic theory of types: predicative part. In eds. H. E. Rose and J. C. Shepherdson, *Logic Colloquium 1973*, North–Holland, Amsterdam (1975).

[41] P. Martin-Löf, *Intuitionistic Type Theory.* Bibliopolis, Naples (1984).

[42] J. Myhill. The undefinability of the set of natural numbers in the ramified Principia. In ed. G. Nakhnikian, *Bertrand Russell's Philosophy*, pp. 19–27. Duckworth, London (1974).

[43] J. Myhill, Constructive set theory, *Journal of Symbolic Logic.* **40**, 347–382 (1975).

[44] E. Nelson, *Predicative arithmetic.* Princeton University Press, Princeton (1986).

[45] C. Parsons. The impredicativity of induction. In ed. M. Detlefsen, *Proof, Logic, and Formalization*, pp. 139–161, Routledge, London (1992).

[46] C. Parsons. Realism and the debate on impredicativity, 1917–1944. In eds. W. Sieg, R. Sommer, and C. Talcott, *Reflections on the Foundations of Mathematics: Essays in Honor of Solomon Feferman*, Association for Symbolic Logic (2002).

[47] C. Parsons, *Mathematical Thought and Its Objects.* Cambridge University Press (2008).

[48] G. Peano, Additione, *Revista de Matematica.* **8**, 143–157 (1902–1906).

[49] W. Pohlers, *Proof Theory: The First Step into Impredicativity.* Universitext, Springer Berlin Heidelberg (2009).

[50] H. Poincaré, Les mathématiques et la logique, *Revue de Métaphysique et Morale.* **1**, 815–835 (1905).

[51] H. Poincaré, Les mathématiques et la logique, *Revue de Métaphysique et de Morale.* **2**, 17–34 (1906).

[52] H. Poincaré, Les mathématiques et la logique, *Revue de Métaphysique et de Morale.* **14**, 294–317 (1906).

[53] H. Poincaré, La logique de l'infini, *Revue de Métaphysique et Morale.* **17**, 461–482 (1909).

[54] H. Poincaré, La logique de l'infini, *Scientia.* **12**, 1–11 (1912).

[55] F. P. Ramsey, Foundations of mathematics, *Proceedings of the London Mathematical Society.* **25** (1926). Reprinted in [56].

[56] F. P. Ramsey, *Foundations of Mathematics and Other Logical Essays.* Routledge and Kegan Paul (1931).

[57] J. Richard, Les principes des mathmatiques et le problme des ensembles, *Revue générale des sciences pures et appliquées.* **16** (12), 541–543 (1905).

[58] B. Russell, On Some Difficulties in the Theory of Transfinite Numbers and Order Types, *Proceedings of the London Mathematical Society.* **4**, 29–53 (1906).

[59] B. Russell, Les paradoxes de la logique, *Revue de métaphysique et de morale.* **14**, 627–650 (1906).

[60] B. Russell, Mathematical logic as based on the theory of types, *American Journal of Mathematics.* **30**, 222–262 (1908).

[61] B. Russell, *Essays in Analysis.* George Braziller, New York (1973). Edited by D. Lackey.

[62] K. Schütte. Predicative well-orderings. In eds. J. Crossley and M. Dummett, *Formal Systems and Recursive Functions*, North–Holland, Amsterdam (1965).

[63] K. Schütte, Eine Grenze für die Beweisbarkeit der Transfiniten Induktion in der verzweigten Typenlogik, *Archiv für mathematische Logik und Grundlagenforschung.* **7**, 45–60 (1965).

[64] K. Schütte, *Proof Theory.* vol. 225, *Grundlehren der mathematischen Wissenschaften* (1977). Translated by J.N. Crossley.

[65] H. Schwichtenberg and S. S. Wainer, *Proofs and Computations*, 1st edn. Cambridge University Press, New York, NY, USA (2012).

[66] S. G. Simpson, *Subsystems of Second Order Arithmetic.* Perspectives in Mathematical Logic, Springer-Verlag 1999).

[67] S. G. Simpson. Predicativity: the outer limits. In *Reflections on the foundations of mathematics (Stanford, CA, 1998)*, vol. 15, *Lect. Notes Log.*, pp. 130–136. Assoc. Symbol. Logic, Urbana, IL (2002).

[68] S. G. Simpson, *Subsystems of second order arithmetic.* Perspectives in Logic, Cambridge University Press (2009). 2nd edition.

[69] H. Stein. Logos, logic and logistiké. In eds. W. Asprey and P. Kitcher, *History and Philosophy of Modern Mathematics*, pp. 238–59, Minneapolis: University of Minnesota (1988).

[70] A. S. Troelstra and D. van Dalen, *Constructivism in Mathematics: an Introduction.* vol. I and II, North-Holland, Amsterdam (1988).

[71] S. S. Wainer. Provable (and unprovable) computability. This volume.

[72] H. Wang, The formalization of mathematics, *The Journal of Symbolic Logic.* **19** (4), pp. 241–266 (1954). ISSN 00224812.

[73] H. Weyl, *Das Kontinuum. Kritische Untersuchungen über die Grundlagen der Analysis.* Veit, Leipzig (1918).

[74] H. Weyl, Über die neue Grundlagenkrise der Mathematik, *Mathematische Zeitschrift.* **10**, 39–79 (1921).

Laura Crosilla

[75] H. Weyl, *Philosophy of Mathematics and Natural Science.* Princeton University Press (1949). An expanded English version of Philosophie der Mathematik und Naturwissenschaft, Mńchen, Leibniz Verlag, 1927.

[76] A. N. Whitehead and B. Russell, *Principia Mathematica, 3 Vols.* vol. 1, Cambridge: Cambridge University Press (1910, 1912, 1913). Second edition, 1925 (Vol 1), 1927 (Vols 2, 3); abridged as Principia Mathematica to *56, Cambridge: Cambridge University Press, 1962.

[77] E. Zermelo, Neuer Beweis für die Möglichkeit einer Wohlordnung, *Mathematische Annalen.* **65**, 107–128 (1908). Translated in [30], pages 183–198. (References are to the English translation).

Chapter 4

Constructive Functional Analysis: An Introduction

Hajime Ishihara

School of Information Science
Japan Advanced Institute of Science and Technology
Nomi, Ishikawa 923-1292, Japan
ishihara@jaist.ac.jp

Contents

1. Introduction

Fifty years ago, in the Preface of his book "Foundations of Constructive Analysis" [8], Errett Bishop wrote

> The book has a threefold purpose: first, to present the constructive point of view; second, to show that the constructive program can succeed; third, to lay a foundation for further work. These immediate ends tend to an ultimate goal — to hasten the inevitable day when constructive mathematics will be the accepted norm.

Fifty years after, has constructive mathematics been the accepted norm, or still a religion [24, pp. 161–162]? This article has a threefold purpose: first, to present a piece of constructive propaganda; second, to show that constructive mathematics is a satisfactory alternative with a *firm foundation*; third, to invite young (and old) people to constructive mathematics. For the reader interested in constructive mathematics, we refer him/her to an excellent article by Bauer [6].

Although the book by Bishop changed the landscape of constructive mathematic, it has a long history [64, 1.4] and varieties. One of the important varieties, which greatly influenced Bishop, is Brouwer's intuitionistic mathematics [19], [25], [21], and [16, Chapter 5], and another is the constructive recursive mathematics of the Russian school of Markov [46] and [16, Chapter 3].

In Section 2, we quickly review the *Brouwer-Heyting-Kolmogorov interpretation* of the logical operators and the natural deduction systems for *intuitionistic logic* and classical logic to highlight the difference between these logics.

In Section 3, we present axioms of *Aczel's constructive Zermelo-Fraenkel set theory* **CZF** [4, 5] as a foundation of the material in the following sections and of constructive mathematics. Then we look into major *omniscience principles* which are underivable, and hence unacceptable in constructive mathematics. A theorem is underivable in constructive mathematics if it implies an omniscience principle.

In Section 4, we present a construction of the real numbers and its basic properties, and we see correspondence between omniscience principles and statements on the relations between real numbers. Then we look into the completeness of the reals, and give, as an application of the completeness, the *constructive supremum principle*.

In Section 5, after giving fundamental definitions on a metric space, we look into complete and compact metric spaces; especially, we see the constructive version of the Baire theorem. Then we introduce various continuity notions (including sequential continuity) of a mapping between metric spaces which have little or no classical significance, and show constructive relationship among them.

In Section 6, after giving basic definitions of normed and Banach spaces, and bounded linear mappings, we introduce the notion of a *normable* linear mapping which also has no classical significance, and look into finite-dimensional normed spaces, uniformly convex spaces, and Hilbert spaces. We see that a constructive version of the Riesz theorem for normable linear functionals.

In Section 7, we give a constructive characterisation of the existence of a Minkowski functional and its applications; note that a Minkowski functional classically always exists. Then, after showing that the classical Hahn-Banach theorem derives an omniscience principle, we see a constructive and approximate version of the theorem, given by Bishop, for normable linear functionals on separable spaces. Furthermore, we show, with a help of geometric properties of a Banach space (and a Hilbert space) such as convexity and differentiability of a norm, an exact version of the Hahn-Banach theorem for normable linear functionals on inseparable spaces without invoking Zorn's lemma.

In Section 8, we show that the classical Banach-Steinhaus theorem implies a nonconstructive boundedness principle BD-N, and give a constructive version of the uniform boundedness theorem. Then we constructively look into the open mapping theorem, the closed graph theorem and Banach's inverse mapping theorem, and see that classical Banach's inverse mapping theorem implies BD-N too. We prove a constructive version of Banach's inverse mapping theorem for sequentially continuous linear mappings, and, as its applications, constructive versions of the open mapping theorem and the closed graph theorem.

In Section 9, we show, although every (bounded) operator on a Hilbert space classically has an adjoint operator, the statement implies an omniscience principle. We introduce the notion of a *weakly compact* operator, and show that an operator has an adjoint if and only if it is weakly compact. We see that the notion of a weakly compact operator which also has no classical significance, plays an important role in compactness of the composition of operators; for each compact operator A and operator C, although the composition CA is always compact, the composition AC is compact if C is weakly compact.

Although omniscience principles have been rejected in constructive framework, some proofs in constructive (forward) mathematics have made use of such nonconstructive principles: see Lemma 7 and Theorem 20 below. Many theorems in classical, intuitionistic and constructive recursive mathematics have been classified using such principles within the framework of constructive mathematics, and such activity is now called *constructive reverse mathematics*; see [36] and [38].

2. Intuitionistic logic

2.1. *The BHK-interpretation*

We use the standard language of (many-sorted) first-order predicate logic based on primitive logical operators $\wedge, \vee, \rightarrow, \perp, \forall, \exists$, and introduce the abbreviations $\neg A \equiv A \rightarrow \perp$ and $A \leftrightarrow B \equiv (A \rightarrow B) \wedge (B \rightarrow A)$.

The *Brouwer-Heyting-Kolmogorov (BHK) interpretation* of the logical operators, which tells us what forms proofs of logically compound statement, is the following.

1. A proof of $A \wedge B$ is given by presenting a proof of A and a proof of B.
2. A proof of $A \vee B$ is given by presenting either a proof of A or a proof of B.
3. A proof of $A \rightarrow B$ is a construction which transform any proof of A into a proof of B.
4. Absurdity $\perp$ has no proof.
5. A proof of $\forall x A(x)$ is a construction which transforms any t into a proof of $A(t)$.
6. A proof of $\exists x A(x)$ is given by presenting a t and a proof of $A(t)$.

Example 1. A proof of $\forall x \exists y A(x, y)$ is a construction which transforms any t into a proof of $\exists y A(t, y)$, and a proof of $\exists y A(t, y)$ is given by presenting an s and a proof of $A(t, s)$. Therefore a proof of $\forall x \exists y A(x, y)$ is a construction which *transforms any t into s* and a proof of $A(t, s)$.

The *principle of excluded middle* (PEM):

$$A \vee \neg A,$$

which is generally valid in classical logic, is not a universally valid principle based on the BHK-interpretation, as it means that we have a universal method of obtaining either a proof of A or a proof of $\neg A$ for any proposition A. If we had such a universal method, we could decide the truth of a statement A whose truth has not been decided.

Remark 1. Note also that a proof of $\neg(\neg A \wedge \neg B)$ is *not* a proof of $A \vee B$, and that a proof of $\neg \forall x \neg A(x)$ is *not* a proof of $\exists x A(x)$.

2.2. *Minimal, intuitionistic and classical logic*

In the following, we give a definition of deduction, conclusion and assumptions in the form of simultaneous inductive definition. We shall use $\mathcal{D}$, possibly with a subscript, for arbitrary deduction, and then write

$$\begin{array}{c} \Gamma \\ \mathcal{D} \\ A \end{array}$$

to indicate that $\mathcal{D}$ is deduction with *conclusion* A and *assumptions* Γ.

Definition 1 (Minimal logic).
Basis: For each formula A,

$$A$$

is a deduction with conclusion A and assumptions $\{A\}$.
Induction step: We give the introduction and elimination rules ($\rightarrow$I and $\rightarrow$E) for $\rightarrow$.

1. If

$$\begin{array}{c} \Gamma \\ \mathcal{D} \\ B \end{array}$$

   is a deduction, then

$$\begin{array}{c} \Gamma \\ \mathcal{D} \\ B \\ \hline A \rightarrow B \end{array} \rightarrow\text{I}$$

   is a deduction with conclusion $A \rightarrow B$ and assumptions $\Gamma \setminus \{A\}$. We then write

$$\begin{array}{c} [A] \\ \mathcal{D} \\ B \\ \hline A \rightarrow B \end{array} \rightarrow\text{I}$$

   to indicate that the set $[A]$ of occurrences of an assumption A are canceled.

2. If

$$\begin{array}{cc} \Gamma_1 & \Gamma_2 \\ \mathcal{D}_1 & \mathcal{D}_2 \\ A \to B & A \end{array}$$

are deductions, then

$$\dfrac{\begin{array}{cc} \Gamma_1 & \Gamma_2 \\ \mathcal{D}_1 & \mathcal{D}_2 \\ A \to B & A \end{array}}{B} \ \to\!\mathrm{E}$$

is a deduction with conclusion B and assumptions $\Gamma_1 \cup \Gamma_2$.

For other logical operators, we have the following introduction and elimination rules.

$$\dfrac{\begin{array}{c} [A] \\ \mathcal{D} \\ B \end{array}}{A \to B} \ \to\!\mathrm{I} \qquad\qquad \dfrac{\begin{array}{cc} \mathcal{D}_1 & \mathcal{D}_2 \\ A \to B & A \end{array}}{B} \ \to\!\mathrm{E}$$

$$\dfrac{\begin{array}{cc} \mathcal{D}_1 & \mathcal{D}_2 \\ A & B \end{array}}{A \wedge B} \ \wedge\mathrm{I} \qquad \dfrac{\begin{array}{c} \mathcal{D} \\ A \wedge B \end{array}}{A} \ \wedge\mathrm{E}_r \quad \dfrac{\begin{array}{c} \mathcal{D} \\ A \wedge B \end{array}}{B} \ \wedge\mathrm{E}_l$$

$$\dfrac{\begin{array}{c} \mathcal{D} \\ A \end{array}}{A \vee B} \ \vee\mathrm{I}_r \quad \dfrac{\begin{array}{c} \mathcal{D} \\ B \end{array}}{A \vee B} \ \vee\mathrm{I}_l \quad \dfrac{\begin{array}{ccc} & [A] & [B] \\ \mathcal{D}_1 & \mathcal{D}_2 & \mathcal{D}_3 \\ A \vee B & C & C \end{array}}{C} \ \vee\mathrm{E}$$

$$\dfrac{\begin{array}{c} \mathcal{D} \\ A \end{array}}{\forall y A[x/y]} \ \forall\mathrm{I} \qquad\qquad \dfrac{\begin{array}{c} \mathcal{D} \\ \forall x A \end{array}}{A[x/t]} \ \forall\mathrm{E}$$

$$\dfrac{\begin{array}{c} \mathcal{D} \\ A[x/t] \end{array}}{\exists x A} \ \exists\mathrm{I} \qquad\qquad \dfrac{\begin{array}{cc} & [A] \\ \mathcal{D}_1 & \mathcal{D}_2 \\ \exists y A[x/y] & C \end{array}}{C} \ \exists\mathrm{E}$$

The rules for the quantifiers are subject to the following restrictions.

1. In $\forall\mathrm{E}$ and $\exists\mathrm{I}$, t must be free for x in A.
2. In $\forall\mathrm{I}$, $\mathcal{D}$ must not contain assumptions containing x free, and $y \equiv x$ or $y \notin \mathrm{FV}(A)$.
3. In $\exists\mathrm{E}$, $\mathcal{D}_2$ must not contain assumptions containing x free except A, $x \notin \mathrm{FV}(C)$, and $y \equiv x$ or $y \notin \mathrm{FV}(A)$.

Example 2. The following is a deduction of $\neg\neg(A \to B) \to (\neg\neg A \to \neg\neg B)$.

$$
\cfrac{
\cfrac{
[\neg\neg A] \quad
\cfrac{
\cfrac{
[\neg\neg(A \to B)] \quad
\cfrac{
[\neg B] \quad
\cfrac{
\cfrac{[A \to B] \quad [A]}{B} \to E
}{\bot} \to E
}{\neg(A \to B)} \to I
}{\bot} \to E
}{\neg A} \to I
}{\bot} \to E
}{
\cfrac{
\cfrac{\neg\neg B}{\neg\neg A \to \neg\neg B} \to I
}{\neg\neg(A \to B) \to (\neg\neg A \to \neg\neg B)} \to I
} \to I
$$

Example 3. The following is a deduction of $(A \to B) \wedge (A \to C) \to (A \to B \wedge C)$.

$$
\cfrac{
\cfrac{
\cfrac{
\cfrac{[(A \to B) \wedge (A \to C)]}{A \to B} \wedge E_r \quad [A]
}{B} \to E
\quad
\cfrac{
\cfrac{[(A \to B) \wedge (A \to C)]}{A \to C} \wedge E_l \quad [A]
}{C} \to E
}{B \wedge C} \wedge I
}{
\cfrac{A \to B \wedge C}{(A \to B) \wedge (A \to C) \to (A \to B \wedge C)} \to I
} \to I
$$

Example 4. The following is a deduction of $(A \to C) \wedge (B \to C) \to (A \vee B \to C)$.

$$
\cfrac{
\cfrac{
[A \vee B] \quad
\cfrac{
\cfrac{[(A \to C) \wedge (B \to C)]}{A \to C} \wedge E_r \quad [A]
}{C} \to E
\quad
\cfrac{
\cfrac{[(A \to C) \wedge (B \to C)]}{B \to C} \wedge E_l \quad [B]
}{C} \to E
}{C} \vee E
}{
\cfrac{A \vee B \to C}{(A \to C) \wedge (B \to C) \to (A \vee B \to C)} \to I
} \to I
$$

Example 5. The following is a deduction of $(A \to \forall x B) \to \forall x (A \to B)$.

$$
\cfrac{
\cfrac{
\cfrac{
\cfrac{
\cfrac{[A \to \forall x B] \quad [A]}{\forall x B} \to E
}{B} \forall E
}{A \to B} \to I
}{\forall x (A \to B)} \forall I
}{(A \to \forall x B) \to \forall x (A \to B)} \to I
$$

where $x \notin FV(A)$.

Example 6. The following is a deduction of $\exists x(A \to B) \to (A \to \exists xB)$.

$$
\cfrac{
  [\exists x(A \to B)] \qquad
  \cfrac{
    \cfrac{
      \cfrac{[A \to B] \quad [A]}{B} \to E
    }{\exists xB} \exists I
  }{}
}{}
$$

$$
\cfrac{\cfrac{[\exists x(A \to B)] \qquad \cfrac{\cfrac{[A \to B] \quad [A]}{B} \to E}{\exists xB} \exists I}{\cfrac{\exists xB}{\cfrac{A \to \exists xB}{\exists x(A \to B) \to (A \to \exists xB)} \to I} \to I} \exists E}{}
$$

where $x \notin \mathrm{FV}(A)$.

Definition 2 (Intuitionistic logic). Intuitionistic logic is obtained from minimal logic by adding the *intuitionistic absurdity rule* (*ex falso quodlibet*). If

$$
\begin{array}{c}
\Gamma \\
\mathcal{D} \\
\bot
\end{array}
$$

is a deduction, then

$$
\begin{array}{c}
\Gamma \\
\mathcal{D} \\
\dfrac{\bot}{A} \bot_i
\end{array}
$$

is a deduction with conclusion A and assumptions Γ.

Example 7. The following is a deduction of $(\neg\neg A \to \neg\neg B) \to \neg\neg(A \to B)$ in intuitionistic logic.

$$
\cfrac{
\cfrac{[\neg\neg A \to \neg\neg B] \qquad \cfrac{\cfrac{[\neg(A \to B)] \quad \cfrac{\cfrac{\cfrac{[\neg A] \quad [A]}{\bot} \to E}{B} \bot_i}{A \to B} \to I}{\bot} \to E}{\neg\neg A} \to I}{\neg\neg B} \to E \qquad \cfrac{[\neg(A \to B)] \quad \cfrac{[B]}{A \to B} \to I}{\cfrac{\bot}{\neg B} \to I} \to E}{
\cfrac{\cfrac{\bot}{\neg\neg(A \to B)} \to I}{(\neg\neg A \to \neg\neg B) \to \neg\neg(A \to B)} \to I}
}
$$

Example 8. The following is a deduction of $A \vee B \to (\neg A \to B)$ in intuitionistic logic.

$$
\cfrac{[A \vee B] \quad \cfrac{\cfrac{\cfrac{[\neg A] \quad [A]}{\bot} \to E}{B} \bot_i \quad [B]}{\cfrac{B}{\cfrac{\neg A \to B}{A \vee B \to (\neg A \to B)} \to I}} \vee E}{}
$$

Definition 3 (Classical logic). Classical logic is obtained from intuitionistic logic by strengthening the absurdity rule to the *classical absurdity rule* (*reductio ad absurdum*).
If

$$
\begin{array}{c} \Gamma \\ \mathcal{D} \\ \bot \end{array}
$$

is a deduction, then

$$
\cfrac{\begin{array}{c} \Gamma \\ \mathcal{D} \\ \bot \end{array}}{A} \bot_c
$$

is a deduction with conclusion A and assumption $\Gamma \setminus \{\neg A\}$.

Example 9 (The double negation elimination DNE).
The following is a deduction of $\neg\neg A \to A$ in classical logic.

$$
\cfrac{\cfrac{\cfrac{[\neg\neg A] \quad [\neg A]}{\bot} \to E}{A} \bot_c}{\neg\neg A \to A} \to I
$$

Example 10 (The principle of excluded middle PEM).
The following is a deduction of $A \vee \neg A$ in classical logic.

$$
\cfrac{[\neg(A \vee \neg A)] \quad \cfrac{\cfrac{[\neg(A \vee \neg A)] \quad \cfrac{\cfrac{[A]}{A \vee \neg A} \vee I_r}{}}{\cfrac{\bot}{\neg A} \to I}}{A \vee \neg A} \vee I_l}{A \vee \neg A} \bot_c
$$

Example 11 (De Morgan's law DML**).**

The following is a deduction of $\neg(A \wedge B) \to \neg A \vee \neg B$ in classical logic.

$$
\cfrac{
  [\neg(\neg A \vee \neg B)] \qquad
  \cfrac{
    \cfrac{
      \cfrac{
        [\neg(\neg A \vee \neg B)] \qquad
        \cfrac{
          \cfrac{
            [\neg(A \wedge B)] \quad
            \cfrac{[A] \quad [B]}{A \wedge B} \, \wedge\mathrm{I}
          }{\bot} \to\mathrm{E}
        }{
          \cfrac{\neg A}{\neg A \vee \neg B} \to\mathrm{I} \;\; \vee\mathrm{I}_r
        }
      }{\bot} \to\mathrm{E}
    }{
      \cfrac{\neg B}{\neg A \vee \neg B} \to\mathrm{I} \;\; \vee\mathrm{I}_l
    }
  }{\bot} \to\mathrm{E}
}{
  \cfrac{\neg A \vee \neg B}{\neg(A \wedge B) \to \neg A \vee \neg B} \; \bot_c \;\; \to\mathrm{I}
}
$$

Remark 2. Note that the rule $\bot_c$ derives A by deducing absurdity ($\bot$) from $\neg A$:

$$
\begin{array}{c}
[\neg A] \\
\mathcal{D} \\
\cfrac{\bot}{A} \; \bot_c
\end{array}
$$

On the other hand, the rule $\to\mathrm{I}$ derives $\neg A$ by deducing absurdity ($\bot$) from A:

$$
\begin{array}{c}
[A] \\
\mathcal{D} \\
\cfrac{\bot}{\neg A} \; \to\mathrm{I}
\end{array}
$$

which is allowed even in minimal logic (hence in intuitionistic logic).

Notes

The BHK-interpretation justifies the slogan "Proofs as Programs" which plainly expresses that there is a direct correspondence (the *Curry-Howard correspondence*) between computer programs and proofs in intuitionistic logic. Hence, in principle, we are able to extract programs from proofs; see [62].

We have chosen natural deduction for our presentation of minimal, intuitionistic and classical logic following [66], [64, Chapter 2], and [62]. For other formalisations of them, see [54, 63, 65].

3. Constructive set theory

3.1. *The constructive set theory* CZF

The constructive set theory **CZF**, founded by Aczel [1–3], grew out of Myhill's constructive set theory [53] as a formal system for Bishop's constructive mathematics, and permits a quite natural interpretation in Martin-Löf type theory [50].

Definition 4. The language of **CZF** contains variables for sets, a constant $\mathbb{N}$, and the binary predicates $=$ and $\in$. The axioms and rules are those of intuitionistic predicate logic with equality, and the following set theoretic axioms:

1. **Extensionality:** $\forall a \forall b (\forall x (x \in a \leftrightarrow x \in b) \to a = b)$.
2. **Pairing:** $\forall a \forall b \exists c \forall x (x \in c \leftrightarrow x = a \vee x = b)$.
3. **Union:** $\forall a \exists b \forall x (x \in b \leftrightarrow \exists y \in a (x \in y))$.
4. **Restricted Separation:**

$$\forall a \exists b \forall x (x \in b \leftrightarrow x \in a \wedge \varphi(x))$$

for every *restricted* formula $\varphi(x)$, where a formula $\varphi(x)$ is restricted, or Δ_0, if all the quantifiers occurring in it are bounded, i.e. of the form $\forall x \in c$ or $\exists x \in c$.

5. **Strong Collection:**

$$\forall a (\forall x \in a \exists y \varphi(x,y) \to \exists b (\forall x \in a \exists y \in b \varphi(x,y) \wedge \forall y \in b \exists x \in a \varphi(x,y)))$$

for every formula $\varphi(x,y)$.

6. **Subset Collection:**

$$\forall a \forall b \exists c \forall u (\forall x \in a \exists y \in b \varphi(x,y,u) \to$$
$$\exists d \in c (\forall x \in a \exists y \in d \varphi(x,y,u) \wedge \forall y \in d \exists x \in a \varphi(x,y,u)))$$

for every formula $\varphi(x,y,u)$.

7. **Infinity:**

(N1) $0 \in \mathbb{N} \wedge \forall x (x \in \mathbb{N} \to x+1 \in \mathbb{N})$,

(N2) $\forall y (0 \in y \wedge \forall x (x \in y \to x+1 \in y) \to \mathbb{N} \subseteq y)$,

where $x+1$ is $x \cup \{x\}$, and 0 is the empty set $\emptyset = \{x \in \mathbb{N} \mid \bot\}$.

8. **∈-Induction:**

$$(\text{IND}_\in) \quad \forall a (\forall x \in a \varphi(x) \to \varphi(a)) \to \forall a \varphi(a)$$

for every formula $\varphi(a)$.

Remark 3. Note that **CZF** does not have the powerset axiom and the full separation axiom.

Let a and b be sets. Using Strong Collection, the *cartesian product* $a \times b$ of a and b consisting of the ordered pairs $(x, y) = \{\{x\}, \{x, y\}\}$ with $x \in a$ and $y \in b$ can be introduced in **CZF**. A *relation* r between a and b is a subset of $a \times b$. A relation $r \subseteq a \times b$ is *total* (or is a *multivalued function*) if for every $x \in a$ there exists $y \in b$ such that $(x, y) \in r$, or $x \, r \, y$. The class of total relations between a and b is denoted by $\mathrm{mv}(a, b)$, or more formally

$$r \in \mathrm{mv}(a, b) \Leftrightarrow r \subseteq a \times b \wedge \forall x \in a \exists y \in b(x \, r \, y).$$

A *function* from a to b is a total relation $f \subseteq a \times b$ such that for every $x \in a$ there is exactly one $y \in b$ with $(x, y) \in f$. The class of functions from a to b is denoted by b^a, or more formally

$$f \in b^a \Leftrightarrow f \in \mathrm{mv}(a, b) \wedge \forall x \in a \forall y, z \in b((x, y) \in f \wedge (x, z) \in f \to y = z).$$

In **CZF**, we can prove

Fullness: $\forall a \forall b \exists c(c \subseteq \mathrm{mv}(a, b) \wedge \forall r \in \mathrm{mv}(a, b) \exists s \in c(s \subseteq r))$,

and, as a corollary, we see that b^a is a set, that is

Exponentiation: $\forall a \forall b \exists c \forall f (f \in c \leftrightarrow f \in b^a)$;

see [4, 5], for more details of **CZF**.

The material in the following could be formalized in **CZF** together with the following choice axioms.

The axiom of countable choice: If R is a relation between $\mathbb{N}$ and a set a, then

$$\forall n \in \mathbb{N} \exists y \in a(n \, R \, y) \to \exists f \in a^{\mathbb{N}} \forall n \in \mathbb{N}(n \, R \, f(n)).$$

The axiom of dependent choice: If R is a binary relation on a set a, that is, $R \subseteq a \times a$, then

$$\forall x \in a \exists y \in a(x \, R \, y) \to \forall x \in a \exists f \in a^{\mathbb{N}}[f(0) = x \wedge \forall n \in \mathbb{N}(f(n) \, R \, f(n+1))].$$

3.2. *Omniscience principles*

Principles of omniscience are principles which are provable in classical logic, but unprovable in intuitionistic logic and hence constructively unacceptable. A theorem is underivable in constructive mathematics if it implies an omniscience principle. Most of them are instances of classical principles, the double negation elimination DNE, the principle of excluded middle PEM and De Morgan's law DML, and we list them below. Here and in the following, we follow the notational conventions in [64]: $\alpha, \beta, \gamma, \delta$ are supposed to range over $\mathbb{N}^{\mathbb{N}}$; $\mathbf{0} = \lambda n.0$ and $\alpha \# \beta \Leftrightarrow \exists n(\alpha(n) \neq \beta(n))$.

1. The limited principle of omniscience (LPO, Σ_1^0-PEM):

$$\forall \alpha[\alpha \# \mathbf{0} \vee \neg \alpha \# \mathbf{0}]$$

2. The weak limited principle of omniscience (WLPO, Π_1^0-PEM):

$$\forall \alpha[\neg\neg\alpha \# \mathbf{0} \vee \neg \alpha \# \mathbf{0}]$$

3. The lesser limited principle of omniscience (LLPO, Σ_1^0-DML):

$$\forall \alpha\beta[\neg(\alpha \# \mathbf{0} \wedge \beta \# \mathbf{0}) \to \neg\alpha \# \mathbf{0} \vee \neg\beta \# \mathbf{0}]$$

4. Markov's principle (MP, Σ_1^0-DNE):

$$\forall \alpha[\neg\neg\alpha \# \mathbf{0} \to \alpha \# \mathbf{0}]$$

5. Markov's principle for disjunction (MP$^\vee$, Π_1^0-DML):

$$\forall \alpha\beta[\neg(\neg\alpha \# \mathbf{0} \wedge \neg\beta \# \mathbf{0}) \to \neg\neg\alpha \# \mathbf{0} \vee \neg\neg\beta \# \mathbf{0}]$$

6. Weak Markov's principle (WMP):

$$\forall \alpha\beta[\forall\gamma(\neg\neg\alpha \# \gamma \vee \neg\neg\gamma \# \beta) \to \alpha \# \beta].$$

Remark 4. In the above, we may assume without loss of generality that α (and β) are ranging over binary sequences, nondecreasing sequences, sequences with at most one nonzero term, or sequences with $\alpha(0) = 0$.

Note that LPO and WLPO are instances of PEM, LLPO and MP$^\vee$ are instances of DML, and MP is an instance of DNE. We have the following diagram which shows implications among them.

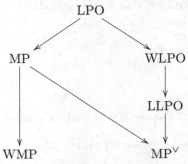

We also know that LPO is equivalent to WLPO + MP, and that MP is equivalent to WMP + MP$^\vee$;

Notes

We refer the reader to [4, 5] for basic set constructions in **CZF** which are used to develop fundamental part of mathematics. For other formalisations of constructive mathematics, see [7, 64, 65].

The principles LPO and LLPO were introduced by Brouwer; the names, the limited principle of omniscience and the lesser limited principle of omniscience, were given by Bishop. In [64], LPO, WLPO and LLPO are called, ∃-PEM, ∀-PEM and SEP, respectively.

In constructive recursive and intuitionistic mathematics, LPO, WLPO and LLPO are provably false; see [64, 4.3.4, 4.3.6, 4.6.4, and 4.6.5]. Markov's principle MP (and hence WMP and MP$^\vee$) is accepted in constructive recursive mathematics; see [64, 4.5].

Mandelkern first considered weak Markov's principle under the names ASP and WLPE in [48] and [49], respectively. Originally, it was stated as follows.

Every pseudo-positive real number is positive,

where a real number a is pseudo-positive if $\forall x \in \mathbf{R}[\neg\neg(0 < x) \vee \neg\neg(x < a)]$; and positive if $0 < a$. Here we consider the above form which is equivalent to its special case in [23, Proposition 6]:

$$\forall\alpha[\forall\gamma(\neg\neg\mathbf{0} \# \gamma \vee \neg\neg\gamma \# \alpha) \to \alpha \# \mathbf{0}].$$

Note that WMP easily follows from Markov's principle MP. In [32], it was shown that WMP is derivable from Church's thesis for disjunctions, and it is known that WMP is a principle which holds in intuitionistic, constructive recursive and classical mathematics. However, Kohlenbach [44] showed, using monotone realisability, that WMP is underivable in $\mathbf{E\text{-}HA}^\omega + AC + CA_{\neg}^\omega$; see also [45, 7.1].

4. Real numbers

4.1. *Cauchy reals*

The set $\mathbb{Z}$ of *integers* is the set $\mathbb{N} \times \mathbb{N}$ with the equality

$$(n, m) =_{\mathbb{z}} (n', m') \Leftrightarrow n + m' = n' + m.$$

The arithmetical relations and operations are defined on $\mathbb{Z}$ in a straightforward way; natural numbers are embedded into $\mathbb{Z}$ by the mapping

$n \mapsto (n, 0)$. The set $\mathbb{Q}_d$ of *dyadic rationals* is the set $\mathbb{Z} \times \mathbb{N}$ with the equality

$$(a, m) =_{\mathbb{Q}_d} (b, n) \Leftrightarrow a \cdot 2^n = zb \cdot 2^m.$$

The arithmetical relations and operations are defined on $\mathbb{Q}_d$ in a straightforward way; integers are embedded into $\mathbb{Q}_d$ by the mapping $a \mapsto (a, 0)$.

Definition 5. A *real number* is a sequence $(p_n)_n$ of dyadic rationals such that

$$\forall mn \left(|p_m - p_n| < 2^{-m} + 2^{-n} \right).$$

We shall write $\mathbb{R}$ for the set of real numbers as usual. Note that dyadic rationals are embedded into $\mathbb{R}$ by the mapping $p \mapsto p^* = \lambda n.p$.

Definition 6. Let $<$ be the *ordering relation* between real numbers $x = (p_n)_n$ and $y = (q_n)_n$ defined by

$$x < y \Leftrightarrow \exists n \left(2^{-n+2} < q_n - p_n \right).$$

Proposition 1. *Let $x, y, z \in \mathbb{R}$. Then*

1. $\neg(x < y \wedge y < x)$,
2. $x < y \rightarrow x < z \vee z < y$.

Proof. (1): Trivial.

(2): Let $x = (p_n)_n$, $y = (q_n)_n$ and $z = (r_n)_n$, and suppose that $x < y$. Then there exists n such that $2^{-n+2} < q_n - p_n$. Setting $N = n + 3$, either $(p_n + q_n)/2 < r_N$ or $r_N \leq (p_n + q_n)/2$. In the former case, we have

$$2^{-N+2} < 2^{-n+1} - (2^{-(n+3)} + 2^{-n}) < (q_n - p_n)/2 - (p_N - p_n)$$
$$= (p_n + q_n)/2 - p_N < r_N - p_N,$$

and hence $x < z$. In the latter case, we have

$$2^{-N+2} < -(2^{-(n+3)} + 2^{-n}) + 2^{-n+1} < (q_N - q_n) + (q_n - p_n)/2$$
$$= q_N - (p_n + q_n)/2 \leq q_N - r_N,$$

and hence $z < y$. □

Definition 7. We define the *apartness* #, the *equality* =, and the ordering relation $\leq$ between real numbers x and y by

1. $x \# y \Leftrightarrow (x < y \vee y < x)$,
2. $x = y \Leftrightarrow \neg(x \# y)$,
3. $x \leq y \Leftrightarrow \neg(y < x)$.

Lemma 1. *Let $x, y, z \in \mathbb{R}$. Then*

1. $x \# y \leftrightarrow y \# x$,
2. $x \# y \rightarrow x \# z \vee z \# y$.

Proof. Straightforward by Proposition 1. □

Proposition 2. *Let $x, y, z \in \mathbb{R}$. Then*

1. $x = x$,
2. $x = y \rightarrow y = x$,
3. $x = y \wedge y = z \rightarrow x = z$.

Proof. Straightforward by Lemma 1. □

Lemma 2. *For each binary sequence α, there exists $x \in \mathbb{R}$ such that*

$$\alpha \# \mathbf{0} \leftrightarrow x \# 0.$$

Conversely, for each $x, y \in \mathbb{R}$, there exists a binary sequence α such that

$$x \# y \leftrightarrow \alpha \# \mathbf{0}.$$

Proof. Suppose that α is a binary sequence; we may assume without loss of generality that α has at most one nonzero term. Define a sequence $(p_n)_n$ of rationals by

$$p_n = \sum_{k=0}^{n} \alpha(k) \cdot 2^{-k}.$$

Then $x = (p_n)_n \in \mathbb{R}$, and $x \# 0 \leftrightarrow \alpha \# \mathbf{0}$.

Conversely, suppose that $x = (p_n)_n, y = (q_n)_n \in \mathbb{R}$, and define a binary sequence α by

$$\alpha(n) = 1 \leftrightarrow 2^{-n+2} < |q_n - p_n|.$$

Then $\alpha \# \mathbf{0} \leftrightarrow x \# y$. □

We have the following correspondences between relations on real numbers and omniscience principles.

Proposition 3.

1. $\forall xy \in \mathbb{R}(x \# y \vee x = y) \Leftrightarrow \mathrm{LPO}$,
2. $\forall xy \in \mathbb{R}(\neg x = y \vee x = y) \Leftrightarrow \mathrm{WLPO}$,
3. $\forall xy \in \mathbb{R}(x \leq y \vee y \leq x) \Leftrightarrow \mathrm{LLPO}$,
4. $\forall xy \in \mathbb{R}(\neg x = y \rightarrow x \# y) \Leftrightarrow \mathrm{MP}$,

5. $\forall xyz \in \mathbb{R}(\neg x = y \to \neg x = z \lor \neg z = y) \Leftrightarrow \mathrm{MP}^{\lor}$,
6. $\forall xy \in \mathbb{R}(\forall z \in \mathbb{R}(\neg x = z \lor \neg z = y) \to x \mathbin{\#} y) \Leftrightarrow \mathrm{WMP}$.

Proof. Straightforward by Lemma 2. $\square$

It is straightforward to define the arithmetical operations on $\mathbb{R}$ in terms of operations on sequences of dyadic rational numbers; see [8, Chapter 2, Proposition 2], [9, Chapter 2, (2.4) and (2.5)], [16, Chapter 1, (5.1)], and [64, Chapter 5] for details.

4.2. *Completeness and total boundedness*

Definition 8. A sequence (x_n) of real numbers is a *Cauchy sequence* if

$$\forall k \exists N \forall mn \geq N[|x_m - x_n| \leq 2^{-k}];$$

converges to a limit $x \in \mathbb{R}$ if

$$\forall k \exists N \forall n \geq N[|x_n - x| \leq 2^{-k}],$$

and we then write $x_n \to x$ as $n \to \infty$.

Theorem 1. *A sequence of real numbers converges if and only if it is a Cauchy sequence.*

Proof. See [8, Chapter 2, Theorem 2], [9, Chapter 2, (3.3)], and [16, Chapter 2, (5.1)]. $\square$

Definition 9. Let S be an inhabited subset of $\mathbb{R}$. A real number z is an *upper bound* of S if $s \leq z$ for each $s \in S$; *least upper bound* of S if z is an upper bound and $z \leq y$ for each upper bound y of S, and we then write $z = \mathrm{lub}\, S$.

A real number z is a *supremum* of S if it is an upper bound of S, and for each k there exists $s \in S$ such that $z < s + 2^{-k}$, and we then write $z = \sup S$. A *lower bound*, *greatest lower bound*, and *infimum* of S are defined similarly.

Classically, the notions of a least upper bound and a supremum coincide, and every subset with an upper bound has a least upper bound. However those notions are constructively distinct, and every subset with an upper bound does not have a least upper bound and a supremum.

Proposition 4. *If every least upper bound of an inhabited bounded subset S of $\mathbb{R}$ is a supremum, then MP holds.*

Proof. Let α be a binary sequence with $\neg\neg\alpha \mathbin{\#} \mathbf{0}$. Then $S = \{\alpha(n) \mid n \in \mathbb{N}\}$ is an inhabited subset of $\mathbb{R}$ with a least upper bound 1. If $\sup S$ exists, then either $0 < \sup S$ or $\sup S < 1$; in the latter case, we have $\neg\alpha \mathbin{\#} \mathbf{0}$, a contradiction. Therefore the former must be the case, and so $\alpha \mathbin{\#} \mathbf{0}$. $\qquad\square$

Proposition 5. *If every inhabited subset S of $\mathbb{R}$ with an upper bound has a least upper bound, then* WLPO *holds.*

Proof. Let α be a binary sequence. Then $S = \{\alpha(n) \mid n \in \mathbb{N}\}$ is an inhabited subset of $\mathbb{R}$ with an upper bound 1. If $\operatorname{lub} S$ exists, then either $0 < \operatorname{lub} S$ or $\operatorname{lub} S < 1$; in the former case, we have $\neg\neg\alpha \mathbin{\#} \mathbf{0}$; in the latter case, we have $\neg\alpha \mathbin{\#} \mathbf{0}$. $\qquad\square$

Proposition 6. *If every inhabited subset S of $\mathbb{R}$ with an upper bound has a supremum, then* LPO *holds.*

Proof. Let α be a binary sequence. Then $S = \{\alpha(n) \mid n \in \mathbb{N}\}$ is an inhabited subset of $\mathbb{R}$ with an upper bound 1. If $\sup S$ exists, then either $0 < \sup S$ or $\sup S < 1$; in the former case, we have $\alpha \mathbin{\#} \mathbf{0}$; in the latter case, we have $\neg\alpha \mathbin{\#} \mathbf{0}$. $\qquad\square$

Remark 5. Note that, in Proposition 4, 5, and 6, taking

$$S = \{0\} \cup \{1 \mid \varphi\}$$

for a restricted formula φ, we can show the stronger consequences: DNE, WPEM ($\neg\varphi \vee \neg\neg\varphi$) and PEM for φ, respectively.

An application of the completeness of the real numbers, we have the following *constructive supremum principle.*

Theorem 2. *Let S be an inhabited subset of $\mathbb{R}$ with an upper bound. If either $\exists s \in S(a < s)$ or $\forall s \in S(s < b)$ for each $a, b \in \mathbb{R}$ with $a < b$, then a supremum of S exists.*

Proof. Let $s_0 \in S$, and let t_0 be an upper bound of S with $s_0 < t_0$. Define sequences $(s_n)_n$ and $(t_n)_n$ of real numbers by

$$s_{n+1} = (2s_n + t_n)/3, t_{n+1} = t_n \text{ if } \exists s \in S[(2s_n + t_n)/3 < s];$$
$$s_{n+1} = s_n, t_{n+1} = (s_n + 2t_n)/3 \text{ if } \forall s \in S[s < (s_n + 2t_n)/3].$$

Note that $s_n < t_n$, $\exists s \in S(s_n \leq s)$ and $\forall s \in S(s \leq t_n)$ for each n. Then $(s_n)_n$ and $(t_n)_n$ converge to the same limit which is a supremum of S. $\qquad\square$

Definition 10. A set S of real numbers is *totally bounded* if for each k there exist $s_0, \ldots, s_{n-1} \in S$ such that

$$\forall y \in S \exists m < n[|s_m - y| < 2^{-k}].$$

Proposition 7. *An inhabited totally bounded set S of real numbers has a supremum.*

Proof. Let $a, b \in \mathbb{R}$ with $a < b$, and let k be such that $2^{-k} < (b - a)/2$. Then there exists $s_0, \ldots, s_{n-1} \in S$ such that

$$\forall y \in S \exists m < n[|s_m - y| < 2^{-k}].$$

Either $a < \max\{s_m \mid m < n\}$ or $\max\{s_m \mid m < n\} < (a + b)/2$; in the former case, there exists $s \in S$ such that $a < s$; in the latter case, for each $s \in S$ there exists m such that $|s - s_m| < 2^{-k}$, and hence

$$s < s_m + |s - s_m| < (a + b)/2 + (b - a)/2 = b.$$

$\square$

Notes

We adopt dyadic rationals instead of standard rationals. However there is no difficulty at all to define arithmetical operations and to establish basic theorems of arithmetic (on real numbers) as in [8, Chapter 2.2], [9, Chapter 2.2], [16, Chapter 1.5], and [64, Chapter 5].

A proof of the Cauchy completeness without countable axiom of choice is given in [64, 5.4.2]. A treatment of Dedekind reals and the difference between the Cauchy completion and the order completion can be found in [64, 5.5]. A construction of the real numbers using a form of interval arithmetic is given in [18].

For other treatments of real numbers in constructive mathematics, see [55], [61], and [57]

5. Metric spaces

5.1. *Fundamental definitions*

Definition 11. A *metric space* is a set X equipped with a *metric* $d :$ $X \times X \to \mathbb{R}$ such that

1. $d(x, y) = 0 \leftrightarrow x = y$,

2. $d(x, y) = d(y, x)$,

3. $d(x, y) \leq d(x, z) + d(z, y)$,

for each $x, y, z \in X$. For $x, y \in X$, we write $x \mathbin{\#} y$ for $0 < d(x, y)$.

Definition 12. A subset S of a metric space X is *open* if

$$\forall x \in S \exists k \forall y \in X[d(x, y) < 2^{-k} \to y \in S];$$

closed if

$$\forall x \in X[\forall k \exists y \in S(d(x, y) < 2^{-k}) \to x \in S].$$

Let S be a subset of a metric space X. Then the *interior* S° of S is defined by

$$S^\circ = \{x \in S \mid \exists k \forall y \in X[d(x, y) < 2^{-k} \to y \in S]\};$$

the *closure* $\overline{S}$ of S is defined by

$$\overline{S} = \{x \in X \mid \forall k \exists y \in S[d(x, y) < 2^{-k}]\}.$$

A subset S of a metric space X is *dense* in X if $\overline{S} = X$; a metric space is *separable* if there exists a countable dense subset.

The *metric complement* $-S$ of a subset S of a metric space X is the set of all points $x \in X$ that are *bounded away* from S, that is,

$$-S = \{x \in X \mid \forall s \in S(d(x, s) \geq a) \text{ for some } a > 0\}.$$

Note that $-S$ is an open subset of X.

Definition 13. A sequence $(x_n)_n$ of X *converges to a limit* $x \in X$ if

$$\forall k \exists N \forall n \geq N[d(x_n, x) \leq 2^{-k}],$$

and we then write $x_n \to x$ as $n \to \infty$.

Remark 6. Note that a subset S of a metric space X is closed if and only if $x \in S$ whenever there exists a sequence $(x_n)_n$ of S converging to x.

5.2. *Complete and compact metric spaces*

Definition 14. A sequence $(x_n)_n$ of a metric space is a *Cauchy sequence* if

$$\forall k \exists N \forall mn \geq N[d(x_m, x_n) \leq 2^{-k}].$$

A metric space is *complete* if every Cauchy sequence converse.

Example 12. The set $\mathbb{R}$ of real numbers is a complete metric space with the metric $d(x, y) = |x - y|$; the n-dimensional Euclidean space $\mathbb{R}^n$ is a compete metric space with the metric

$$d((x_0, \ldots, x_{n-1}), (y_0, \ldots, y_{n-1})) = |x_0 - y_0| + \cdots + |x_{n-1} - y_{n-1}|.$$

We have the following constructive version of the Baire theorem.

Theorem 3. *If* $(U_n)_n$ *is a sequence of open dense subsets of a complete metric space* X, *then* $\cap_{n=0}^{\infty} U_n$ *is dense in* X.

Proof. See [8, Chapter 4, Theorem 4], [9, Chapter 4, (3.9)], and [16, Chapter 2, (1.3)]. □

Corollary 1. *If* $(x_n)_n$ *is a sequence of real numbers, then there exists* $a \in \mathbb{R}$ *such that* $x_n \mathbin{\#} a$ *for each* n.

Proof. Let $U_n = \{x \in \mathbb{R} \mid x \mathbin{\#} x_n\}$ for each n. Then U_n is open and dense. Therefore there exists $a \in \cap_{n=0}^{\infty} U_n$. □

Definition 15. A metric space X is *bounded* if there exists $K > 0$ such that $d(x, y) < K$ for each $x, y \in X$; a subset S of an inhabited metric space X is *bounded* if there exists $K > 0$ and $x \in X$ such that $d(x, y) < K$ for each $y \in S$.

Remark 7. Note that, for a binary sequence α with at most one nonzero term, a subset $S = \{n \mid \alpha(n) = 1\}$ of $\mathbb{N}$ is bounded as a metric space, but if it is bounded as a subset of $\mathbb{N}$, then LPO holds.

Definition 16. A metric space X is *totally bounded* if for each k there exist $x_0, \ldots, x_{n-1} \in X$ such that

$$\forall y \in X \exists m < n[d(x_m, y) < 2^{-k}].$$

Note that every totally bounded metric space is separable.

Definition 17. A mapping f between metric spaces X and Y is *uniformly continuous* if

$$\forall k \exists N \forall xy \in X[d(x, y) < 2^{-N} \to d(f(x), f(y)) \leq 2^{-k}].$$

Proposition 8. *If* f *is a uniformly continuous mapping from a totally bounded metric space* X *into a metric space, then*

$$f(X) = \{f(x) \mid x \in X\}$$

is totally bounded.

Proof. See [8, Chapter 4, Proposition 8], [9, Chapter 4, (4.2)], and [16, Chapter 2, (4.3)]. □

Proposition 9. *If every subset S of a totally bounded metric space is totally bounded, then* LPO *holds.*

Proof. Let α be a binary sequence. Then $S = \{\alpha(n) \mid n \in \mathbb{N}\}$ is a subset of a totally bounded metric space $\{0,1\}$. If S is totally bounded, then there exist $s_0, \ldots, s_{n-1} \in S$ such that

$$\forall y \in S \exists m < n[|s_m - y| < 2^{-1}],$$

and therefore either $0 < \max\{s_m \mid m < n\}$ or $\max\{s_m \mid m < n\} < 2^{-1}$; in the former case, we have $\alpha \,\#\, \mathbf{0}$; in the later case, we have $\neg\alpha \,\#\, \mathbf{0}$. □

Definition 18. A subset S of a metric space X is *located* if

$$d(x, S) = \inf\{d(x,y) \mid y \in S\}$$

exists for each $x \in X$.

Proposition 10. *Every located subset S of a totally bounded metric space X is totally bounded.*

Proof. For given a k, there exist $x_0, \ldots, x_{n-1} \in X$ such that

$$\forall y \in X \exists m < n[d(x_m, y) < 2^{-(k+2)}].$$

Let $s_0, \ldots, s_{n-1} \in S$ be such that $d(x_m, s_m) < d(x_m, S) + 2^{-(k+1)}$ for each $m < n$. Then for each $y \in S$ there exists $m < n$ such that $d(x_m, y) < 2^{-(k+2)}$, and therefore

$$d(s_m, y) \leq d(s_m, x_m) + d(x_m, y) < d(x_m, S) + 2^{-(k+1)} + d(x_m, y)$$
$$\leq d(x_m, y) + 2^{-(k+1)} + d(x_m, y) < 2^{-k}.$$

□

Proposition 11. *A totally bounded subset S of a metric space X is located.*

Proof. Let $x \in X$. Then the mapping $y \mapsto d(x,y)$ is a uniformly continuous mapping from S into $\mathbb{R}$, and hence $\{d(x,y) \mid y \in S\}$ is totally bounded, by Proposition 8. Therefore $\inf\{d(x,y) \mid y \in S\}$ exists, by Theorem 2. □

Definition 19. A metric space is *compact* if it is totally bounded and complete.

Classically, a metric space is compact if and only if the Cantor intersection theorem holds: *every sequence $(F_n)_n$ of closed subsets with the finite intersection property has an inhabited intersection.* However the Cantor intersection theorem does not hold constructively for every compact metric space.

Proposition 12. *If every sequence $(F_n)_n$ of closed subsets of a compact metric space X with the finite intersection property has an inhabited intersection, then* LLPO *holds.*

Proof. Let α and β be sequences of natural numbers, and suppose that $\neg(\alpha \mathbin{\#} \mathbf{0} \wedge \beta \mathbin{\#} \mathbf{0})$. Define a sequence $(F_n)_n$ of closed subsets of the compact metric space $\{0, 1\}$ by

$$F_n = \begin{cases} \{0, 1\} & \text{if } \forall k \le n(\alpha(k) = 0 \wedge \beta(k) = 0), \\ \{0\} & \text{if } \exists k \le n(\alpha(k) \ne 0), \\ \{1\} & \text{if } \exists k \le n(\beta(k) \ne 0). \end{cases}$$

Then $(F_n)_n$ has the finite intersection property. If $0 \in \bigcap_{n=0}^{\infty} F_n$, then $\neg\beta \mathbin{\#} \mathbf{0}$; if $1 \in \bigcap_{n=0}^{\infty} F_n$, then $\neg\alpha \mathbin{\#} \mathbf{0}$. $\square$

5.3. *Continuity properties*

Definition 20. A mapping f between metric spaces X and Y is *strongly extensional* if

$$f(x) \mathbin{\#} f(y) \to x \mathbin{\#} y$$

for each $x, y \in X$; (pointwise) *continuous* if

$$\forall x \in X \forall k \exists N \forall y \in X[d(x, y) < 2^{-N} \to d(f(x), f(y)) \le 2^{-k}].$$

Note that, trivially, every continuous mapping is strongly extensional, and every uniformly continuous mapping is continuous.

The following proposition shows that we can construct a mapping which is *not* strongly extensional.

Proposition 13. *If every mapping is strongly extensional, then* MP *holds.*

Proof. Let α be a binary sequence with $\neg\neg\alpha \mathbin{\#} \mathbf{0}$. Then there exists $a \in \mathbb{R}$ such that $\alpha \mathbin{\#} \mathbf{0} \leftrightarrow a \mathbin{\#} 0$, and define a (linear) mapping f of $a\mathbb{R}$ into $\mathbb{R}$ by $f(ax) = x$; in fact f is a mapping; for if $ax = ay$ and $x \mathbin{\#} y$, then $a = 0$, and hence $\neg\alpha \mathbin{\#} \mathbf{0}$, a contradiction. If f is strongly extensional, then $f(a0) = 0 \mathbin{\#} 1 = f(a1)$, and hence $0 \mathbin{\#} a$. $\square$

Definition 21. A mapping f between metric spaces is *sequentially continuous* if

$$f(x_n) \to f(x) \text{ as } n \to \infty$$

for each sequence $(x_n)_n$ converging to a limit x; *nondiscontinuous* if

$$\forall n[d(f(x_n), f(x)) \geq \delta] \to \delta \leq 0$$

for each sequence $(x_n)_n$ converging to a limit x. Note that, trivially, every
sequentially continuous mapping is nondiscontinuous.

Lemma 3. *Every sequentially continuous mapping f of a metric space X
into a metric space Y is strongly extensional.*

Proof. Let x, y be points of X with $f(x) \# f(y)$. Construct a nondecreasing binary sequence α such that

$$\alpha(n) = 0 \Rightarrow d(x, y) < 2^{-n},$$
$$\alpha(n) = 1 \Rightarrow x \# y.$$

Define a sequence $(z_n)_n$ in X as follows: if $\alpha(n) = 0$, set $z_n = x$; if $\alpha(n) = 1$,
set $z_n = y$. Then $(z_n)_n$ converges to y. Then, since f is sequentially
continuous, the sequence $(f(z_n))_n$ in Y converges to the limit $f(y)$. Choose
a natural number N such that $d(f(z_N), f(y)) < d(f(x), f(y))$. If $\alpha(N) = 0$,
then $z_N = x$; whence $d(f(x), f(y)) = d(f(z_n), f(y)) < d(f(x), f(y))$, a
contradiction. Hence $\alpha(N) = 1$, and so $x \# y$. $\square$

We will see a relationship among sequential continuity, nondiscontinuity
and strong extensionality, and to this end, we need a couple of Lemmata.

Lemma 4. *Let f be a strongly extensional mapping of a complete metric
space X into a metric space Y, and let $(x_n)_n$ be a sequence in X converging to a limit x. Then for each positive numbers a, b with $a < b$, either
$d(f(x_n), f(x)) > a$ for some n or $d(f(x_n), f(x)) < b$ for each n.*

Proof. Let γ be a sequence of natural numbers such that $\gamma(0) = 0$
and $d(x_k, x) < 2^{-(n+1)}$ for each $k \geq \gamma(n)$ and $n \geq 1$, and set $s_n =
\max\{d(f(x_k), f(x)) : \gamma(n-1) \leq k < \gamma(n)\}$ for each n. Construct a nonde-
creasing binary sequence α such that

$$\alpha(n) = 0 \Rightarrow \forall k \leq n(s_k < b),$$
$$\alpha(n) = 1 \Rightarrow \exists k \leq n(s_k > a).$$

We may assume that $\alpha(0) = 0$. Define a sequence $(y_n)_n$ in X as follows: if $\alpha(n) = 0$, set $y_n = x$; if $\alpha(n) = 1 - \alpha(n-1)$, choose k_n with

$\gamma(n-1) \leq k_n < \gamma(n)$ and $d(f(x_{k_n}), f(x)) > a$ and set $y_i = x_{k_n}$ for all $i \geq n$. Then $(y_n)_n$ is a Cauchy sequence: in fact, $d(y_m, y_n) \leq 2^{-n}$ whenever $m \geq n$. So $(y_n)_n$ converges to a limit y in X. Either $d(f(x), f(y)) < a$ or $d(f(x), f(y)) > 0$. In the former case, if $\alpha(n) = 1 - \alpha(n-1)$, then $y = x_{k_n}$ with $d(f(x_{k_n}), f(x)) > a$, a contradiction; whence $\alpha(n) = 0$ for each n. In the latter case, as f is strongly extensional, $x \# y$. Choose a positive integer N such that $x \# y_N$. If $\alpha(N) = 0$, then $x \# y_N = x$, a contradiction; whence $\alpha(N) = 1$. $\qquad\square$

Lemma 5. *Let f be a strongly extensional mapping of a complete metric space X into a metric space Y, and let $(x_n)_n$ be a sequence in X converging to a limit x. Then for each positive numbers a, b with $a < b$, either $d(f(x_n), f(x)) > a$ for infinitely many n or $d(f(x_n), f(x)) < b$ for all sufficiently large n.*

Proof. Let γ be a sequence natural numbers such that $d(x_n, x) < 2^{-(k+1)}$ for each $n \geq \gamma(k)$. Construct a nondecreasing binary sequence α, by successively applying Lemma 4 to the subsequence $(x_n)_{n \geq \gamma(k)}$, such that

$$\alpha(k) = 0 \Rightarrow \exists n \geq \gamma(k)[d(f(x_n), f(x)) > a],$$
$$\alpha(k) = 1 \Rightarrow \forall n \geq \gamma(k)[d(f(x_n), f(x)) < b].$$

We may assume that $\alpha(0) = 0$. Define a sequence $(y_k)_k$ in X as follows: if $\alpha(k) = 0$, choose $n_k \geq \gamma(k)$ so that $d(f(x_{n_k}), f(x)) > a$ and set $y_k = x_{n_k}$; if $\alpha(k) = 1 - \alpha(k-1)$, set $y_i = y_{k-1}$ for each $i \geq k$. Then $(y_k)_k$ is a Cauchy sequence: in fact, $d(y_m, y_n) \leq 2^{-n}$ whenever $m \geq n$. Let y be the limit of $(y_k)_k$ in X. Either $0 < d(f(x), f(y))$ or $d(f(x), f(y)) < a$. In the former case, since f is strongly extensional, $x \# y$. Choose a positive integer N so that $d(x, y_N) > 2^{-(N+1)}$. If $\alpha(N) = 0$, then $y_N = x_{n_N}$ for some $n_N \geq \gamma(N)$; whence $2^{-(N+1)} < d(x, y_N) = d(x, x_{n_N}) < 2^{-(N+1)}$, a contradiction, and therefore $\alpha(N) = 1$. In the latter case, if $\alpha(k+1) = 1 - \alpha(k)$ for some k, then $y = x_{n_k}$ with $d(f(x_{n_k}), f(x)) > a$, a contradiction; whence $\alpha(k) = 0$ for each k. $\qquad\square$

Theorem 4. *Let f be a mapping of a complete metric space X into a metric space Y. Then f is sequentially continuous if and only if f is nondiscontinuous and strongly extensional.*

Proof. Suppose that f is sequentially continuous. Then, trivially, f is nondiscontinuous, and f is strongly extensional, by Lemma 4.

Conversely, suppose that f is nondiscontinuous and strongly extensional. Let $(x_n)_n$ be a sequence in X converging to a limit x and let

$\epsilon > 0$. Then either there exists a subsequence $(x_{n_k})_k$ of (x_n) such that $d(f(x_{n_k}), f(x)) > \epsilon/2$ for all k or there exists a positive number N such that $d(f(x_n), f(x)) < \epsilon$ for all $n \geq N$, by Lemma 5. In the former case, as f is nondiscontinuous, $\epsilon \leq 0$, a contradiction. Therefore the latter must be the case. $\qquad\square$

Lemma 6. *If there exists a mapping f of a complete metric space into a metric space such that f is strongly extensional and discontinuous, in the sense that $\forall n[d(f(x_n), f(x)) > \delta]$ for some sequence $(x_n)_n$ converging to a limit x and $\delta > 0$, then LPO holds.*

Proof. Suppose such an $f : X \to Y$ exists, where X is complete. Then there exist a sequence $(x_n)_n$ in X converging to a limit x, and a positive number δ, such that $d(f(x_n), f(x)) > \delta$ for each n. Given α, define a sequence $(y_n)_n$ in X as follows: if $\forall k \leq n(\alpha k = 0)$, set $y_n = x$; if $\exists k \leq n(\alpha k \neq 0)$, set $y_n = x_j$ where $j = \min\{k : \alpha k \neq 0\}$. Then $(y_n)_n$ is a Cauchy sequence. Let y be the limit of $(y_n)_n$ in X. Either $0 < d(f(y), f(x))$ or $d(f(y), f(x)) < \delta$. In the former case, since f is strongly extensional, $y \mathrel{\#} x$. Hence there exists n such that $x_n \mathrel{\#} x$, and so $\alpha \mathrel{\#} \mathbf{0}$. In the latter case, if $\exists n(\alpha n \neq 0)$, then $\delta < d(f(y), f(x)) < \delta$, a contradiction; thus $\neg \alpha \mathrel{\#} \mathbf{0}$. $\qquad\square$

The limited principle of omniscience LPO holds in classical logic, and so the above lemma classically says nothing. Since we cannot accept LPO as a general constructive principle, the lemma is of constructive worth as we will see in Section 8.2. Many well-known theorems in classical analysis could be proved with LPO. For example, one can see that the following lemma is immediate consequence of Theorem 2 with LPO.

Lemma 7. *Assuming LPO, every separable subset of a metric space is located.*

Notes

A detailed analysis of the constructive Baire theorem [8, Chapter 4, Theorem 4] and [9, Chapter 4, (3.9)] is given in [16, Chapter 2.2], and its applications can be found in [18, 6.6].

A treatment of function spaces and the Stone-Weierstraß theorem are given in [8, Chapter 4.4] and [9, Chapter 4.5]. For locally compact spaces, a one-point compactification and the Tietze extension theorem, see [8, Chapter 4.5], [9, Chapter 4.6], and [65, Chapter 7].

The definition of a located set in Definition 18 is the notion of a *metrically* located set, and the notion of a *topologically* located set can be found in [65, Chapter 7]. The Lindelöf theorem and the Heine-Borel theorem are also dealt in [65, Chapter 7]; see also [32].

Lemma 4, Lemma 5 and Theorem 4 were first given in [29]; see also [17], [18, 3.2], and [22]. A treatment of them in a formal system can be found in [38]. Lemma 3 and Lemma 6 were given respectively in [39] and in [33].

6. Normed spaces

6.1. *Normed and Banach spaces*

Definition 22. A *normed space* is a linear space E equipped with a *norm* $\|\cdot\| : E \to \mathbb{R}$ such that

1. $\|x\| = 0 \leftrightarrow x = 0$,
2. $\|ax\| = |a|\|x\|$,
3. $\|x + y\| \leq \|x\| + \|y\|$,

for each $x, y \in E$ and $a \in \mathbb{R}$. Note that a normed space E is a metric space with the metric

$$d(x, y) = \|x - y\|.$$

A *Banach space* is a normed space which is complete with respect to the metric.

Example 13. Let ξ be a sequence of real numbers. Then we write ξ_n for $\xi(n)$. For $1 \leq p < \infty$, let

$$l_p = \{\xi \in \mathbb{R}^{\mathbb{N}} \mid \sum_{n=0}^{\infty} |\xi_n|^p < \infty\}$$

and define a norm by $\|\xi\| = (\sum_{n=0}^{\infty} |\xi_n|^p)^{1/p}$. Then l_p is a (separable) Banach space. Classically the normed space

$$l_\infty = \{\xi \in \mathbb{R}^{\mathbb{N}} \mid (\xi_n) \text{ is bounded}\}$$

with the norm $\|\xi\| = \sup_n |\xi_n|$ is an *inseparable* Banach space. However, constructively, it is *not* a normed space.

Definition 23. A *subspace* M of a normed space E is a linear subset of E; for a subset S of E, we write

$$\text{span}(S) = \{t_0 x_0 + \cdots + t_{n-1} x_{n-1} \mid t_0, \ldots, t_{n-1} \in \mathbb{R}, x_0, \ldots, x_{n-1} \in S\}$$

for the smallest subspace of E containing S.

Proposition 14. *Let* Y *be a located subspace of a normed space* E, *and define*

$$\|x\|_{E/Y} = d(x, Y).$$

Then E *with the equality relation* $x =_{E/Y} y \Leftrightarrow \|x - y\|_{E/Y} = 0$ *is a normed space with the norm* $\| \cdot \|_{E/Y}$, *called the quotient space of* E *by* Y, *and written* E/Y. *Moreover, if* E *is a Banach space, then* E/Y *is a Banach space.*

Proof. It is straightforward to see that E/Y is a normed space with a norm $\| \cdot \|_{E/Y}$. Suppose that E is a Banach space, and $(x_n)_n$ is a Cauchy sequence in E/Y. Then there exists a subsequence $(x_{n_k})_k$ of $(x_n)_n$ such that $\|x_{n_k} - x_{n_{k+1}}\|_{E/Y} < 2^{-(k+1)}$, and we can then inductively choose a sequence $(y_k)_k$ of Y such that $\|(x_{n_k} - y_k) - (x_{n_{k+1}} - y_{k+1})\| < 2^{-k}$. Since E is complete, the Cauchy sequence $(x_{n_k} - y_k)_k$ converges to a limit $z \in E$, and we have

$$\|x_{n_k} - z\|_{E/Y} \leq \|(x_{n_k} - y_k) - z\| \to 0$$

as $k \to \infty$. Therefore, since $(x_n)_n$ is a Cauchy sequence and has a convergent subsequence, $(x_n)_n$ must converge. $\square$

Definition 24. Let E be a normed space. Then we write $B_E(x, r)$ to denote the *open ball* with *centre* x and *radius* r, that is,

$$B_E(x, r) = \{y \in E \mid \|x - y\| < r\};$$

$B_E(r)$ to denote $B_E(0, r)$; and B_E to denote $B_E(1)$. For subsets S and S' of E and $a \in \mathbb{R}$, we write $S + S'$ and aS respectively for the subsets

$$S + S' = \{x + y \mid x \in S, y \in S'\}$$

and

$$aS = \{ax \mid x \in S\};$$

for $z \in E$, we write $S + z$ for $S + \{z\} = \{z\} + S$.

Remark 8. Note that if S is located and $a > 0$, then $S + z$ and aS are located.

Definition 25. A subset C of a linear space E is *convex* if $\lambda x + (1 - \lambda)y \in C$ for each $x, y \in C$ and $\lambda \in [0, 1]$; *absorbing* if every x in E lies in rC for some $r > 0$.

Example 14. The open ball $B_E(x, r)$ is a convex subset, and $B_E(r)$ is a convex absorbing subset for $r > 0$.

Remark 9. Note that if C and D are convex subsets and $a \in \mathbb{R}$, then $C + D$ and aC are convex subsets of E; if C is a convex subset of E and $\lambda_0, \ldots, \lambda_{n-1}$ are nonnegative real numbers such that $\sum_{i=0}^{n-1} \lambda_i = 1$, then $\sum_{i=0}^{n-1} \lambda_i C \subseteq C$.

If C is absorbing, then $0 \in C$; if C is a convex absorbing subset, then $\lambda C \subseteq C$ for each $\lambda \in [0, 1]$, and hence $sC \subseteq tC$ if $0 \le s < t$. More generally, if C is a convex absorbing subset of E and $\lambda_0, \ldots, \lambda_{n-1}$ are nonnegative real numbers such that $\sum_{i=0}^{n-1} \lambda_i \le 1$, then $\sum_{i=0}^{n-1} \lambda_i C \subseteq C$; for, setting $\lambda_n = 1 - \sum_{i=1}^{n-1} \lambda_i$, we have $\sum_{i=0}^{n} \lambda_i = 1$, and hence $\sum_{i=0}^{n-1} \lambda_i C \subseteq \sum_{i=0}^{n} \lambda_i C \subseteq C$.

Lemma 8. *Let C be a convex absorbing subset of a Banach space E. Then $0 \notin \overline{-C}$, where $-C$ is the metric complement of C.*

Proof. Suppose that $0 \in \overline{-C}$. Then there exists a sequence (x_n) in $-C$ converging to 0. We shall prove that $-nC$ is dense in E for each n. To this end, fix n, let $y \in E$, and let $\epsilon > 0$. Then there exist $r > 0$, k and $a > 0$ such that $-y \in rC$, $\|x_k\| < (n + r)^{-1}\epsilon$, and $\|x_k - w\| \ge a$ for each $w \in C$. Set $y' = y + (n + r)x_k$. Then $\|y - y'\| = (n + r)\|x_k\| < \epsilon$. Moreover if $z \in nC$, then $(n + r)^{-1}(z - y) \in (n + r)^{-1}(nC + rC) = C$; whence

$$\|y' - z\| = \|y + (n + r)x_k - z\| = (n + r)\|x_k - (n + r)^{-1}(z - y)\|$$
$$\ge (n + r)a;$$

so $y' \in -nC$. Hence $-nC$ is dense in E; clearly, it is open in E. Applying Theorem 3, construct a point x in $\bigcap_{n=1}^{\infty} -nC$. Choose m such that $x \in mC$. Then, as $x \in -mC$, we have a contradiction. Thus $0 \notin \overline{-C}$. $\square$

6.2. Bounded and normable linear mappings

Definition 26. A mapping T between linear spaces E and F is *linear* if

1. $T(ax) = aTx$,
2. $T(x + y) = Tx + Ty$

for each $x, y \in E$ and $a \in \mathbb{R}$. A *linear functional* f on a linear space E is a linear mapping from E into $\mathbb{R}$. The *kernel* $\ker(T)$ of a linear mapping T between linear spaces E and F is defined by

$$\ker(T) = \{x \in E \mid Tx = 0\}.$$

Definition 27. A linear mapping T between normed spaces E and F is *bounded* if $T(B_E) = \{Tx \mid x \in B_E\}$ is bounded; *compact* if $T(B_E)$ is totally bounded; *open* if $T(B_E)$ has inhabited interior.

Proposition 15. *Let T be a linear mapping between normed spaces E and F. Then the following are equivalent.*

1. *T is continuous,*
2. *T is uniformly continuous,*
3. *T is bounded.*

Proof. See [8, Chapter 9, Proposition 2], [9, Chapter 7, (1.5)], and [18, Proposition 2.3.3]. □

Definition 28. A linear mapping T between normed spaces E and F is *normable* if
$$\|T\| = \sup\{\|Tx\| \mid x \in B_E\}$$
exists. Note that every compact mapping is normable.

Classically, the following theorem holds.

Theorem 5 (classical). *Every bounded linear functional is normable, and the set of bounded linear functionals on a normed space forms a Banach space.*

However, constructively, we can construct a bounded linear functional which is *not* normable.

Proposition 16. *If every bounded linear functional on l_2 is normable, then LPO holds.*

Proof. Let α be a binary sequence with at most one nonzero term, and define a linear functional f on l_2 by
$$f(\xi) = \sum_{k=0}^{\infty} \alpha(k)\xi_k.$$
Then f is bounded. If f is normable, then either $0 < \|f\|$ or $\|f\| < 1$; in the former case, we have $\alpha \# \mathbf{0}$; in the latter case, we have $\neg\alpha \# \mathbf{0}$. □

Let E^* be the set of normable linear functionals on a normed space E. The following proposition shows that E^* of a Banach space E is not always linear constructively.

Proposition 17. *If the set $(l_1)^*$ of normable linear functionals on l_1 is linear, then LPO holds.*

Proof. Let α be a binary sequence with $\alpha(0) = 0$, and define linear functionals on l_1 by

$$f(\xi) = \sum_{k=0}^{\infty} \xi_k, \quad g(\xi) = \sum_{k=0}^{\infty} (\alpha(k) - 1)\xi_k.$$

Then f and g are normable with $\|f\| = \|g\| = 1$. If $f + g$ is normable, then either $0 < \|f + g\|$ or $\|f + g\| < 1$; in the former case, we have $\alpha \mathbin{\#} \mathbf{0}$; in the latter case, we have $\neg\alpha \mathbin{\#} \mathbf{0}$. $\square$

Remark 10. Note that, as classically, $(l_p)^*$ is a linear space for $1 < p < \infty$.

Proposition 18. *A nonzero bounded linear functional f on a normed space E is normable if and only if its kernel $\ker(f)$ is located.*

Proof. See [8, Chapter 9, Proposition 8], [9, Chapter 7, (1.10)], and [18, Proposition 2.3.6]. $\square$

6.3. *Finite-dimensional normed spaces*

Definition 29. A normed space E if *finite-dimensional* if there exists a finitely enumerable subset $\{e_0, \ldots, e_{n-1}\}$ of E, called a *metric basis* of E, such that the mapping

$$(\lambda_0, \ldots, \lambda_{n-1}) \mapsto \sum_{i=0}^{n-1} \lambda_i e_i$$

of $\mathbb{R}^n$ into E has a uniformly continuous inverse. Here we assume that $\mathbb{R}^n$ is equipped with the standard norm

$$\|(\lambda_0, \ldots, \lambda_{n-1})\| = |\lambda_0| + \cdots + |\lambda_{n-1}|.$$

Proposition 19. *An open ball $B_E(x, r)$ of a finite-dimensional normed space E is totally bounded.*

Proof. See [8, Chapter 9, Proposition 5], [9, Chapter 7, (2.3)], and [18, Corollary 4.1.7]. $\square$

Corollary 2. *A finite-dimensional subspace Y of a normed space E is located.*

Proposition 20. *Every linear mapping of a finite-dimensional normed space in to a normed space is bounded, and hence normable.*

Proof. See [18, Corollary 4.1.4 and Corollary 4.1.8]. $\square$

Proposition 21. *Let* $\{x_0, \ldots, x_{n-1}\}$ *be a finitely enumerable subset of a normed space* E. *Then for each* k *there exists a finite-dimensional subspace* Y *of* $\mathrm{span}\{x_0, \ldots, x_{n-1}\}$ *such that* $d(x_i, Y) < 2^{-k}$ *for each* $i < n$.

Proof. See [9, Chapter 7, (2.5)] and [18, Lemma 4.1.11]. $\square$

Lemma 9. *Let* C *be a convex absorbing subset of a finite-dimensional normed space* E. *Then* C *has inhabited interior.*

Proof. Let $\{e_0, \ldots, e_{n-1}\}$ be a metric base for E, and let $r > 0$ be such that $e_i \in rC$ and $-e_i \in rC$ for each $i < n$. It suffices to show that there exists $\delta > 0$ such that $B_E(\delta) \subseteq rC$. Choose $\delta > 0$ so that if $\|\sum_{i=0}^{n-1} \lambda_i e_i\| < \delta$, then $\sum_{i=0}^{n-1} |\lambda_i| \leq 1$. Let $x = \sum_{i=0}^{n-1} \lambda_i e_i$ be such that $\|x\| < \delta$, and set $\lambda_i^+ = \max\{0, \lambda_i\}$ and $\lambda_i^- = \max\{0, -\lambda_i\}$ for each $i < n$. Then $\lambda_i = \lambda_i^+ - \lambda_i^-$ and $|\lambda_i| = \lambda_i^+ + \lambda_i^-$ for each $i < n$. Since $\sum_{i=0}^{n-1} |\lambda_i| = \sum_{i=0}^{n-1} \lambda_i^+ + \sum_{i=0}^{n-1} \lambda_i^- \leq 1$, we have

$$x = \sum_{i=0}^{n-1} \lambda_i e_i = \sum_{i=0}^{n-1} \lambda_i^+ e_i + \sum_{i=0}^{n-1} \lambda_i^- (-e_i) \in rC.$$

$\square$

6.4. *Uniformly convex spaces*

Definition 30. A normed space E is *uniformly convex* if for each $\epsilon > 0$ there exists $\delta > 0$ such that

$$\|x - y\| > \epsilon \to \|(x+y)/2\| \leq 1 - \delta$$

for each $x, y \in B_E$.

It is straightforward to see that if E is a uniformly convex normed space, then for each $\epsilon > 0$ there exists $\delta > 0$ such that

$$\|x - y\| > d\epsilon \to \|(x+y)/2\| \leq d(1 - \delta)$$

for each $d > 0$ and $x, y \in B_E(d)$.

Remark 11. l_p is uniformly convex for $1 < p < \infty$; see [8, Chapter 9, Theorem 1 and Corollary] and [9, Chapter 7, (3.22)].

Lemma 10. *Let* C *be a convex subset of a uniformly convex normed space* E, *and let* x *be an element of* E *such that*

$$d(x, C) = \inf\{\|x - z\| \mid z \in C\}$$

exists. Then there exists strongly at most one element y *of* C *such that* $d(x, C) = \|x - y\|$, *in the sense that for each* $y, z \in C$, *if* $y \mathbin{\#} z$, *then* $d(x, C) < \|x - y\|$ *or* $d(x, C) < \|x - z\|$.

Proof. Note that, since

$$0 < \|y - z\| \leq \|x - y\| + \|x - z\| \leq 2\max\{\|x - y\|, \|x - z\|\},$$

we have $0 < r = \max\{\|x-y\|, \|x-z\|\}$. Choose $\epsilon > 0$ so that $r\epsilon < \|y - z\|$. Then there exists $\delta > 0$ such that

$$d(x, C) \leq \|x - (y + z)/2\| \leq r(1 - \delta) < r,$$

and hence either $d(x, C) < \|x - y\|$ or $d(x, C) < \|x - z\|$. $\square$

Proposition 22. *Let C be a closed convex subset of a uniformly convex Banach space E, and let x be an element of E such that*

$$d(x, C) = \inf\{\|x - z\| \mid z \in C\}$$

exists. Then there exists an element y of C such that $d(x, C) = \|x - y\|$.

Proof. Let $d = d(x, C)$, and let $(z_n)_n$ be a sequence of C such that $\|x - z_n\| \leq d + 2^{-n}$. Then for each $\epsilon > 0$, either $0 < d$ or $d < \epsilon/4$. In the former case, for $\epsilon' = \epsilon/(d+1)$, there exists $\delta > 0$ such that if $(d+2^{-n})\epsilon' < \|z_m - z_n\|$, then

$$d \leq \|x - (z_m + z_n)/2\| \leq (d + 2^{-n})(1 - \delta)$$

for each n and $m \geq n$, and hence for sufficiently large n with $(d+2^{-n})(1 - \delta) < d$, we have $\|z_m - z_n\| \leq (d + 2^{-n})\epsilon' \leq (d + 1)\epsilon' = \epsilon$ for $m \geq n$. In the latter case, for sufficiently large n with $2^{-n} < \epsilon/4$, we have $\|z_m - z_n\| \leq \|x - z_m\| + \|x - z_n\| < \epsilon$ for $m \geq n$. Therefore $(z_n)_n$ is a Cauchy sequence, and so it converges to a limit $y \in C$. Thus we have $d(x, C) = \|x - y\|$. $\square$

Proposition 23. *Let f be a nonzero normable linear functional on a uniformly convex Banach space E. Then there exists $x \in E$ such that $f(x) = \|f\|$ and $\|x\| = 1$.*

Proof. See [9, Chapter 7, (3.23)]. $\square$

6.5. *Hilbert spaces*

Definition 31. An *inner product space* is a linear space E equipped with an *inner product* $\langle \cdot, \cdot \rangle : E \times E \to \mathbb{R}$ such that

1. $\langle x, x \rangle \geq 0$ and $\langle x, x \rangle = 0 \leftrightarrow x = 0$,
2. $\langle x, y \rangle = \langle y, x \rangle$,
3. $\langle ax, y \rangle = a\langle x, y \rangle$,

4. $\langle x + y, z \rangle = \langle x, z \rangle + \langle y, z \rangle$

for each $x, y, z \in E$ and $a \in \mathbb{R}$. Note that an inner product space E is a normed space with the norm

$$\|x\| = \langle x, x \rangle^{1/2}.$$

A *Hilbert space* is an inner product space which is a Banach space.

Example 15. Let

$$l_2 = \{\xi \in \mathbb{R}^\mathbb{N} \mid \sum_{n=0}^\infty |\xi_n|^2 < \infty\}$$

and define an inner product by $\langle \xi, \zeta \rangle = \sum_{n=0}^\infty \xi_n \zeta_n$. Then l_2 is a Hilbert space.

Proposition 24. *Let E be an inner product space. Then for each $x, y \in E$, the Cauchy-Schwarz inequality*

$$|\langle x, y \rangle| \leq \|x\| \|y\|$$

and the parallelogram identity

$$\|x + y\|^2 + \|x - y\|^2 = 2\|x\|^2 + 2\|y\|^2$$

hold.

Proof. See [8, Chapter 9, Theorem 5] and [9, Chapter 7, (8.4)]. $\quad\square$

Remark 12. Note that if E is an inner product space, then, by the parallelogram identity, we have

$$\|(x + y)/2\|^2 = \|x\|^2/2 + \|y\|^2/2 - \|x - y\|^2/4 \leq 1 - \|x - y\|^2/4$$

for each $x, y \in B_E$, and hence E is uniformly convex.

The following proposition characterises the closest point of a convex subset of an inner product space; the classical proof of [10, Theorem 5.2] works constructively.

Proposition 25. *Let C be a convex subset of an inner product space E, and let $x \in E$ and $y \in C$. Then $\|x - y\| \leq \|x - z\|$ for each $z \in C$ if and only if $0 \leq \langle x - y, y - z \rangle$ for each $z \in C$.*

Proof. Suppose that $\|x - y\| \leq \|x - z\|$ for each $z \in C$. Then, for λ with $0 < \lambda < 1$ and $z \in C$, we have

$$\|x - y\|^2 \leq \|x - (\lambda y + (1 - \lambda)z)\|^2 = \|x - y - \lambda(z - y)\|^2$$
$$\leq \|x - y\|^2 - 2\lambda\langle x - y, z - y \rangle + \lambda^2\|z - y\|^2.$$

Therefore $-\lambda^2\|z-y\|^2 \leq 2\lambda\langle x-y, y-z\rangle$, and so $-\lambda/2\|z-y\|^2 \leq \langle x-y, y-z\rangle$. Thus, letting $\lambda \to 0$, we have $0 \leq \langle x-y, y-z\rangle$.

Conversely, suppose that $0 \leq \langle x-y, y-z\rangle$ for each $z \in C$. Then, for each $z \in C$, we have

$$\|x-z\|^2 = \|x-y+y-z\|^2$$
$$= \|x-y\|^2 + 2\langle x-y, y-z\rangle + \|y-z\|^2 \geq \|x-y\|^2.$$

$\square$

Theorem 6. *Let M be a closed located subspace of a Hilbert space H. Then for each $x \in H$ there exists an element Px of M such that $d(x, M) = \|x - Px\|$, and hence $\langle x - Px, z\rangle = 0$ for each $z \in M$.*

Proof. Since M is a closed located convex subset of a uniformly convex Banach space H, there exists an element Px of M such that $d(x, M) = \|x - Px\|$, by Proposition 22. Since M is subspace, $Px - z, Px + z \in M$ for each $z \in M$, and hence we have $0 \leq \langle x - Px, Px - (Px - z)\rangle = \langle x - Px, z\rangle$ and $0 \leq \langle x - Px, Px - (Px + z)\rangle = -\langle x - Px, z\rangle$ for each $z \in M$. $\square$

Remark 13. Note that for each $x \in H$, Px is strongly unique, in the sense that for each $z \in M$, if $Px \neq z$, then $\|x - Px\| < \|x - z\|$, by Lemma 10; the mapping $P : H \to M$ is called the *projection of H onto M*. Note also that P is linear; P is *idempotent*, in the sense that $P^2 x = P(Px) = Px$ for each $x \in H$; and $\|Px\| \leq \|x\|$ for each $x \in H$; see [8, Chapter 9, Theorem 6], [9, Chapter 7, (8.7)], and [18, 4.3] for details.

We have the following constructive version of the Riesz theorem for normable linear functionals.

Theorem 7. *Let f be a bounded linear functional on a Hilbert space H. Then f is normable if and only if there exists $x_0 \in H$ such that*

$$f(x) = \langle x, x_0\rangle$$

for each $x \in H$.

Proof. See [9, Chapter 8, (2.3)] and [18, Theorem 4.3.6]. $\square$

The following corollary shows that if H is a Hilbert space, then H^* is linear always.

Corollary 3. *If f and g are normable linear functionals of a Hilbert space H, then $f + g$ is normable.*

Proof. By Theorem 7, there exist $x_0, y_0 \in H$ such that

$$f(x) = \langle x, x_0 \rangle \quad \text{and} \quad g(x) = \langle x, y_0 \rangle$$

for each $x \in H$. Therefore

$$(f + g)(x) = f(x) + g(x) = \langle x, x_0 \rangle + \langle x, y_0 \rangle = \langle x, x_0 + y_0 \rangle$$

for each $x \in H$, and so $f + g$ is normable, by Theorem 7. $\square$

Notes

Treatments of the L_p spaces and the Radon-Nikodym theorem, and the L_∞ space as a quasinormed space are given respectively in [8, Chapter 9.2] and [9, Chapter 7.3], and in [9, Chapter 7.5]. Dual spaces and extreme points are dealt in [9, Chapters 7.6 and 7.7], and Hilbert spaces and the spectral theorem are dealt in [8, Chapter 9.4] and [9, Chapter 7.8]. A treatment of locally convex spaces can be found in [8, Chapter 9.5] and [18, Chapter 5.4].

Proposition 14 and Lemma 8 were given respectively in [18, Proposition 2.3.8] and in [33]. Proposition 22 was given as a problem in [8, Chapter 9, Problem 5] and [9, Chapter 7, Problem 11], and Theorem 6 was given in [8, Chapter 9, Theorem 6], [9, Chapter 7, (8.7)], and [18, Theorem 4.3.1].

7. Convexity

7.1. *Minkowski functionals*

Definition 32. A *Minkowski functional* μ of a convex absorbing subset C of a linear space E is defined by

$$\mu(x) = \inf\{r > 0 \mid x \in rC\}$$

for each $x \in E$.

Classically, a Minkowski functional of a convex absorbing subset of a linear space exists. However, constructively, it does not always exist.

Proposition 26. *If every convex absorbing subset of* $\mathbb{R}$ *has a Minkowski functional, then* LPO *holds.*

Proof. Consider a convex absorbing subset

$$C = \bigcup_{n=1}^{\infty} [-1 - \alpha n, 1 + \alpha n]$$

of $\mathbb{R}$, where α is a binary sequence. If a Minkowski functional μ of C exists, then either $2 > \mu(2)$ or $\mu(2) > 1$. In the former case, we have $2 = rx$ for some $x \in C$ and r with $0 < r < 2$; whence there exists n such that $x \in [-1 - \alpha n, 1 + \alpha n]$; if $\alpha n = 0$, then $1 < 2/r = x \leq 1$, a contradiction; so $\alpha \# \mathbf{0}$. In the latter case, if $\alpha n = 1$ for some n, then $C = [-2, 2]$; whence $1 \geq \mu(2) > 1$, a contradiction; so $\neg \alpha \# \mathbf{0}$. $\qquad \square$

Remark 14. If a Minkowski functional μ of a convex absorbing subset C of a normed space E exists, then $\mu(x+y) \leq \mu(x) + \mu(y)$ and $\mu(ax) = a\mu(x)$ for each $x, y \in E$ and $a \geq 0$.

Lemma 11. *Let C be a convex absorbing subset of a linear space E. Then C has a Minkowski functional if and only if for each x in E and positive real numbers s and t with $s < t$, either $x \notin sC$ or $x \in tC$.*

Proof. Suppose that $\mu(x) \doteq \inf\{r > 0 \mid x \in rC\}$ exists for each $x \in E$. Then either $s < \mu(x)$ or $\mu(x) < t$. In the former case, if $x \in sC$ then we have $\mu(x) \leq s < \mu(x)$, and a contradiction; whence $x \notin sC$. In the latter case, clearly $x \in tC$.

Conversely, suppose that for each positive real numbers s and t with $s < t$, either $x \notin sC$ or $x \in tC$. Let a and b be real numbers with $a < b$. Then either $a < 0$ or $0 < b$. In the former case, a is a lower bound of the set $S = \{r > 0 \mid x \in rC\}$. In the latter case, choose a' so that $\max\{0, a\} < a' < b$. Then either $x \notin a'C$ or $x \in bC$; in the former case, a', and hence a, is a lower bound of S; in the latter case, we have $b \in S$. Therefore $\inf\{r > 0 \mid x \in rC\}$ exists, by Theorem 2. $\qquad \square$

Theorem 8. *Let C be a located convex absorbing subset of a normed space E with inhabited interior. Then C has a Minkowski functional.*

Proof. We may assume that there exists $\delta > 0$ such that $B_E(\delta) \subseteq C$. Consider positive real numbers s and t with $s < t$, and set $\epsilon = \delta(t - s)$. For each x in E, either $d(x, sC) > 0$ or $d(x, sC) < \epsilon$. In the former case, $x \notin sC$. In the latter case, there exists $y \in sC$ such that $\|x - y\| < \epsilon$, and hence we have

$$x = y + (x - y) \in sC + (\epsilon/\delta)C \subseteq tC.$$

Thus $\inf\{r > 0 \mid x \in rC\}$ exists, by Lemma 11. $\qquad \square$

Remark 15. A Minkowski functional μ of a convex absorbing subset C of a normed space E with inhabited interior, if it exists, is uniformly continuous: for, since $B_E(\delta) \subseteq C$ for some $\delta > 0$, if $\|x - y\| < \delta$, then $\mu(x - y) \leq 1$.

Proposition 27. *Let T be a linear mapping of a normed space E onto a linear space F. Then $\ker(T)$ is located in E if and only if $T(B_E)$ has a Minkowski functional.*

Proof. Note that, for $x \in E$, we have

$$d(x, \ker(T)) = \inf\{t > 0 \mid \|x - y\| < t \text{ for some } y \in \ker(T)\}$$
$$= \inf\{t > 0 \mid T(x - tz) = 0 \text{ for some } z \in E \text{ with } \|z\| < 1\}$$
$$= \inf\{t > 0 \mid Tx \in tT(B_E)\}$$
$$= \mu(Tx).$$

$\square$

Corollary 4. *Let T be an open linear mapping T of a normed space E onto a normed space F such that $T(B_E)$ is located in F. Then $\ker(T)$ is located in E.*

Proof. By Theorem 8 and Proposition 27. $\square$

Theorem 9. *Let C be a convex absorbing subset of a finite-dimensional normed space E. Then C has a Minkowski functional if and only if C is located in E.*

Proof. Suppose that C is located. Then, since C has inhabited interior, by Lemma 9, C has a Minkowski functional by Proposition 8.

Conversely, suppose that $\mu(x) = \inf\{r > 0 \mid x \in rC\}$ exists for each $x \in E$. Let a and b be real numbers with $a < b$. Then either $a < 0$ or $0 < b$; in the latter case, since $c = \max\{0, a\} < b$, either $\|x\| < b$ or $(b+c)/2 < \|x\|$; so $a < 0$, or $\|x\| < b$, or $(b+c)/2 < \|x\|$. In the first case, a is a lower bound of the set $\{\|x - y\| \mid y \in C\}$; in the second case, $0 \in C$ and $\|x - 0\| < b$. In the last case, let $\epsilon = (b - c)/(2\|x\| - (b+c))$. Then since the map $x \mapsto \mu(x)$ is uniformly continuous, $d = \inf\{\mu(y) \mid y \in B_E(x, c)\}$ exists, by Proposition 8, Proposition 19, and Proposition 7; whence either $d < 1 + \epsilon$ or $d > 1 + \epsilon/2$. In the former case, there exists $y \in B_E(x, c)$ such that $\mu(y) < 1 + \epsilon$; hence, $y/(1 + \epsilon) \in C$ and

$$\|x - y/(1 + \epsilon)\| = \|x + \epsilon x - y\|/(1 + \epsilon) \leq (\|x - y\| + \epsilon\|x\|)/(1 + \epsilon)$$

$$\leq (c + \epsilon\|x\|)/(1 + \epsilon) = \frac{2c\|x\| - c(b + c) + (b - c)\|x\|}{2\|x\| - (b + c) + (b - c)}$$

$$= (b + c)/2 < b.$$

In the latter case, if $\|x - y\| < c$ for some $y \in C$, then $1 + \epsilon/2 < d \leq \mu(y) \leq 1$; this contradiction ensures that $a \leq c \leq \|x - y\|$ for each $y \in C$. So C is located by Theorem 2. $\square$

We have the following corollary which is a generalisation of Proposition 18.

Corollary 5. *Let T be a linear mapping of a normed space E onto a finite-dimensional normed space F. Then $\ker(T)$ is located in E if and only if $T(B_E)$ is located in F.*

Proof. By Theorem 9 and Proposition 27. $\qquad\square$

7.2. *The Hahn-Banach theorem*

Classically, the following Hahn-Banach theorem (the continuous extension theorem) and its corollary hold.

Theorem 10 (classical). *Let M be a subspace of a normed space E, and let f be a bounded linear functional on M. Then there exists a bounded linear functional g on E such that $g(x) = f(x)$ for each $x \in M$ and $\|g\| = \|f\|$.*

Corollary 6 (classical). *Let x be a nonzero element of a normed space E. Then there exists a bounded linear functional f on E such that $f(x) = \|x\|$ and $\|f\| = 1$.*

However, constructively, the corollary implies LLPO.

Proposition 28. *The classical Hahn-Banach theorem implies LLPO.*

Proof. Let α and β be sequences of natural numbers, and suppose that $\neg(\alpha \mathbin{\#} \mathbf{0} \wedge \beta \mathbin{\#} \mathbf{0})$. Construct real numbers a and b such that $\alpha \mathbin{\#} \mathbf{0} \leftrightarrow a \mathbin{\#} 0$ $\beta \mathbin{\#} \mathbf{0} \leftrightarrow b \mathbin{\#} 0$, and set $c = |a| - |b|$. Then $(1, c)$ is a nonzero element of the normed space $\mathbb{R}^2$ with a norm $\|(x, y)\| = |x| + |y|$. Hence there exists a bounded linear functional f such that $f(1, c) = 1 + |c|$ and $\|f\| = 1$. Since $|f(1, 0)| \le 1$ and $|f(0, 1)| \le 1$, we have

$$1 + |c| = f(1, c) = f(1, 0) + cf(0, 1) \le f(1, 0) + |c|,$$

and therefore $f(1, 0) = 1$ and $cf(0, 1) = |c|$. Either $-1 < f(0, 1)$ or $f(0, 1) < 1$; in the former case, we have $0 \le c$, and hence $\neg\beta \mathbin{\#} \mathbf{0}$; in the latter case, we have $c \le 0$, and hence $\neg\alpha \mathbin{\#} \mathbf{0}$. $\qquad\square$

Bishop [8] first gave a constructive proof of the following approximate version of Hahn-Banach theorem (the separation theorem).

Theorem 11. *Let A and B be bounded convex subsets of a separable normed space E such that the algebraic difference*

$$B - A = \{y - x \mid x \in A, y \in B\}$$

is located and $d = d(0, B - A) > 0$. Then for each $\epsilon > 0$ there exists a normable linear functional f on E such that $\|f\| = 1$ and

$$f(x) + d \leq f(y) + \epsilon$$

for each $x \in A$ and $y \in B$.

Proof. See [8, Chapter 9, Theorem 3], [9, Chapter 7, (4.3)], or [18, Theorem 5.2.9]. □

Corollary 7. *Let x_0 be a nonzero element of a separable normed space E. Then for each $\epsilon > 0$ there exists a normable linear functional f on E such that $\|f\| = 1$ and*

$$\|x_0\| \leq f(x_0) + \epsilon.$$

Proof. Apply Theorem 11 for $A = \{0\}$ and $B = \{x_0\}$; see [8, Chapter 9, Corollary of Theorem 3] and [9, Chapter 7, (4.5)] for details. □

Theorem 12. *Let M be a subspace of a separable normed space E, and let f be a nonzero linear functional on M such that the kernel $\ker(f)$ is located in E. Then for each $\epsilon > 0$ there exists a normable linear functional g on E such that $g(x) = f(x)$ for each $x \in M$ and*

$$\|g\| \leq \|f\| + \epsilon.$$

Proof. Apply Corollary 7 for $x_0 \in E/\ker(f)$ with $f(x_0) = 1$; see [8, Chapter 9, Theorem 4] and [9, Chapter 7, (4.6)] for details. □

In the following, we show, with a help of geometric properties of a Banach space such as uniform convexity and Gâteaux differentiability of a norm, an exact version of the separation theorem and the continuous extension theorem for normable linear functionals on inseparable spaces without invoking Zorn's lemma.

Definition 33. The norm of a normed space E is *Gâteaux differentiable at* $x \in E$ with the derivative $f : E \to \mathbb{R}$ if for each $y \in E$ with $\|y\| = 1$ and $\epsilon > 0$ there exists $\delta > 0$ such that

$$\forall t \in \mathbb{R}(|t| < \delta \to |\|x + ty\| - \|x\| - tf(y)| \leq \epsilon|t|).$$

Note that the derivative f is linear; see [18, Lemma 5.3.5, Proposition 5.3.6]. The norm of a normed space E is *Gâteaux differentiable* if it is Gâteaux differentiable at each $x \in E$ with $\|x\| = 1$.

Remark 16. The norm of l_p for $1 < p < \infty$ and the norm of a Hilbert space are Gâteaux differentiable at each $x \in E$ with $x \# 0$; see [43] and [14, Proposition 3.3].

Proposition 29. *Let x be a nonzero element of a normed linear space E whose norm is Gâteaux differentiable at x. Then there exists a unique normable linear functional f on E such that $f(x) = \|x\|$ and $\|f\| = 1$.*

Proof. Take the derivative f of the norm at x; see [26, Lemma 1] for details. □

Theorem 13. *Let M be a subspace of a uniformly convex Banach space E with a Gâteaux differentiable norm, and let f be a normable linear functional on M. Then there exists a unique normable linear functional g on E such that $g(x) = f(x)$ for each $x \in M$ and $\|g\| = \|f\|$.*

Proof. We may assume without loss of generality that $\|f\| = 1$. Let $\overline{M}$ be the closure of M. Then there exists a normable extension $\overline{f}$ of f on $\overline{M}$. Since $\overline{M}$ is a uniformly convex Banach, there exists $x \in \overline{M}$ such that $\overline{f}(x) = \|x\| = 1$. Take the derivative g of the norm at x; see [26, Theorem 1] for details. □

Theorem 14. *Let A and B be convex subsets of a uniformly convex Banach space E with a Gâteaux differentiable norm such that*

$$d = d(0, B - A) = \inf\{y - x \mid x \in A, y \in B\}$$

exists. Then there exists a normable linear functional f on E such that $\|f\| = 1$ and

$$f(x) + d \leq f(y)$$

for each $x \in A$ and $y \in B$.

Proof. Let $z \in \overline{B - A}$ be such that $\|z\| = d(0, B - A)$, and let f be the derivative of the norm at z. Then $f(x) + d \leq f(y)$ for each $x \in A$ and $y \in B$. See [26, Theorem 2] for details. □

Since Hilbert spaces have very good geometric properties, we have simple and direct proofs of the separation theorem and the continuous extension theorem as follows.

Theorem 15. *Let A and B be convex subsets of a Hilbert space H such that*

$$d = d(0, B - A) = \inf\{\|y - x\| \mid x \in A, y \in B\}$$

exists. Then there exists a normable linear functional f on H such that $\|f\| = d$ *and*

$$f(x) + d^2 \leq f(y)$$

for each $x \in A$ and $y \in B$.

Proof. Let $z \in \overline{B - A}$ be such that $\|z\| = d$, and let $f(u) = \langle u, z \rangle$. Then, since $0 \leq \langle -z, z - (y - x) \rangle$ for each $x \in A$ and $y \in B$, we have $f(x) + d^2 \leq f(y)$ for each $x \in A$ and $y \in B$. $\qquad\square$

Corollary 8. *Let x_0 be a nonzero element of a Hilbert space H. Then there exists a normable linear functional f on H such that $f(x_0) = \|x_0\|$ and $\|f\| = 1$.*

Proof. Take $f(x) = \langle x, x_0/\|x_0\| \rangle$. $\qquad\square$

Theorem 16. *Let M be a subspace of a Hilbert space H, and let f be a normable linear functional on M. Then there exists a normable linear functional g on H such that $g(x) = f(x)$ for each $x \in M$ and $\|g\| = \|f\|$.*

Proof. Let $\overline{M}$ be the closure of M. Then there exists a normable extension $\overline{f}$ of f on $\overline{M}$. Since $\overline{M}$ is a Hilbert space, there exists $x_0 \in \overline{M}$ such that

$$\overline{f}(x) = \langle x, x_0 \rangle$$

for each $x \in \overline{M}$. Let $g(x) = \langle x, x_0 \rangle$ for each $x \in H$. Then it is straightforward to show that $g(x) = f(x)$ for each $x \in M$ and $\|g\| = \|f\|$. $\qquad\square$

Notes

The material of Section 7.1 is drawn from [30]. A slightly different version of Corollary 5 was given in [12]; see also [16, Chapter 2, (5.4)].

A constructive proof of Theorem 12 (the continuous extension theorem) along by a standard classical proof, for example, [59, 3.2], was given in [16, Chapter 2, (5.9) and (5.10)]. Proposition 29, Theorem 13, and Theorem 14 are drawn from [26]. A treatment of the Hahn-Banach theorem in the setting of formal topology [60] can be found in [20]. A recursive counterexample of the classical Hahn-Banach theorem was given in [51].

8. Completeness

8.1. *The Banach-Steinhaus theorem*

Classically, the following Banach-Steinhaus theorem and its corollary hold.

Theorem 17 (classical). *Let* $(T_m)_m$ *be a sequence of bounded linear mappings from a Banach space E into a normed space F such that the set*

$$\{T_m x \mid m \in \mathbb{N}\}$$

is bounded in F for each $x \in E$. Then $(T_m)_m$ is equicontinuous, that is, $\{T_m x \mid m \in \mathbb{N}, x \in B_E\}$ *is bounded.*

Corollary 9 (classical). *Let* $(T_m)_m$ *be a sequence of bounded linear mappings from a Banach space E into a normed space F such that the limit*

$$T x = \lim_{m \to \infty} T_m x$$

exists for each $x \in E$. Then (being obviously linear) T is bounded.

However, we will see that the corollary implies a nonconstructive boundedness principle BD-N.

Definition 34. A subset S of $\mathbb{N}$ is *pseudo-bounded* if, for each sequence $(s_n)_n$ of S, $s_n < n$ for sufficiently large n. A *boundedness principle* BD-N is stated as follows.

BD-N: Every countable pseudo-bounded subset of $\mathbb{N}$ is bounded.

Lemma 12. *Let S be a countable pseudo-bounded subset of $\mathbb{N}$. Then there there exists a sequence $(T_m)_m$ of bounded linear mappings of l_2 into itself such that*

$$T \xi = \lim_{m \to \infty} T_m \xi_n$$

exists for each $\xi \in l_2$, and T is a one-one linear mapping of l_2 onto itself with a bounded inverse. Moreover, if T is bounded, then S is bounded.

Proof. Let $S = \{s_n \mid n \in \mathbb{N}\}$ be a countable pseudo-bounded subset of $\mathbb{N}$. Define a sequence $(T_m)_m$ of continuous linear mapping of l_2 by

$$T_m \xi = (2^{s_0} \xi_0, \ldots, 2^{s_m} \xi_m, 0, 0, \ldots).$$

We show that

$$T \xi = \lim_{m \to \infty} T_m \xi$$

exists for each $\xi \in l_2$. To this end, given $\xi \in l_2$, let γ be a nondecreasing sequence of natural numbers such that

$$\sum_{k=\gamma(n)}^{\infty} |\xi_k|^2 < 2^{-(3n+1)},$$

and construct a binary sequence α such that

$$\alpha(n) = 0 \Rightarrow \sum_{k=\gamma(n)}^{\gamma(n+1)-1} 2^{2s_k} |\xi_k|^2 < 2^{-n},$$

$$\alpha(n) = 1 \Rightarrow \sum_{k=\gamma(n)}^{\gamma(n+1)-1} 2^{2s_k} |\xi_k|^2 > 2^{-(n+1)}.$$

Define a sequence $(s'_n)_n$ in S as follows: if $\alpha(n) = 0$, set $s'_n = s_0$; if $\alpha(n) = 1$, then, noting

$$2^{-(n+1)} < \sum_{k=\gamma(n)}^{\gamma(n+1)-1} 2^{2s_k} |\xi_k|^2 \leq \max\{2^{2s_k} \mid \gamma(n) \leq k < \gamma(n+1)\} \cdot 2^{-(3n+1)},$$

choose k with $\gamma(n) \leq k < \gamma(n+1)$ and $n < s_k$, and set $s'_n = s_k$. Since S is pseudo-bounded, there exists N such that $s'_n < n$ for each $n \geq N$. If $\alpha(n) = 1$ for some $n \geq N$, then $n < s'_n < n$, a contradiction. Hence $\alpha(n) = 0$ for each $n \geq N$, and therefore $\lim_{m\to\infty} T_m \xi$ exists.

It is clear that T is a one-one linear mapping of l_2 onto itself with the bounded inverse

$$T^{-1}\xi = (2^{-s_0}\xi_0, 2^{-s_1}\xi_1, \dots).$$

$\square$

Proposition 30. *The classical uniform boundedness theorem implies* BD-N.

Proof. Consider T in Lemma 12. $\square$

We have the following version of the uniform boundedness theorem.

Theorem 18. *Let* $(T_m)_m$ *be a sequence of bounded linear mappings from a Banach space E into a normed space F. If (x_m) is a sequence of B_E such that $\{T_m x_m \mid m \in \mathbb{N}\}$ is unbounded, then there exists $x \in E$ such that*

$$\{T_m x \mid m \in \mathbb{N}\}$$

is unbounded.

Proof. Let (x_m) be a sequence of B_E such that $\{T_m x_m \mid m \in \mathbb{N}\}$ is unbounded, and, given n, let

$$U_n = \{x \in E \mid n < \|T_m x\| \text{ for some } m\}.$$

Then, trivially, U_n is open. Consider $y \in E$ and $\epsilon > 0$, and choose m so that $(2n+1)/\epsilon < \|T_m x_m\|$. Then either $n < \|T_m y\|$ or $\|T_m y\| < n + 1$; in the former case, we have $y \in U_n$; in the latter case, we have $\|\epsilon x_m\| < \epsilon$ and

$$\|T_m(y + \epsilon x_m)\| \geq \epsilon\|T_m x_m\| - \|T_m y\| > (2n+1) - (n+1) > n.$$

Therefore U_n is dense in E. Thus there exists $x \in \bigcap_{n=0}^{\infty} U_n$, by Theorem 3, and so $\{T_m x \mid m \in \mathbb{N}\}$ is unbounded. $\square$

8.2. *The open mapping theorem and the closed graph theorem*

Classically, the open mapping theorem, and, as its corollaries, the closed graph theorem and the Banach inverse mapping theorem hold.

Theorem 19 (classical). *Let T be a bounded linear mapping between Banach spaces. Then T is an open mapping.*

Corollary 10 (classical). *Let T be a linear mapping between Banach spaces. Then T is bounded if and only if its graph is closed.*

Corollary 11 (classical). *Let T be a bounded one-to-one linear mapping from a Banach space onto a Banach space. Then its inverse T^{-1} is bounded.*

However, the Banach inverse mapping theorem implies a nonconstructive boundedness principle BD-N.

Proposition 31. *The classical Banach inverse mapping theorem implies BD-N.*

Proof. For a countable pseudo-bounded subset S of $\mathbb{N}$, consider T^{-1} in Lemma 12. Then T^{-1} is continuous bijection of l_2 into itself. If its inverse $T = (T^{-1})^{-1}$ is bounded, then S is bounded. $\qquad\square$

The following lemma shows that a linear mapping of a Banach space into a normed space is well-behaved.

Lemma 13. *Let T be a linear mapping of a Banach space E into a normed space F. Then T is strongly extensional.*

Proof. Suppose that $0 < \|Tx\|$, and construct a sequence α of natural numbers such that

$$\alpha n = 0 \Rightarrow \|x\| < 2^{-2n},$$
$$\alpha n \neq 0 \Rightarrow 0 < \|x\|.$$

Since if $\alpha 0 \neq 0$, then we have $0 < \|x\|$, we may assume that $\alpha 0 = 0$, and define a sequence $(y_n)_n$ of E by $y_n = 2^n x$ if $\forall k \leq n(\alpha k = 0)$; $y_n = y_{n-1}$, otherwise. Then $(y_n)_n$ is a Cauchy sequence, and hence converges to a limit $y \in E$. Choose N so that $\|Ty\| < 2^N \|Tx\|$. Either $\forall k \leq N(\alpha k = 0)$ or $\exists k \leq N(\alpha k \neq 0)$. In the former case, if $\alpha n \neq 0$ for some $n > N$, then $y = 2^m x$ for some $m \geq N$, and hence

$$2^N \|Tx\| < 2^m \|Tx\| = \|Ty\| < 2^N \|Tx\|,$$

a contradiction. Therefore $\forall n(\alpha n = 0)$, and so $\|x\| = 0$; whence $\|Tx\| = 0$, a contradiction. Thus the latter must be the case, and so $0 < \|x\|$. □

Corollary 12. *Let T be a linear mapping of a Banach space E into a normed linear space F. Then T is continuous if and only if T is sequentially nondiscontinuous.*

Proof. By Theorem 4 and Lemma 13. □

We will show that the Banach inverse mapping theorem constructively holds for sequentially continuous bijections, and, as its corollaries sequential versions of the open mapping theorem and the closed graph theorem. To this end, we need the following technical lemma whose original form was proved in [15, Lemma 4.4].

Lemma 14. *Let T be a sequentially continuous linear mapping of a Banach space E onto a normed linear space F such that $T(B_E)$ is located, and let ϵ be a positive number. Then for each y in $B_F(\epsilon)$ there exists an x in B_E such that if $y \neq Tx$, then $d(y', T(B_E)) > 0$ for some y' in $B_F(2\epsilon)$.*

Proof. Let $\epsilon > 0$, and given y in $B_F(\epsilon)$, construct a nondecreasing binary sequence α and a sequence $(x_k)_k$ in B_E such that for each k,

1. if $\alpha(k) = 0$, then $\| \sum_{i=0}^{k} 2^{-i}(y - Tx_i)\| < \epsilon/2^{k+1}$,
2. if $N = \min\{k \mid \alpha(k) = 1\}$, then $d(2^N \sum_{i=0}^{N-1} 2^{-i}(y - Tx_i) + y, T(B_E)) > 0$.

We proceed by induction. For $k = 0$, either $d(y, T(B_E)) > 0$, in which case we set $\alpha(0) = 1$ and $x_0 = 0$; or $d(y, T(B_E)) < \epsilon/2$. In the latter case, set $\alpha(0) = 0$ and choose x_0 in B_E such that $\|y - Tx_0\| < \epsilon/2$. Hence (1) and (2) hold for $k = 0$. Assume that we have constructed $\alpha(0), \ldots, \alpha(k)$ and $x_0, \ldots, x_k$. If $\alpha(k) = 1$, set $\alpha(k + 1) = 1$ and $x_{k+1} = 0$. Otherwise, $\alpha(k) = 0$ and either

$$d(2^{k+1} \sum_{i=0}^{k} 2^{-i}(y - Tx_i) + y, T(B_E)) > 0$$

or

$$d(2^{k+1} \sum_{i=0}^{k} 2^{-i}(y - Tx_i) + y, T(B_E)) < \epsilon/2.$$

In the former case, define $\alpha(k + 1) = 1$ and $x_{k+1} = 0$. In the latter case, set $\alpha(k + 1) = 0$ and choose x_{k+1} in B_E such that

$$\|2^{k+1} \sum_{i=0}^{k} 2^{-i}(y - Tx_i) + y - Tx_{k+1}\| < \epsilon/2.$$

Then $\| \sum_{i=0}^{k+1} 2^{-i}(y - Tx_i) \| < \epsilon/2^{k+2}$. Hence (1) and (2) hold for $k + 1$. This completes the induction.

As E is complete, the series $\sum_{k=0}^{\infty} 2^{-k} x_k$ converges to an element x in B_E. Suppose that $y \neq Tx$. Since T is sequentially continuous, there exists a number n such that $3\epsilon/2^{n+1} < \| y - \sum_{k=0}^{n} 2^{-k} Tx_k \|$. If $\alpha(n) = 0$, then

$$\| y - \sum_{k=0}^{n} 2^{-k} Tx_k \| \leq \| \sum_{k=0}^{n} 2^{-k}(y - Tx_k) \| + \| \sum_{k=n+1}^{\infty} 2^{-k} y \|$$
$$< \epsilon/2^{n+1} + \epsilon/2^n = 3\epsilon/2^{n+1},$$

which contradicts our choice of n. Hence $\alpha(n) = 1$. Let $N = \min\{k \leq n \mid \alpha(k) = 1\}$, and let

$$y' = 2^N \sum_{k=0}^{N-1} 2^{-k}(y - Tx_k) + y.$$

Then $d(y', T(B_E)) > 0$ and $\| y - y' \| < \epsilon$. $\qquad\square$

Theorem 20. *Let T be a one-one sequentially continuous linear mapping of a separable Banach space E onto a Banach space F. Then T^{-1} is sequentially continuous.*

Proof. Since F is complete, $T^{-1} : F \to E$ is strongly extensional, by Lemma 13. Let $(Tx_n)_n$ be a sequence in F converging to 0. Then, by Lemma 5, for each $\epsilon > 0$, either $\| x_n \| > \epsilon/2$ for infinitely many n or $\| x_n \| < \epsilon$ for all sufficiently large n. In the former case, taking a subsequence, we may assume that $\| x_n \| > \epsilon/2$ for each n. Hence LPO holds, by Lemma 6. Let $(y_k)_k$ be a dense sequence in B_E. Then since T is sequentially continuous, $(Ty_k)_k$ is dense in $T(B_E)$. Hence, by Lemma 7, $T(B_E)$ is located. Let $z_n = 2\epsilon^{-1} x_n$. Then $Tz_n \to 0$ and $\| z_n \| > 1$. It follows from the strong extensionality of T^{-1} that $Tz_n \neq Tx$ for each x in B_E. Applying Lemma 14 to each Tz_n, construct a sequence $(Tz_n')_n$ such that $Tz_n' \to 0$ and $d(Tz_n', T(B_E)) > 0$. Then $0 \in \overline{-T(B_E)}$. Since $T(B_E)$ is convex and absorbing, we have a contradiction, by Lemma 8. Thus $\| x_n \| < \epsilon$ for all sufficiently large n. Since $\epsilon > 0$ is arbitrary, it follows that T^{-1} is sequentially continuous. $\qquad\square$

Definition 35. A linear mapping T between normed spaces is *sequentially open* if, whenever $Tx_n \to 0$ as $n \to \infty$, there exists a sequence (y_n) in $\ker(T)$ such that $x_n + y_n \to 0$.

Corollary 13. *Let T be a sequentially continuous linear mapping of a separable Banach space E onto a Banach space F such that $\ker(T)$ is located. Then T is sequentially open.*

Proof. Replacing E by $E/\ker(T)$, we may assume T is one-one. By Theorem 20, T^{-1} is sequentially continuous; so T is sequentially open.   □

Corollary 14. *Let T be a linear mapping of a Banach space E into a Banach space F such that $\mathrm{graph}(T)$ is closed and separable. Then T is sequentially continuous.*

Proof. The projection $p : (x, Tx) \mapsto x$ is a bounded one-one linear mapping of $\mathrm{graph}(T)$ onto E. By Theorem 20, p^{-1} is sequentially continuous; so T is sequentially continuous.   □

Notes

Lemma 12 was given in [34], and Theorem 18 was given as a problem in [8, Chapter 9, Problem 6] and [9, Chapter 7, Problem 20]. The material of Section 8.2 is drawn from [33], and its extension to an F-space can be found in [41]. Lemma 13 was proved in [13].

A boundedness principle BD-N was first recognised in [31] as follows.

Every countable pseudo-bounded set of natural numbers is bounded,

where a set S of natural numbers is pseudo-bounded if $\lim_{n\to\infty} s_n/n = 0$ for each sequence $(s_n)_n$ in A; bounded if there exist n such that $s < n$ for each $s \in A$. Then, in [42, Lemma 3], it was shown that a set S of natural numbers is pseudo-bounded if and only if for each sequence $(s_n)_n$ in S, $s_n < n$ for all sufficiently large n; see also [58]. We have known that BD-N is a principle, like WMP, which holds in intuitionistic, constructive recursive and classical mathematics, and is equivalent to various theorems in analysis (see, for example, [31, 37, 42]). However, Lietz and Streicher [47] showed, using realisability, that BD-N is underivable in $\mathbf{E\text{-}HA}^\omega + \mathrm{AC}$, by invoking the fact that $\mathbf{E\text{-}HA}^\omega + \mathrm{AC} + \mathrm{WC\text{-}N}$ is inconsistent [65, 9.6.10].

9. Adjoint and compact operators

9.1. *Adjoint operators*

Definition 36. Let H be a Hilbert space. Then an *operator* A on H is a bounded linear mapping of H into itself; an operator A^* on H is an *adjoint* of an operator A on H if

$$\langle Ax, y \rangle \doteq \langle x, A^*y \rangle$$

for each $x, y \in H$; an operator A on H is *selfadjoint* if

$$\langle Ax, y \rangle = \langle x, Ay \rangle$$

for each $x, y \in H$.

Example 16. The projection P of a Hilbert space onto a closed located subspace is a selfadjoint operator: in fact, since $\langle x - Px, Py \rangle = \langle Px, y - Py \rangle = 0$, by Theorem 6, we have

$$\langle Px, y \rangle = \langle Px, Py \rangle + \langle Px, y - Py \rangle = \langle Px, Py \rangle + \langle x - Px, Py \rangle = \langle x, Py \rangle;$$

see [8, Chapter 9.4], [9, Chapter 7.8], and [18, Chapter 4.3].

Classically, every operator has an adjoint. However, constructively, we have the following proposition.

Proposition 32. *If every operator on l_2 has an adjoint, then* LPO *holds.*

Proof. Let α be a binary sequence with at most one nonzero term, and define a linear mapping C from l_2 into itself by

$$C\xi = \left(\sum_{k=0}^{\infty} \alpha(k)\xi_k / \sqrt{2^{n+1}} \right)_n.$$

Then C is an operator. Note that

$$\langle C\xi, \zeta \rangle = \sum_{k=0}^{\infty} \alpha(k)\xi_k$$

for $\zeta = (1/\sqrt{2^{n+1}})_n$. If C has an adjoint, then, the linear functional $f : \xi \mapsto \langle C\xi, \zeta \rangle$ is normable, by Theorem 7, and therefore either $0 < \|f\|$ or $\|f\| < 1$; in the former case, we have $\alpha \,\#\, \mathbf{0}$; in the latter case, we have $\neg \alpha \,\#\, \mathbf{0}$. $\qquad\square$

Definition 37. An operator A on a Hilbert space H is *weakly compact* if

$$\{ \langle Ax, y \rangle \mid x \in B_H \}$$

is a totally bounded subset of $\mathbb{R}$ for each $y \in H$.

Remark 17. Note that every compact operator is weakly compact, and the identity operator $I : x \mapsto x$ on a Hilbert space is not compact in general, but weakly compact.

Proposition 33. *An operator A has an adjoint if and only if A is weakly compact.*

Proof. By Theorem 7, A has an adjoint if and only if the linear functional $x \mapsto \langle Ax, y \rangle$ is normable for each $y \in H$ if and only if $\{ \langle Ax, y \rangle \mid x \in B_H \}$ is totally bounded for each $y \in H$. $\qquad\square$

9.2. *Compact operators*

We have the following classical properties of compact operators on a Hilbert space.

Theorem 21 (classical). *Let A and B be compact operators on a Hilbert space H, let C be an operator on H, and let $a \in \mathbb{R}$. Then*

1. *aA, $A + B$ and A^* are compact,*
2. *CA and AC are compact.*

However, we can construct a compact operator A and a bounded operator C such that AC is *not* compact.

Proposition 34. *If AC is compact for each compact operator A and bounded operator C on l_2, then LPO holds.*

Proof. Let α be a binary sequence with at most one nonzero term, and define linear mappings A and C from l_2 into itself by

$$A\xi = \left(\xi_n / \sqrt{2^{n+1}} \right)_n ,$$
$$C\xi = \left(\sum_{k=0}^{\infty} \alpha(k)\xi_k / \sqrt{2^{n+1}} \right)_n .$$

Then A is compact and C is bounded, and

$$\|AC\xi\|^2 = |\textstyle\sum_{k=0}^{\infty} \alpha(k)\xi_k|^2 / 3.$$

Therefore either $0 < \|AC\|$ or $\|AC\| < 1/3$; in the former case, we have $\alpha \mathrel{\#} \mathbf{0}$; in the latter case, we have $\neg \alpha \mathrel{\#} \mathbf{0}$. $\qquad\square$

Proposition 35. *Let u be a bounded linear mapping of a Hilbert space H into $\mathbb{R}^n$, and for $i < n$, let $\pi_i : \mathbb{R}^n \to \mathbb{R}$ be a mapping defined by $\pi_i(x_0, \dots, x_{n-1}) = x_i$. Then u is compact if and only if $\pi_i \circ u$ is normable for each $i < n$.*

Proof. Let u be compact. Then $\pi_i \circ u$ is compact, and therefore normable, for each $i < n$.

Conversely, suppose that $\pi_i \circ u$ is normable for each $i < n$. Then for each $i < n$ there exist $y_i \in H$ such that $(\pi_i \circ u)(x) = \langle x, y_i \rangle$, by Theorem 7. Then, given N, there exists a finite-dimensional subspace Y of H such that $d(y_i, Y) < 2^{-(N+n+1)}$ for each $i < n$, by Proposition 21. Since $u(B_Y)$ is totally bounded, by Proposition 19 and Proposition 8, there exist $x_0, \dots, x_{m-1}$ in B_Y such that

$$\forall y \in B_Y \exists k < m \left(\|u(y) - u(x_k)\| < 2^{-(N+1)} \right).$$

Therefore for each $x \in B_H$, since $Px \in B_Y$, there exists $k < m$ such that $\|u(Px) - u(x_k)\| < 2^{-(N+1)}$, where P is the projection onto Y. Since P is selfadjoint and $\|y_i - Py_i\| < 2^{-(N+n+1)}$ for each $i < n$, we have

$$
\begin{aligned}
\|u(x) - u(x_k)\| &\leq \|u(x - Px)\| + \|u(Px) - u(x_k)\| \\
&< \textstyle\sum_{i=0}^{n-1} |(\pi_i \circ u)(x - Px)| + 2^{-(N+1)} \\
&= \textstyle\sum_{i=0}^{n-1} |\langle x - Px, y_i \rangle| + 2^{-(N+1)} \\
&= \textstyle\sum_{i=0}^{n-1} |\langle x, y_i - Py_i \rangle| + 2^{-(N+1)} \\
&\leq \textstyle\sum_{i=0}^{n-1} \|x\| \|y_i - Py_i\| + 2^{-(N+1)} \\
&< \textstyle\sum_{i=0}^{n-1} 2^{-(N+n+1)} + 2^{-(N+1)} \leq 2^{-N}.
\end{aligned}
$$

$\square$

Theorem 22. *Let A and B be compact operators on a Hilbert space H, let C be an operator on H and let $\alpha \in \mathbb{R}$. Then*

1. *αA is compact;*
2. *$A + B$ is compact;*
3. *A^* exists and is compact;*
4. *CA is compact;*
5. *if C is weakly compact, then AC is compact.*

Proof. (1) and (4) are trivial.

(2): Given N, there exist $x_0, \ldots, x_{n-1}$ and $y_0, \ldots, y_{m-1}$ in B_H such that

$$
\forall x \in B_H \exists i < n \exists j < m \left[\|Ax - Ax_i\| < 2^{-(N+5)} \wedge \|Bx - By_j\| < 2^{-(N+5)} \right].
$$

By applying Proposition 21, construct a finite-dimensional subspace Y of H such that

$$
\forall i < n \forall j < m \left[d(Ax_i, Y) < 2^{-(N+4)} \wedge d(By_j, Y) < 2^{-(N+4)} \right],
$$

and let P be the projection onto Y. Then for each $x \in B_H$, since there exists $i < n$ such that $\|Ax - Ax_i\| < 2^{-(N+5)}$, we have

$$
\begin{aligned}
\|Ax - PAx\| &\leq \|Ax - Ax_i\| + \|Ax_i - PAx_i\| + \|PAx_i - PAx\| \\
&< 2^{-(N+5)} + 2^{-(N+4)} + 2^{-(N+5)} = 2^{-(N+3)}.
\end{aligned}
$$

Similarly, we have $\|Bx - PBx\| < 2^{-(N+3)}$. Hence

$$
\|(A + B)x - (PA + PB)x\| \leq \|Ax - PAx\| + \|Bx - PBx\| < 2^{-(N+2)}
$$

for each $x \in B_H$. Let $\{e_0, \ldots, e_{n-1}\}$ be a metric basis of Y. Then we can write PA and PB in the forms

$$PA(x) = f_0(x)e_0 + \cdots + f_{n-1}(x)e_{n-1},$$
$$PB(x) = g_1(x)e_0 + \cdots + g_{n-1}(x)e_{n-1}.$$

Since PA and PB are compact, by (4), f_i and g_i are normable linear functionals on H for each $i < n$, by Proposition 35, and hence $f_i + g_i$ is normable, by Corollary 3. Therefore the linear mapping

$$(PA + PB)(x) = (f_0 + g_0)(x)e_0 + \cdots + (f_{n-1} + g_{n-1})(x)e_{n-1}$$

of H into Y is compact, by Proposition 35. Thus there exists $z_0, \ldots, z_{l-1}$ in B_H such that $\forall x \in B_H \exists k < l[\|(PA + PB)x - (PA + PB)z_k\| < 2^{-(N+1)}]$, and so for each $x \in B_H$ there exists $k < l$ such that

$$\begin{aligned}
\|(A+B)x - (A+B)z_k\| &\leq \|(A+B)x - (PA+PB)x\| \\
&\quad + \|(PA+PB)x - (PA+PB)z_k\| \\
&\quad + \|(PA+PB)z_k - (A+B)z_k\| \\
&< 2^{-(N+2)} + 2^{-(N+1)} + 2^{-(N+2)} = 2^{-N}.
\end{aligned}$$

(3): Since A is compact, $\{\langle Ax, y\rangle \mid x \in B_H\}$ is totally bounded for each $y \in H$; whence A^* exists, by Proposition 33. Let $x_0, \ldots, x_{n-1}$ be elements of B_H such that $\forall x \in B_H \exists i < n[\|Ax - Ax_i\| < 2^{-(N+2)}]$, and let $u : H \to \mathbb{R}^n$ be the bounded linear mapping defined by

$$u(y) = (\langle Ax_0, y\rangle, \ldots, \langle Ax_{n-1}, y\rangle)$$

for each $y \in H$. Then u is compact, by Proposition 35, and hence there exists $y_0, \ldots, y_{m-1}$ in B_H such that $\forall x \in B_H \exists j < m[\|u(x) - u(y_j)\| < 2^{-(N+1)}]$. Therefore for each $x \in B_H$ there exists $j < m$ such that

$$|\langle Ax_i, x\rangle - \langle Ax_i, y_j\rangle| \leq \|u(x) - u(y_j)\| < 2^{-(N+1)}$$

for each $i < n$. Either $0 < \|A^*x - A^*y_j\|$ or $\|A^*x - A^*y_j\| < 2^{-N}$. In the former case, setting $z = (A^*x - A^*y_j)/\|A^*x - A^*y_j\|$, there exists $i < n$ such that $\|Az - Ax_i\| < 2^{-(N+2)}$, and hence we have

$$\begin{aligned}
\|A^*x - A^*y_j\| &= \langle z, A^*x - A^*y_j\rangle \\
&\leq |\langle Ax_i, x\rangle - \langle Ax_i, y_j\rangle| + |\langle Az - Ax_i, x\rangle| + |\langle Az - Ax_i, y_j\rangle| \\
&< 2^{-(N+1)} + 2^{-(N+2)} + 2^{-(N+2)} = 2^{-N}.
\end{aligned}$$

(5): By (3), A^* is compact, and hence C^*A^* is compact. Therefore $AC = (C^*A^*)^*$ is compact, by (3). $\qquad\square$

Notes

The material of Section 9 is drawn from [28], and its extension to a mapping of a normed space can be found in [27]. The notion of a weakly compact set is constructively useful, as an inhabited bounded convex subset of a Hilbert space is located if and only if it is weakly compact; see [35, Corollary 5]. Further results on weakly compact convex sets are given in [14, 35, 40, 56].

Acknowledgments

The author heartily thanks the organisers, Klaus Mainzer, Peter Schuster and Helmut Schwichtenberg, of the autumn school "Proof and Computation" held from 3rd to 8th October 2016 in Fischbachau, for giving him an opportunity of three lectures on constructive (functional) analysis. The author also thanks the Japan Society for the Promotion of Science (JSPS), Core-to-Core Program (A. Advanced Research Networks), and Grant-in-Aid for Scientific Research (C) No. 16K05251 for supporting the research.

References

[1] Peter Aczel, *The type theoretic interpretation of constructive set theory*, Logic Colloquium '77 (Proc. Conf., Wroclaw, 1977), pp. 55–66, Stud. Logic Foundations Math., 96, North-Holland, Amsterdam-New York, 1978.

[2] Peter Aczel, *The type theoretic interpretation of constructive set theory: choice principles*, The L. E. J. Brouwer Centenary Symposium (Noordwijkerhout, 1981), 1–40, Stud. Logic Found. Math., 110, North-Holland, Amsterdam, 1982.

[3] Peter Aczel, *The type theoretic interpretation of constructive set theory: inductive definitions*, Logic, methodology and philosophy of science, VII (Salzburg, 1983), 17–49, Stud. Logic Found. Math., 114, North-Holland, Amsterdam, 1986.

[4] Peter Aczel and Michael Rathjen, *Notes on constructive set theory*, Report No. 40, Institut Mittag-Leffler, The Royal Swedish Academy of Sciences, 2001.

[5] Peter Aczel and Michael Rathjen, *CST Book draft*, August 19, 2010, http://www1.maths.leeds.ac.uk/~rathjen/book.pdf.

[6] Andrej Bauer, *Five stages of accepting constructive mathematics*, Bull. Amer. Math. Soc. (N.S.) **54** (2017), 481–498.

[7] Michael J. Beeson, *Foundations of constructive mathematics*, Metamathematical studies, Ergebnisse der Mathematik und ihrer Grenzgebiete (3) [Results in Mathematics and Related Areas (3)], 6. Springer-Verlag, Berlin, 1985.

[8] Errett Bishop, *Foundations of constructive analysis*, McGraw-Hill Book Co., New York-Toronto, Ont.-London, 1967.

[9] Errett Bishop and Douglas Bridges, *Constructive analysis*, Grundlehren der Mathematischen Wissenschaften [Fundamental Principles of Mathematical Sciences], 279. Springer-Verlag, Berlin, 1985.

[10] Haim Brezis, *Functional analysis, Sobolev spaces and partial differential equations*, Universitext, Springer, New York, 2011.

[11] Douglas S. Bridges, *Constructive functional analysis*, Research Notes in Mathematics, 28. Pitman (Advanced Publishing Program), Boston, Mass.-London, 1979.

[12] Douglas Bridges, Allan Calder, William Julian, Ray Mines and Fred Richman, *Fred. Bounded linear mappings of finite rank*, J. Funct. Anal. **43** (1981), 143–148.

[13] Douglas Bridges and Hajime Ishihara, *Linear mappings are fairly well-behaved*, Arch. Math. (Basel) **54** (1990), 558–562.

[14] Douglas Bridges, Hajime Ishihara and Luminiţa Vîţă, *Computing infima on convex sets, with applications in Hilbert spaces*, Proc. Amer. Math. Soc. **132** (2004), 2723–2732.

[15] Douglas Bridges, William Julian and Ray Mines, *A constructive treatment of open and unopen mapping theorems*, Z. Math. Logik Grundlag. Math. **35** (1989), 29–43.

[16] Douglas Bridges and Fred Richman, *Varieties of constructive mathematics*, London Mathematical Society Lecture Note Series, 97. Cambridge University Press, Cambridge, 1987.

[17] Douglas Bridges, Dirk van Dalen and Hajime Ishihara, *Ishihara's proof technique in constructive analysis*, Indag. Math. (N.S.) **14** (2003), 163–168.

[18] Douglas S. Bridges and Luminiţa Simona Vîţă, *Techniques of constructive analysis*, Universitext, Springer, New York, 2006.

[19] L.E.J. Brouwer, *Brouwer's Cambridge lectures on intuitionism*, Edited and with a preface by D. van Dalen, Cambridge University Press, Cambridge-New York, 1981.

[20] Jan Cederquist, Thierry Coquand and Sara Negri, *The Hahn-Banach theorem in type theory*, Twenty-five years of constructive type theory (Venice, 1995), 57–72, Oxford Logic Guides, 36, Oxford Univ. Press, New York, 1998.

[21] Michael Dummett, *Elements of intuitionism*, Second edition, Oxford Logic Guides, 39. The Clarendon Press, Oxford University Press, New York, 2000.

[22] Martín H. Escardó, *Infinite sets that satisfy the principle of omniscience in any variety of constructive mathematics*, J. Symbolic Logic **78** (2013), 764–784.

[23] Makoto Fujiwara, Hajime Ishihara and Takako Nemoto, *Some principles weaker than Markov's principle*, Arch. Math. Logic **54** (2015), 861–870.

[24] Paul R. Halmos, *I want to be a mathematician. An automathography*, Springer-Verlag, New York, 1985.

[25] A. Heyting, *Intuitionism: An introduction*, Second revised edition North-Holland Publishing Co., Amsterdam 1966.

[26] Hajime Ishihara, *On the constructive Hahn-Banach theorem*, Bull. London Math. Soc. **21** (1989), 79–81.

[27] Hajime Ishihara, *Constructive compact linear mappings*, Bull. London Math. Soc. **21** (1989), 577–584.

[28] Hajime Ishihara, *Constructive compact operators on a Hilbert space*, International Symposium on Mathematical Logic and its Applications (Nagoya, 1988), Ann. Pure Appl. Logic **52** (1991), 31–37.

[29] Hajime Ishihara, *Continuity and nondiscontinuity in constructive mathematics*, J. Symbolic Logic **56** (1991), 1349–1354.

[30] Hajime Ishihara, *Constructive existence of Minkowski functionals*, Proc. Amer. Math. Soc. **116** (1992), 79–84.

[31] Hajime Ishihara, *Continuity properties in constructive mathematics*, J. Symbolic Logic **57** (1992), 557–565.

[32] Hajime Ishihara, *Markov's principle, Church's thesis and Lindelöf's theorem*, Indag. Math. (N.S.) **4** (1993), 321–325.

[33] Hajime Ishihara, *A constructive version of Banach's inverse mapping theorem*, New Zealand J. Math. **23** (1994), 71–75.

[34] Hajime Ishihara, *Sequential continuity in constructive mathematics*, Combinatorics, computability and logic (Constanţa, 2001), 5–12, Springer Ser. Discrete Math. Theor. Comput. Sci., Springer, London, 2001.

[35] Hajime Ishihara, *Locating subsets of a Hilbert space*, Proc. Amer. Math. Soc. **129** (2001), 1385–1390.

[36] Hajime Ishihara, *Constructive reverse mathematics: compactness properties*, From sets and types to topology and analysis, 245–267, Oxford Logic Guides, 48, Oxford Sci. Publ., Oxford Univ. Press, Oxford, 2005.

[37] Hajime Ishihara, *The uniform boundedness theorem and a boundedness principle*, Ann. Pure Appl. Logic **163** (2012), 1057–1061.

[38] Hajime Ishihara, *On Brouwer's continuity principle*, Indag. Math. (N.S.), to appear.

[39] Hajime Ishihara and Ray Mines, *Various continuity properties in constructive analysis*, Reuniting the antipodes — constructive and nonstandard views of the continuum (Venice, 1999), 103–110, Synthese Lib., 306, Kluwer Acad. Publ., Dordrecht, 2001.

[40] Hajime Ishihara and Luminiţa Vîţă, *Locating subsets of a normed space*, Proc. Amer. Math. Soc. **131** (2003), 3231–3239.

[41] Hajime Ishihara and Luminiţa Vîţă, *A constructive Banach inverse mapping theorem in F-spaces*, New Zealand J. Math. **35** (2006), 183–187.

[42] Hajime Ishihara and Satoru Yoshida, *A constructive look at the completeness of the space $\mathcal{D}(\mathbf{R})$*, J. Symbolic Logic **67** (2002), 1511–1519.

[43] D. L. Johns and C. G. Gibson, *A constructive approach to the duality theorem for certain Orlicz spaces*, Math. Proc. Cambridge Philos. Soc. **89** (1981), 49–69.

[44] Ulrich Kohlenbach, *On weak Markov's principle*, Dagstuhl Seminar on Computability and Complexity in Analysis, 2001. MLQ Math. Log. Q. **48** (2002), 59–65.

[45] Ulrich Kohlenbach, *Applied proof theory: proof interpretations and their use in mathematics*, Springer Monographs in Mathematics, Springer-Verlag, Berlin, 2008.

[46] B.A. Kushner, *Lectures on constructive mathematical analysis*, Translated from the Russian by E. Mendelson, Translation edited by Lev J. Leifman, Translations of Mathematical Monographs, 60. American Mathematical Society, Providence, RI, 1984.

[47] Peter Lietz and Thomas Streicher, *Realizability models refuting Ishihara's boundedness principle*, Ann. Pure Appl. Logic **163** (2012), 1803–1807.

[48] Mark Mandelkern, *Constructive continuity*, Mem. Amer. Math. Soc. **42** (1983), no. 277,

[49] Mark Mandelkern, *Constructively complete finite sets*, Z. Math. Logik Grundlag. Math. **34** (1988), 97–103.

[50] Per Martin-Löf, *Intuitionistic type theory*, Notes by Giovanni Sambin. Studies in Proof Theory. Lecture Notes, 1. Bibliopolis, Naples, 1984.

[51] G. Metakides, A. Nerode and R.A. Shore, *Recursive limits on the Hahn-Banach theorem*, Errett Bishop: reflections on him and his research (San Diego, Calif., 1983), 85–91, Contemp. Math., **39**, Amer. Math. Soc., Providence, RI, 1985.

[52] Ray Mines, Fred Richman and Wim Ruitenburg, *A course in constructive algebra*, Universitext, Springer-Verlag, New York, 1988.

[53] John Myhill, *Constructive set theory*, J. Symbolic Logic **40** (1975), 347–382.

[54] Sara Negri and Jan von Plato, *Structural proof theory*, Appendix C by Aarne Ranta, Cambridge University Press, Cambridge, 2001.

[55] Fred Richman, *Generalized real numbers in constructive mathematics*, Indag. Math. (N.S.) **9** (1998), 595–606.

[56] Fred Richman, *Adjoints and the image of the ball*, Proc. Amer. Math. Soc. **129** (2001), 1189–1193.

[57] Fred Richman, *Real numbers and other completions*, MLQ Math. Log. Q. **54** (2008), 98–108.

[58] Fred Richman, *Intuitionistic notions of boundedness in* N, MLQ Math. Log. Q. **55** (2009), 31–36.

[59] Walter Rudin, *Functional analysis*, Second edition. International Series in Pure and Applied Mathematics. McGraw-Hill, Inc., New York, 1991.

[60] Giovanni Sambin, *Some points in formal topology*, Topology in computer science (Schloß Dagstuhl, 2000). Theoret. Comput. Sci. 305 (2003), 347–408.

[61] Peter M. Schuster, *Real numbers as black boxes*, New Zealand J. Math. **31** (2002), 189–202.

[62] Helmut Schwichtenberg and Stanley S. Wainer, *Proofs and computations*, Perspectives in Logic, Cambridge University Press, Cambridge; Association for Symbolic Logic, Chicago, IL, 2012.

[63] A.S. Troelstra and H. Schwichtenberg, *Basic proof theory*, Second edition, Cambridge Tracts in Theoretical Computer Science, 43. Cambridge University Press, Cambridge, 2000.

[64] Anne S. Troelstra and Dirk van Dalen, *Constructivism in mathematics. Vol. I. An introduction*, Studies in Logic and the Foundations of Mathematics, 121, North-Holland Publishing Co., Amsterdam, 1988.

[65] Anne S. Troelstra and Dirk van Dalen, *Constructivism in mathematics. Vol. II. An introduction*, Studies in Logic and the Foundations of Mathematics, 123. North-Holland Publishing Co., Amsterdam, 1988.

[66] Dirk van Dalen, *Logic and structure*, Fifth edition, Universitext, Springer, London, 2013.

Chapter 5

Program Extraction

Kenji Miyamoto

Institute for Informatics
University of Innsbruck,
Technikerstraße 21a, A-6020 Innsbruck, Austria
Kenji.Miyamoto@uibk.ac.at

This note provides a concise introduction to the theory of program extraction and case studies.

Contents

1. Introduction

An outcome of studying mathematics in a constructive setting is that proofs are more informative than classical ones. They tell us not only truth but also computational insight which leads us to come to see the truth. Program extraction is a method to make such a computational content hidden in a proof be an explicit computational object. The obtained object is an algorithmic entity, so that it is supposed to be executable as a program, and moreover it is an evidence claiming that the corresponding formula is true through a computational mean. The theoretical foundation of program extraction goes back to the BHK-interpretation and the realizability interpretation.

In this note we describe TCF, the theory for program extraction, which makes use of the notion of realizability, and see a couple of case studies of program extraction. As we are interested in extracting human-readable programs, our theory of program extraction is based on Kreisel's modified realizability rather than Kleene's realizability. We translate through a mapping called program extraction a proof into a program which is provably a computational solution of the original claim. As the size of proof is large, we need computer's assistance to perform program extraction correctly and quickly. The proof assistant Minlog is an implementation of TCF which offers necessary features to practice program extraction in computer. It is an open source software and available for free. The case studies in this note are also available, so that one can use Minlog to see the extracted programs and execute them.

The rest of this note consists of three sections. Section 2 describes the theory of program extraction. Section 3 describes case studies of program extraction within the theory presented in Section 2. Section 4 gives notes on related topics.

2. Realizability logic TCF

This section describes the theory TCF [15]. The theory TCF is first-order minimal logic with inductive definitions. The main objective of this theory is to provide a constructive framework for program extraction based on Kreisel's modified realizability. The term calculus of TCF is an extension of Gödel's T. Our term calculus is a language of first-order objects in TCF, which is also used as the target language of program extraction. Therefore, TCF can express that an extracted term is provably a realizer of the corresponding formula within TCF. The soundness theorem guarantees that our program extraction finds a desired realizer.

2.1. *Terms*

We focus on its syntactic aspects in this note. Semantics of this term calculus is given by Scott-Ershov domain of partial continuous functionals.

Concerning the development of the semantics, see [15].

Definition 1 (Algebras and types). Let α, β, ξ be variables ranging over types, and simultaneously define *algebras* ι and *types* τ, σ as

$$\iota := \mu_\xi(((\rho_{ij})_{j<n_i} \to \xi)_{i<k}), \qquad \tau, \sigma := \alpha \mid \iota \mid \tau \to \sigma,$$

where $n_0 = 0$, $k > 0$, and ρ_{ij} is a type in which ξ occurs at most strictly positive, namely, $SP(\xi, \rho_{ij})$ holds. Let $TV(\sigma)$ denote free variables in σ taking μ as a binder, then SP is defined as follows.

$$SP(\alpha, \beta), \qquad \frac{\alpha \notin TV(\rho) \qquad SP(\alpha, \sigma)}{SP(\alpha, \rho \to \sigma)}.$$

The arrow type constructor $\to$ is right associative. The vector notation $\vec{\tau}$ and the indexed notations $(\tau_i)_{i<k}$ and $(\tau)_k$ are used to express a (finite) list of types. We abbreviate $\tau_0 \to \dots \to \tau_{k-1} \to \sigma$ as $(\tau_i)_{i<k} \to \sigma$, as $(\tau)_k \to \sigma$ in case $\tau_i = \tau$ for each $i < k$, and as $\vec{\tau} \to \sigma$. In case k is 0, $(\tau_i)_{i<k} \to \sigma$ denotes σ. We allow the extensions of the indexed notation $(\alpha_i, \beta_i)_{i<k}$ and $(\alpha, \beta)_k$ so that $\alpha_0 \to \beta_0 \to \dots \to \alpha_{k-1} \to \beta_{k-1} \to \tau$ is abbreviated as $(\alpha_i, \beta_i)_{i<k} \to \tau$ and as $(\alpha, \beta)_k \to \tau$ in case $\alpha_i = \alpha$ and $\beta_i = \beta$ for each $i < k$. We can omit the outermost parentheses of expressions given to μ. We can also omit some indices for brevity.

We allow to write algebra definitions as $\mu_\xi(C_i^{(\rho_{ij}(\xi))_{j<n_i} \to \xi})_{i<k}$, so that constructor names are specified as well. When the definition of ι contains free variables $\vec{\alpha}$, these variables are called *type parameters*. Such algebras can be written as $\iota_{\vec{\alpha}}$ and as $\iota(\vec{\alpha})$, and its instantiation by $\vec{\sigma}$ is written as $\iota_{\vec{\sigma}}$ and as $\iota(\vec{\sigma})$. We also write $\rho_{ij}(\xi)$ for a type ρ_{ij} with possible occurrences of ξ.

Example 1. The algebras of unit, product, sum, natural numbers, lists of type α, binary trees, and ordinals are defined as follows:

$$\mathbf{U} := \mu_\xi(\cdot^\xi), \quad \alpha \times \beta := \mu_\xi(\langle -, - \rangle^{\alpha \to \beta \to \xi}), \quad \alpha + \beta := \mu_\xi(\mathsf{Inl}^{\alpha \to \xi}, \mathsf{Inr}^{\beta \to \xi}),$$

$$\mathbf{N} := \mu_\xi(0^\xi, \mathsf{S}^{\xi \to \xi}), \quad \mathbf{L}_\alpha := \mu_\xi(\mathsf{Nil}^\xi, ::^{\alpha \to \xi \to \xi}), \quad \mathbf{T} := \mu_\xi(\mathsf{I}^\xi, \mathsf{C}^{\xi \to \xi \to \xi}),$$

$$\mathbf{O} := \mu_\xi(\mathsf{Zer}^\xi, \mathsf{Suc}^{\xi \to \xi}, \mathsf{Sup}^{(\mathbf{N} \to \xi) \to \xi}).$$

We introduce two sorts of constants, constructors and program constants, which are supposed to be disjoint.

Definition 2 (Constructors). For $\iota = \mu_\xi((\rho_{ij}(\xi))_{j<n_i} \to \xi)_{i<k}$, its *constructors* are defined to be symbols $C_0, C_1, \dots, C_{k-1}$ of the *constructor types* $(\rho_{0j}(\iota))_{j<n_0} \to \iota, (\rho_{1j}(\iota))_{j<n_1} \to \iota, \dots, (\rho_{(k-1)j}(\iota))_{j<n_{k-1}} \to \iota$.

If ξ occurs in $\rho_{ij}(\xi)$, it is a *recursive* argument type, and otherwise it is a *parameter* argument type.

Definition 3 (Program constants). A program constant is defined to be a symbol D with a type assigned to it.

Definition 4 (Terms). Let C and D range over constructors and program constants. Define terms t, s by the following grammar.

$$t, s := x^\rho \mid \lambda_{x^\rho} t^\sigma \mid t^{\rho \to \sigma} s^\rho \mid C^\rho \mid D^\rho.$$

We omit types when there is no confusion. We write "_" for an unused abstracted variable (e.g. λ_0). Due to the vector notation or the indexed notation, $ts_0 \ldots s_{n-1}$ is abbreviated as $t\vec{s}$, $t(s_i)_{i<n}$, and so on. We also abbreviate $r^{(\rho_i, \sigma_i)_{i<n} \to \tau} s_0^{\rho_0} t_0^{\sigma_0} \ldots s_{n-1}^{\rho_{n-1}} t_{n-1}^{\sigma_{n-1}}$ as $r(s_i, t_i)_{i<n}$. Our terms are ones of typed lambda calculus extended with algebras as base types, constructors, and program constants. The conversion relation is the union of the β-reduction and left-to-right conversions yielded by computation rules, which will shortly be introduced. Note that there are non-terminating programs in our term calculus due to the following computation rules:

Definition 5 (Computation rules). Let C range over constructors. Define constructor patterns P, Q to be either a variable x^ρ or $(C\vec{P})^\iota$ where ι is an algebra. We denote by $P(\vec{x})$, $\vec{P}(\vec{x})$ that the constructor patterns P and $\vec{P}$ have free variables $\vec{x}$. Computation rules are a system of finite equations of the form $D\vec{P}_i(\vec{x}_i) = M_i$, where free variables of M_i are among $\vec{x}_i$. We require that for $i \neq j$ $\vec{P}_i$ and $\vec{P}_j$ are non-unifiable, or otherwise the free variables of $\vec{P}_i$ and $\vec{P}_j$ are disjoint and M_i and M_j are unified by the most general unifier of $\vec{P}_i$ and $\vec{P}_j$.

A constructor pattern can be viewed as a singleton list of the constructor pattern. For each algebra the (primitive) recursion operator is defined as an instance of program constants and computation rules.

Definition 6 (Recursion operators).
Let ι be an algebra $\mu_\xi((\rho_{ij}(\xi))_{j<n_i} \to \xi)_{i<k}$, τ a type, δ_i the type $((\rho_{ij}((\iota, \tau)))_{j<n_i} \to \tau)_{j<n_i} \to \tau$ for each $i < k$. The recursion operator $\mathcal{R}_\iota^\tau$ is a program constant of type $\iota \to (\delta_i)_{i<k} \to \tau$ with the computation rules $\mathcal{R}_\iota^\tau(C_i \vec{s}\vec{t})\vec{M} = M_i \vec{s}(t_j, \lambda_{\vec{x}}(\mathcal{R}_\iota^\tau(t_j \vec{x})\vec{M}))_{j<n_i}$ for $i < k$, where $\vec{s}$ and $\vec{t}$ are corresponding to parameter and recursive argument types, respectively.

Example 2. The recursion operators $\mathcal{R}_\mathbf{N}^\tau$, $\mathcal{R}_{\mathbf{L}_\alpha}^\tau$, $\mathcal{R}_\mathbf{T}^\tau$, and $\mathcal{R}_\mathbf{O}^\tau$ are as follows.

$$\mathcal{R}_\mathbf{N}^\tau : \mathbf{N} \to \tau \to (\mathbf{N} \to \tau \to \tau) \to \tau,$$

$$\mathcal{R}_\mathbf{N}^\tau 0 M_1 M_2 = M_1, \qquad \mathcal{R}_\mathbf{N}^\tau (\mathrm{S}n) M_1 M_2 = M_2 n (\mathcal{R}_\mathbf{N}^\tau n M_1 M_2),$$

$$\mathcal{R}_{\mathbf{L}_\alpha}^\tau : \mathbf{L}_\alpha \to \tau \to (\alpha \to \mathbf{L}_\alpha \to \tau \to \tau) \to \tau,$$

$$\mathcal{R}_{\mathbf{L}_\alpha}^\tau \mathrm{Nil} M_1 M_2 = M_1, \qquad \mathcal{R}_{\mathbf{L}_\alpha}^\tau (x{::}xs) M_1 M_2 = M_2 x xs (\mathcal{R}_\mathbf{L}^\tau xs M_1 M_2),$$

$$\mathcal{R}_\mathbf{T}^\tau : \mathbf{T} \to \tau \to (\mathbf{T} \to \tau \to \mathbf{T} \to \tau \to \tau) \to \tau,$$

$$\mathcal{R}_\mathbf{T}^\tau \mathrm{I} M_1 M_2 = M_1, \qquad \mathcal{R}_\mathbf{T}^\tau (\mathrm{C}ts) M_1 M_2 = M_2 t (\mathcal{R}_\mathbf{T}^\tau t M_1 M_2) s (\mathcal{R}_\mathbf{T}^\tau s M_1 M_2),$$

$$\mathcal{R}_\mathbf{O}^\tau : \mathbf{O} \to \tau \to (\mathbf{O} \to \tau \to \tau) \to ((\mathbf{N} \to \mathbf{O}) \to (\mathbf{N} \to \tau) \to \tau) \to \tau,$$

$$\mathcal{R}_\mathbf{O}^\tau \mathrm{Zer} M_1 M_2 M_3 = M_1, \qquad \mathcal{R}_\mathbf{O}^\tau (\mathrm{Suc}t) M_1 M_2 M_3 = M_2 t (\mathcal{R}_\mathbf{O}^\tau t M_1 M_2 M_3),$$

$$\mathcal{R}_\mathbf{O}^\tau (\mathrm{Sup}f) M_1 M_2 M_3 = M_3 f \lambda_n (\mathcal{R}_\mathbf{O}^\tau (fn) M_1 M_2 M_3).$$

Definition 7 (Case distinctions). The case distinction or the if-construct $\mathcal{C}_\iota^\tau$ is defined for an algebra ι as a program constant as follows.

$$\mathcal{C}_\iota^\tau : \iota \to ((\rho_{ij}(\iota))_{j<n_i} \to \tau)_{i<k} \to \tau, \qquad \mathcal{C}_\iota^\tau (C_i \vec{t})(M_i)_{i<k} = M_i \vec{t}.$$

2.2. *Formulas*

Based on predicates and formulas of first-order logic we define the ones of TCF. There are decorated implications and quantifiers in order to fine-tune computational content of proofs. Implication has two variants $\to^c$ and $\to^{nc}$ which are computational implication and non-computational implication, respectively. The computational implication is a usual implication in constructive mathematics, following the traditional BHK-interpretation. The other one is non-computational implication which is intended to express an implication whose premise is not computationally used. A proof of non-computational implication does not involve any computation depending on the realizer of the premise, hence we can safely discard anything concerning the premise during the program extraction procedure.

Definition 8 (Extended types). We define *extended types* to be our types enriched with the nulltype $\circ$, which satisfies

$$\circ = \circ \to \circ, \qquad \circ = \tau \to \circ, \qquad \tau = \circ \to \tau.$$

Definition 9 (Predicates and formulas). An arity and an associated type are a list of types and an extended type, respectively. A·predicate variable X is a symbol with a pair of an arity $(\vec{\tau})$ and an associated type τ_X. Predicates and formulas are simultaneously defined as follows:

$$P := X \mid \{\vec{x} \mid A(\vec{x})\} \mid \mu_X^{c/nc}((\forall_{\vec{x}_i}^{c/nc}((A_{ij}(X))_{j<n_i} \to^{c/nc} X\vec{t}_i))_{i<k}),$$

$$A, B := P\vec{t} \mid A \to^{c/nc} B \mid \forall_{x^\rho}^{c/nc} A,$$

where formulas given to $\mu^{c/nc}$ are called *clauses*, which must be closed formulas and $n_0 = 0$. The set of free (object) variables $\mathrm{FV}(A)$ is defined as usual. A predicate $\{\vec{x} \mid A(\vec{x})\}$ is called a *comprehension term*, such as $\{\vec{x} \mid A(\vec{x})\}\vec{t}$ denotes $A(\vec{t})$ for $\vec{x}$ and $\vec{t}$ of the same length. The decoration $^{c/nc}$ is intended to be an arbitrary one between ^{c} and ^{nc}. Predicates defined by μ^c and μ^{nc} are called (computational) inductively defined predicates and non-computational inductively defined predicates, respectively. In premises, only strictly positive occurrences is allowed for X. We can omit the outermost parentheses in formulas given to $\mu^{c/nc}$ and indices for brevity. We follow the same convention in algebra definitions.

The vector and the indexed notation are used here as well. The expressions $\forall_{\vec{x}}^c B$ and $\forall_{\vec{x}}^{nc} B$ abbreviate $\forall_{x_0}^c \forall_{x_1}^c \cdots \forall_{x_{m-1}}^c B$ and $\forall_{x_0}^{nc} \forall_{x_1}^{nc} \cdots \forall_{x_{m-1}}^{nc} B$, respectively. In order to indicate occurrences of x in A, we can write $A(x)$ instead, and $A(t)$ means the results of substituting a term t for x. We use a similar convention for predicate variables. We abbreviate $A_0(x), \ldots, A_{k-1}(x)$ as $\vec{A}(x)$ and $A_0(X), \ldots, A_{k-1}(X)$ as $\vec{A}(X)$.

Example 3. Conjunction, disjunction, the existential quantifier, and Leibniz equality are defined as follows. Assume P is a predicate of arity (ρ).

$$A \wedge B := \mu_X^c(A \to^c B \to^c X), \qquad A \vee B := \mu_X^c(A \to^c X, B \to^c X),$$

$$\exists_{x^\rho} Px := \mu_X^c(\forall_{x^\rho}^c(Px \to^c X)), \qquad \equiv := \mu_X^{nc}(\forall_{x^\rho}^{nc}(x \equiv x)).$$

Here we use the common infix notation. We can explicitly denote type parameters as $\exists_x^\rho Px$ and $t \equiv^\sigma s$ if it is necessary. A boolean term b can be taken as a formula by $b \equiv \mathsf{T}$. We abbreviate $A \to \mathsf{F}$ as $\neg A$.

2.3. *Derivation*

We introduce the notion of *derivations*, which is in principle same as proofs but intended to be formal. We first define axioms, then simultaneously define the syntax of derivation and the sets of free computational variables.

Definition 10 (Axioms). An axiom is a constant with an axiom formula which has to be closed. We write c^A for the axiom constant c with the axiom formula A.

The theory TCF is extended by the corresponding introduction and elimination axioms whenever new inductive predicates are defined. We define $P_0 \cap P_1$ to be $\mu_X(\forall_{\vec{x}}^{nc}(P_0\vec{x} \to^c P_1\vec{x} \to^c X\vec{x}))$, then we can abbreviate $(\vec{A} \to P_0\vec{t}) \to^{c/nc} (\vec{A} \to P_1\vec{t}) \to^{c/nc} B$ as $(\vec{A} \to (P_0 \cap P_1)\vec{t}) \to^{c/nc} B$.

Definition 11 (Introduction and elimination axioms). Assume I is $\mu_X^{c/nc}(K_i(X))_{i<k}$, where $K_i(X) = \forall_{\vec{x}_i}^{c/nc}((A_{ij}(X))_{j<n_i} \to^{c/nc} X\vec{t}_i)$. The introduction axioms of c.r. I are (axiom) constants I_i^+ with the axiom formulas $\forall_{\vec{y}_i}^{nc} K_i(I)$ where $\vec{y}_i$ are chosen to make the formulas closed. We can write I^+ for I_0^+ if $k = 1$. Assume P is a predicate of the same arity as I. The elimination axiom of c.r. I is a constant I^- with axiom formula $\forall_{\vec{x}}^{nc}(I\vec{x} \to^c (S_i(I, P))_{i<k} \to^c P\vec{x})$. The *i-th step formula* $S_i(Y, Z)$ is defined to be $\forall_{\vec{x}_i}^{c/nc}((A_{ij}(Y \cap Z))_{j<n_i} \to^{c/nc} Z\vec{t}_i)$. Here P is called the *competitor predicate*. If $k = 1$ and K_0 is formed by the logical connectives $\forall^{nc}$ and $\to^{nc}$ only, I is a *one-clause non-computational inductively defined predicate*. In this case, n.c. I^+ and n.c. I^- are available in the same way as the computational case. Otherwise, we restrict the competitor predicate for n.c. I^- to be n.c.

The exceptional treatment for one-clause non-computational inductive predicates is possible because we can safely ignore the computational significance of elimination axioms of such predicates.

Example 4. The introduction and elimination axioms for conjunction, disjunction, existential quantifier, and Leibniz equality are as follows.

$$A \to^c B \to^c A \wedge B, \quad A \wedge B \to^c (A \to^c B \to^c P) \to^c P,$$
$$A \to^c A \vee B, \quad B \to^c A \vee B, \quad A \vee B \to^c (A \to^c P) \to^c (B \to^c P) \to^c P,$$
$$\forall_x^c(A(x) \to^c \exists_x A(x)), \quad \exists_x A(x) \to^c \forall_x^c(A(x) \to^c P) \to^c P,$$
$$\forall_x^{nc}(x \equiv x), \quad \forall_{x,y}^{nc}(x \equiv y \to \forall_x^{nc}(Pxx) \to Pxy).$$

Commonly used well-founded induction is the use of the elimination axiom of the corresponding (structural) totality predicate. We first define the structural totality, and then the totality comes as a special case of it.

Definition 12 (Totality). Assume τ is a type. The structural totality predicate ST_τ of type τ is inductively defined predicate of arity (τ) as follows.

$$ST_\xi := X,$$
$$ST_\alpha := \{x^\alpha \mid \top\} \qquad \text{if } \alpha \text{ is a type variable other than } \xi,$$
$$ST_{\rho \to \sigma} := \{f^{\rho \to \sigma} \mid \forall^{\mathrm{nc}}_{x^\rho}(ST_\rho x \to^{\mathrm{c}} ST_\sigma(fx))\},$$
$$ST_{\mu_\xi((\rho_{ij})_{j<n_i} \to \xi)_{i<k}} := \mu^{\mathrm{c}}_X(\forall^{\mathrm{nc}}_{\vec{x}_i}((ST_{\rho_{ij}} x_{ij})_{j<n_i} \to^{\mathrm{c}} X(C_i \vec{x}_i))_{i<k}.$$

Here, C_i is the i-th constructor of the above algebra $\mu_\xi((\rho_{ij})_{j<n_i} \to \xi)_{i<k}$. If τ contains no free type variable, ST_τ is especially called the *totality predicate* and written as T_τ.

We abbreviate $\forall^{\mathrm{nc}}_{x^\iota}(ST_\iota x \to A)$ as $\forall^{\mathrm{c}}_{x^\iota \in ST_\iota} A$, and abuse the wording *induction* for the use of ST^-_ι. We also refer to premises in a step formula by *induction hypotheses*.

Assume disjoint domains for variables of objects and assumptions. Object and assumption variables range over terms and derivations, resp. We write $u : A$ and $c : A$ to denote an assumption variable u and an axiom c of a formula A, respectively.

Definition 13 (Derivation). Let x be an object variable, u an assumption variable of a formula A, and c an axiom constant claiming A_c.

$$u : A \qquad\qquad c : A_c$$

$$\begin{array}{c} \vdots \; K \\ \dfrac{A(x)}{\forall^{\mathrm{c}}_{x^\rho} A(x)} \; \forall^{\mathrm{c}+}, x \end{array} \qquad \begin{array}{c} \vdots \; L \\ \dfrac{A(x)}{\forall^{\mathrm{nc}}_{x^\rho} A(x)} \; \forall^{\mathrm{nc}+}, x \end{array} \qquad \begin{array}{c} \vdots \; M \\ \dfrac{\forall^{\mathrm{c}}_{x^\rho} A(x) \quad t}{A(t)} \; \forall^{\mathrm{c}-} \end{array} \qquad \begin{array}{c} \vdots \; M \\ \dfrac{\forall^{\mathrm{nc}}_{x^\rho} A(x) \quad t}{A(t)} \; \forall^{\mathrm{nc}-} \end{array}$$

$$\begin{array}{c} [u : A] \\ \vdots \; M \\ \dfrac{B}{A \to^{\mathrm{c}} B} \; \to^{\mathrm{c}+}, u \end{array} \qquad \begin{array}{c} [u : A] \\ \vdots \; N \\ \dfrac{B}{A \to^{\mathrm{nc}} B} \; \to^{\mathrm{nc}+}, u \end{array} \qquad \begin{array}{c} \vdots \; M \quad \vdots \; M' \\ \dfrac{A \to^{\mathrm{c}} B \quad A}{B} \; \to^{\mathrm{c}-} \end{array} \qquad \begin{array}{c} \vdots \; M \quad \vdots \; M' \\ \dfrac{A \to^{\mathrm{nc}} B \quad A}{B} \; \to^{\mathrm{nc}-} \end{array}$$

The assumption variables $u : A$ in the square brackets are discharged due to $\to^{\mathrm{c}+}, \to^{\mathrm{nc}+}$, and not free anymore. As side conditions, x is not allowed to be a free variable of open assumptions in K or in L, $x \notin \mathrm{CV}(L)$, and $u \notin \mathrm{CV}(N)$. We also use the following term expression of derivations.

$$u^A, \qquad\qquad c^{A_c},$$
$$(\lambda_x K^{A(x)})^{\forall^{\mathrm{c}}_{x^\rho} A(x)}, \quad (\lambda_x L^{A(x)})^{\forall^{\mathrm{nc}}_{x^\rho} A(x)}, \quad (M^{\forall^{\mathrm{c}}_{x^\rho} A(x)} t)^{A(t)}, \quad (M^{\forall^{\mathrm{nc}}_{x^\rho} A(x)} t)^{A(t)},$$
$$(\lambda_u M^B)^{A \to^{\mathrm{c}} B}, \qquad (\lambda_u N^B)^{A \to^{\mathrm{nc}} B}, \qquad (M^{A \to^{\mathrm{c}} B} M'^A)^B, \quad (M^{A \to^{\mathrm{nc}} B} M'^A)^B.$$

Then the *free computational variables* are defined as follows.

$$\mathrm{CV}(u) := \{u\}, \quad \mathrm{CV}(c) := \emptyset,$$

$$\mathrm{CV}((\lambda_x M)^{\forall_x^c A}) := \mathrm{CV}((\lambda_x M)^{\forall_x^{nc} A}) := \mathrm{CV}(M) \setminus \{x\},$$

$$\mathrm{CV}((M^{\forall_x^c A(x)} t)^{A(t)}) := \mathrm{CV}(M) \cup \mathrm{FV}(t), \quad \mathrm{CV}((M^{\forall_x^{nc} A(x)} t)^{A(t)}) := \mathrm{CV}(M),$$

$$\mathrm{CV}((\lambda_u M)^{A \to^c B}) := \mathrm{CV}((\lambda_u M)^{A \to^c B}) := \mathrm{CV}(M) \setminus \{u\},$$

$$\mathrm{CV}((M^{A \to^c B} N^A)^B) := \mathrm{CV}(M) \cup \mathrm{CV}(N), \quad \mathrm{CV}((M^{A \to^{nc} B} N^A)^B) := \mathrm{CV}(M).$$

Free computational variables in a derivation are object and assumption variables which are supposed to have computational significance in the derivation. We can abstract object variables by $\forall^{nc+}$ if they have no computational significance, namely, no occurrence in $\mathrm{CV}(L)$, and same for $\to^{nc+}$ and $\mathrm{CV}(N)$.

As TCF is based on minimal logic, there is no falsity in the sense of intuitionistic and classical logic. However, we can define a falsity using Leibniz equality and prove ex-falso-quodlibet under some constraint. We say a predicate variable X is *free* if there is no corresponding $\mu_X^{c/nc}$.

Theorem 1 (Ex-falso-quodlibet). *If A does not have any free predicate variable,* $\mathbf{F} \to A$ *holds where $\mathbf{F}$ is defined to be* $\mathsf{T} \equiv \mathsf{F}$.

Proof. First prove that $\mathbf{F} \to t \equiv s$ for any terms t, s of the same type. Consider the formula $A(b) := \mathcal{C}_\mathbf{B} \mathbf{F} s t \equiv \mathcal{C}_\mathbf{B} b t s$, then $A(\mathsf{T})$, i.e. $t \equiv t$, is proved by $\equiv^+$. Using $\equiv^-$, from $A(\mathsf{T})$ and $\mathbf{F}$, i.e. $\mathsf{T} \equiv \mathsf{F}$, we can prove $A(\mathsf{F})$, that is, $t \equiv s$. We do induction on the construction of A. If A is $I\vec{t}$ for an inductive predicate I, we replace $\vec{t}$ so that we can prove this by I_0^+. If A is $A_0 \to^{c/nc} A_1$, assume A_0 and use the induction hypothesis for A_1. The other cases are straightforward. $\qquad\square$

2.4. *Realizability*

Program extraction is a mapping from a proof into a term, which is provably a realizer of the corresponding formula. We first define the mapping, then prove the soundness theorem of our program extraction, stating that the extracted term is a realizer of the corresponding formula.

Definition 14 (Type of predicate and formula). The mapping τ from a predicate and a formula into an extended type is defined as follows.

$$\tau(X) := \xi_X, \qquad \tau(\{\vec{x} \mid A\}) := \tau(A), \qquad \tau(\mu_X^{\mathrm{nc}}(\vec{K})) := \circ,$$
$$\tau(\mu_X^{\mathrm{c}}(K_0, \ldots, K_{k-1})) := \mu_{\xi_X}(\tau(K_0), \ldots, \tau(K_{k-1})),$$
$$\tau(P\vec{t}) := \tau(P), \quad \tau(A \to^{\mathrm{c}} B) := \tau(A) \to \tau(B), \quad \tau(A \to^{\mathrm{nc}} B) := \tau(B),$$
$$\tau(\forall_{x\rho}^{\mathrm{c}} A) := \rho \to \tau(A), \quad \tau(\forall_{x\rho}^{\mathrm{nc}} A) := \tau(A).$$

For an inductively defined predicate I, the type $\tau(I)$ is an algebra and is called the *associated algebra* of I.

Definition 15 (Computational and non-computational formula).

A formula A is computational, computationally relevant or c.r. if $\tau(A) \neq \circ$, otherwise A is non-computational or n.c.

Whenever A is n.c. all the decorations in A do not make any difference, hence we can safely remove decorations in such a case. We simultaneously define witnessing predicates and realizability of formulas.

Definition 16 (Realizability).

Assume a global assignment, so that for each predicate variable X of arity $(\vec{\rho})$ there is $X^{\mathbf{r}}$ of arity $(\tau_X, \vec{\rho})$. For c.r. formulas A we define the realizability $A^{\mathbf{r}}$ as follows.

$$(X\vec{s})^{\mathbf{r}} := \{y \mid X^{\mathbf{r}} y \vec{s}\}, \quad (\{\vec{x} \mid A(\vec{x})\} \vec{s})^{\mathbf{r}} := (A(\vec{s}))^{\mathbf{r}}, \quad (I\vec{s})^{\mathbf{r}} := \{y \mid I^{\mathbf{r}} y \vec{s}\},$$
$$(A \to^{\mathrm{c}} B)^{\mathbf{r}} := \{y \mid \forall_x (A^{\mathbf{r}} x \to B^{\mathbf{r}}(yx))\} \quad \text{where } A \text{ is c.r.,}$$
$$(A \to^{\mathrm{c}} B)^{\mathbf{r}} := \{y \mid A \to B^{\mathbf{r}} y\} \quad \text{where } A \text{ is n.c.,}$$
$$(A \to^{\mathrm{nc}} B)^{\mathbf{r}} := \{y \mid A \to B^{\mathbf{r}} y\},$$
$$(\forall_x^{\mathrm{c}} A)^{\mathbf{r}} := \{y \mid \forall_x (A^{\mathbf{r}}(yx)))\}, \quad (\forall_x^{\mathrm{nc}} A)^{\mathbf{r}} := \{y \mid \forall_x (A^{\mathbf{r}} y)\}.$$

For $K_i(X) = \forall_{\vec{x}_i}^{\mathrm{c}/\mathrm{nc}} ((A_{ij}(X))_{j<n_i} \to^{\mathrm{c}/\mathrm{nc}} X\vec{t}_i)$ let I be $\mu_X^{\mathrm{c}}(K_i(X))_{i<k}$ of arity $(\vec{\sigma})$ and ι_I the associated algebra of I where $(C_i)_{i<k}$ are its constructors. Let the clause formula $K_i^{\mathbf{r}}(X^{\mathbf{r}})$ be $\forall_{\vec{x}_i, \vec{u}_i}((u_{ij} \mathbf{r} A_{ij}(X))_{j<n_i} \to C_i \vec{x}_i \vec{u}_i \mathbf{r} X\vec{t}_i)$, where in case A_{ij} is n.c. $u_{ij} \mathbf{r} A_{ij}(X)$ is replaced by A_{ij} and u_{ij} is missing, and x_{ij} and u_{ij} are fed to C_i if x_{ij} is quantified by $\forall^{\mathrm{c}}$ and if A_{ij} is c.r. and followed by $\to^{\mathrm{c}}$, respectively. Then the *witnessing predicate* $I^{\mathbf{r}}$ of I is defined to be $\mu_{X^{\mathbf{r}}}^{\mathrm{nc}}(K_i^{\mathbf{r}}(X^{\mathbf{r}}))_{i<k}$ of arity $(\iota_I, \vec{\sigma})$. For n.c. formula A we define $A^{\mathbf{r}} := A$.

Conventionally, $t \mathbf{r} A$ means the same as $A^{\mathbf{r}} t$. When $t \mathbf{r} A$, we say that t is a realizer of A. Note that the decoration is safely omitted due to the non-computationality of formulas stating realizability. Each clause formula of I^r is designed so as the original clause formula of I is realized by the corresponding constructor with a consistent treatment with respect to non-computational variables and premises. We admit the following invariance axiom, any instance of which is realized by identity functions.

Definition 17 (Invariance axiom). For c.r. formulas A, Inv_A^+ and Inv_A^- are the axiom constants of $A \to^c \exists_u(u \, \mathbf{r} \, A)$ and $\exists_u(u \, \mathbf{r} \, A) \to^c A$, resp.

The invariance axiom clarifies what we aim to have through the realizability. Due to the invariance axiom, the following formulas are derivable.

Example 5. Assume x does not occur freely in B. The following formulas hold.

$$\forall_{x^\tau}^c \exists_{y^\sigma} A(x, y) \to^c \exists_{f^{\tau \to \sigma}} \forall_x^c A(x, fx),$$
$$(B \to^{nc} \exists_x C(x)) \to^c \exists_x(B \to^{nc} C(x)).$$

We define the extracted term of a derivation, which is used to find a suitable realizer in the proof of Theorem 2.

Definition 18 (Extended terms). An extended term is a term or a symbol ε such that $t\varepsilon := t$ and $\varepsilon t := \varepsilon$.

Definition 19 (Extracted term). For a derivation M^A where A is n.c., $\text{et}(M^A) := \varepsilon$. For M^A where A is c.r., if M is the invariance axiom or I^- of an n.c. inductive predicate with c.r. competitor predicate, $\text{et}(M^A)$ is an identity function of type $\tau(A)$, otherwise:

$$\text{et}(u^A) := x_u, \quad \text{et}(\text{Inv}^A) := \text{id}^{\tau(A) \to \tau(A)}, \quad \text{et}(I_i^+) := C_i, \quad \text{et}(I^-) := \mathcal{R}_{\iota_I}^\rho,$$

$$\text{et}((\lambda_{u^A} M^B)^{A \to^c B}) := \lambda_{x_u} \text{et}(M) \quad \text{where A is c.r.,}$$

$$\text{et}((\lambda_{u^A} M^B)^{A \to^c B}) := \text{et}(M) \quad \text{where A is n.c.,}$$

$$\text{et}((\lambda_{u^A} M^B)^{A \to^{nc} B}) := \text{et}(M),$$

$$\text{et}((M^{A \to^c B} N^A)^B) := \text{et}(M)\text{et}(N), \quad \text{et}((M^{A \to^{nc} B} N^A)^B) := \text{et}(M),$$

$$\text{et}((\lambda_{x^\rho} M^A)^{\forall_{x^\rho}^c A}) := \lambda_{x^\rho} \text{et}(M), \quad \text{et}((M^{\forall_{x^\rho}^c A(x)} t^\rho)^{A(t)}) := \text{et}(M)t,$$

$$\text{et}((\lambda_{x^\rho} M^A)^{\forall_{x^\rho}^{nc} A}) := \text{et}(M), \quad \text{et}((M^{\forall_{x^\rho}^{nc} A(x)} t^\rho)^{A(t)}) := \text{et}(M),$$

where C_i is the i-th constructor of the algebra ι_I and ρ is the type of the corresponding competitor predicate.

We prove that whenever a proof of a formula is given in our theory, there is a term which provably realizes the formula.

Theorem 2 (Soundness). *Let M^A be a derivation from assumptions $(u_i^{B_i})_{i<n}$. For each $i < n$ let v_i be an assumption of a formula $x_{u_i} \, \mathbf{r} \, B_i$ if B_i is c.r., and $u_i^{B_i}$ if B_i is n.c. If A is c.r., there is a realizer t and a derivation $N^{t \, \mathbf{r} \, A}$ from the assumptions $(v_i)_{i<n}$, otherwise a derivation N^A from the assumptions $(v_i)_{i<n}$.*

Proof. By induction on the derivation M^A, we prove that $\mathrm{et}(M)$ is a desired realizer t. Case u^A If A is c.r., $\mathrm{et}(M) = x_u$ and v is an assumption of $x_u \, \mathbf{r} \, A$. If A is n.c., we take a derivation v which is u^A. Case $(\lambda_{u^B} L^C)^{B \to^c C}$ We first deal with the case $B \to^c C$ is c.r. If B is c.r., we prove $\forall_{x_u}(x_u \, \mathbf{r} \, B \to \mathrm{et}(L) \, \mathbf{r} \, C)$. By induction hypothesis there is a derivation $N^{\mathrm{et}(L) \, \mathbf{r} \, C}$ from the assumptions $(v_i)_{i<n}$ and $v^{x_u \, \mathbf{r} \, B}$. Using $\to^+$ and $\forall^+$ we get a derivation $(\lambda_x (\lambda_{v^{x_u \, \mathbf{r} \, B}} N^{\mathrm{et}(L) \, \mathbf{r} \, C})^{x_u \, \mathbf{r} \, B \to \mathrm{et}(L) \, \mathbf{r} \, C})^{\forall_{x_u}(x_u \, \mathbf{r} \, B \to \mathrm{et}(L) \, \mathbf{r} \, C)}$ from the assumptions $(v_i)_{i<n}$. If B is n.c., we prove $B \to \mathrm{et}(L) \, \mathbf{r} \, C$. By induction hypothesis there is a derivation $N^{\mathrm{et}(L) \, \mathbf{r} \, C}$ from the assumptions $(v_i)_{i<n}$ and $v^B = u$. Using $\to^+$ we get a derivation $(\lambda_{u^B} N^{\mathrm{et}(L) \, \mathbf{r} \, C})^{B \to \mathrm{et}(L) \, \mathbf{r} \, C}$ from the assumptions $(v_i)_{i<n}$. In case $B \to^c C$ is n.c., there is a derivation K^C from the assumptions $(v_i)_{i<n}$ and $v^{x_u \, \mathbf{r} \, B}$ or u^B depending on whether B is c.r. or n.c. If B is c.r., we assume $J^{\forall_{x_u}(x_u \, \mathbf{r} \, B \to C) \to \exists_{x_u}(x_u \, \mathbf{r} \, B) \to C}$, which is trivial, and the invariance axiom $(\mathrm{Inv}_B^+)^{B \to^c \exists_{x_u}(x_u \, \mathbf{r} \, B)}$. Let M' be a derivation $(J(\lambda_{x_u}(\lambda_{v^{x_u \, \mathbf{r} \, B}} K^C)^{x_u \, \mathbf{r} \, B \to C})^{\forall_{x_u}(x_u \, \mathbf{r} \, B \to C)})^{\exists_{x_u}(x_u \, \mathbf{r} \, B) \to C}$, then our desired derivation is $(\lambda_{u^B}(M'(\mathrm{Inv}^+ u)^{\exists_{x_u}(x_u \, \mathbf{r} \, B)})^C)^{B \to C}$ from $(v_i)_{i<n}$. If B is n.c., $(\lambda_{u^B} K^C)^{B \to C}$ from $(v_i)_{i<n}$. Case $(\lambda_{u^B} L^C)^{B \to^{nc} C}$ Similar to the previous case for B n.c. Case $(M_0^{B \to^c C} M_1^B)^C$ If C is c.r., we make a case distinction on B. If B is c.r., by induction hypotheses there are derivations $N_0^{\forall_{x_u}(x_u \, \mathbf{r} \, B \to \mathrm{et}(M_0) x_u \, \mathbf{r} \, C))}$ and $N_1^{\mathrm{et}(M_1) \, \mathbf{r} \, B}$ from the assumptions $(v_i)_{i<n}$. Then the desired derivation is $(N_0 \mathrm{et}(M_1) N_1)^{\mathrm{et}(M_0 M_1) \, \mathbf{r} \, C}$ from the assumptions $(v_i)_{i<n}$, where $\mathrm{et}(M_0 M_1) = \mathrm{et}(M_0) \mathrm{et}(M_1)$. If B is n.c., by induction hypotheses there are derivations $N_0^{B \to \mathrm{et}(M_0) \, \mathbf{r} \, C}$ and N_1^B from the assumptions $(v_i)_{i<n}$. The desired derivation is $(N_0 N_1)^{\mathrm{et}(M_0) \, \mathbf{r} \, C}$ from the assumptions $(v_i)_{i<n}$, where $\mathrm{et}(M_0) = \mathrm{et}(M_0 M_1)$. If C is n.c., we make a case distinction on B. If B is c.r., by induction hypotheses there are derivations $N_0^{B \to C}$ and $N_1^{\mathrm{et}(M_1) \, \mathbf{r} \, B}$ from the assumptions $(v_i)_{i<n}$. The desired derivation is $(N_0 (\mathrm{Inv}_B^-(\exists^+ \mathrm{et}(M_1) N_1)^{\exists_x (x \, \mathbf{r} \, B)})^B)^C$ from the assumptions. If B is n.c., by induction hypotheses there are derivations $N_0^{B \to C}$ and N_1^B from the assumptions $(v_i)_{i<n}$. The desired derivation is $(N_0 N_1)^C$ from the assumptions. Case $(M_0^{B \to^{nc} C} M_1^B)^C$ Similar to the previous case for B n.c. Case $(\lambda_x M^{B(x)})^{\forall_x^c B(x)}$ If $B(x)$ is c.r., by induction hypothesis there is a derivation $N^{\mathrm{et}(M) \, \mathbf{r} \, B(x)}$ from the assumptions

$(v_i)_{i<n}$, the formulas of which do not contain x as a free variable. The desired derivation is $(\lambda_x N^{\text{et}(M)\,\mathbf{r}\,B(x)})^{\forall_x((\lambda_x \text{et}(M))x\,\mathbf{r}\,B(x))}$ from the assumptions. If B is n.c., by induction hypothesis there is a derivation $N^{B(x)}$ from the assumptions $(v_i)_{i<n}$, the formulas of which do not contain x as a free variable. The desired derivation is $(\lambda_x N^{B(x)})^{\forall_x B(x)}$ from the assumptions. Case $(\lambda_x M^{B(x)})^{\forall_x^{\text{nc}} B(x)}$ If $B(x)$ is c.r., by induction hypothesis there is a derivation $N^{\text{et}(M)\,\mathbf{r}\,B(x)}$ from the assumptions $(v_i)_{i<n}$, the formulas of which do not contain x as a free variable. The desired derivation is $(\lambda_x N^{\text{et}(M)\,\mathbf{r}\,B(x)})^{\forall_x(\text{et}(M)\,\mathbf{r}\,B(x))}$ from the assumptions. If $B(x)$ is n.c., same as the case of $\forall_x^c B(x)$. Case $(M^{\forall_x^c B(x)}t)^{B(t)}$ If $B(x)$ is c.r., by induction hypothesis there is a derivation $N^{\forall_x^c(\text{et}(M)x\,\mathbf{r}\,B(x))}$ from the assumptions $(v_i)_{i<n}$. The desired proof is $(Nt)^{\text{et}(M)t\,\mathbf{r}\,B(t)}$ from the assumptions. If $B(x)$ is n.c., by induction hypothesis there is a derivation $N^{\forall_x^c B(x)}$ from the assumptions $(v_i)_{i<n}$. The desired proof is $(Nt)^{B(t)}$ from the assumptions. Case $(M^{\forall_x^{\text{nc}} B(x)}t)^{B(t)}$ If $B(x)$ is c.r., by induction hypothesis there is a derivation $N^{\forall_x^{\text{nc}}(\text{et}(M)\,\mathbf{r}\,B(x))}$ from the assumptions $(v_i)_{i<n}$. The desired proof is $(Nt)^{\text{et}(M)\,\mathbf{r}\,B(t)}$ from the assumptions. If $B(x)$ is n.c., it is similar to the case of $(M^{\forall_x^c B(x)}t)^{B(t)}$. Case I_i^+ In case I is c.r., we prove that $\text{et}(M) = C_i$ realizes the formula $\forall_{\vec{x}_i}^{c/\text{nc}}((A_{ij}(I))_{j<n_i} \to^{c/\text{nc}} I\vec{t}_i)$. It is trivial by $I_i^{\mathbf{r}+}$, because due to the definition the formula of $I_i^{\mathbf{r}+}$ is same as our goal. In case I is n.c., our goal is not changed by the realizability. Case I^- Assume I and the competitor predicate P are c.r. We prove that for $\tau = \tau(P)$, $\text{et}(M) = \mathcal{R}_{\iota_I}^\tau$ realizes the formula $\forall_{\vec{x}}^{\text{nc}}(I\vec{x} \to^c (K_i(I,P))_{i<k} \to^c P\vec{x})$, where $K_i(I,P)$ is defined to be $\forall_{\vec{x}_i}^{c/\text{nc}}((A_{ij}(I \cap P))_{j<n_i} \to^{c/\text{nc}} P\vec{t}_i)$. Our goal is $\forall_{\vec{x},u}(I^{\mathbf{r}}u\vec{x} \to \forall_{\vec{w}}((w_i\,\mathbf{r}\,K_i(I,P))_{i<k} \to \mathcal{R}_{\iota_I}^\tau u\vec{w}\,\mathbf{r}\,P\vec{x}))$. Let Q be $\{u,\vec{x} \mid \mathcal{R}_{\iota_I}^\tau u\vec{w}\,\mathbf{r}\,P\vec{x}\}$. Assume $\vec{x}$, u, $I^{\mathbf{r}}u\vec{x}$, $\vec{w}$, and $(w_i\,\mathbf{r}\,K_i(I,P))_{i<k}$, and prove the goal $\mathcal{R}_{\iota_I}^\tau u\vec{w}\,\mathbf{r}\,P\vec{x}$ $(= Qu\vec{x})$ using $I^{\mathbf{r}-}$. It suffices to prove the step formulas of $I^{\mathbf{r}-}$ from the corresponding assumption $w_i\,\mathbf{r}\,K_i(I,P)$. We simplify the matter, assuming $K_i(I,P)$ is of the form $\forall_{\vec{x}_i}^{\text{nc}}(A_i \to \forall_{\vec{y}}^{\text{nc}}(\vec{B}_i \to^c I\vec{s}_i) \to \forall_{\vec{y}}^{\text{nc}}(\vec{B}_i \to^c P\vec{s}_i) \to P\vec{t}_i)$, then $w_i\,\mathbf{r}\,K_i(I,P)$ is

$$\forall_{\vec{x}_i,u_i,f_i,g_i}(u_i\,\mathbf{r}\,A_i \to \forall_{\vec{y},\vec{v}}(\vec{v}\,\mathbf{r}\,\vec{B}_i \to f_i\vec{v}\,\mathbf{r}\,I\vec{s}_i) \to$$
$$\forall_{\vec{y},\vec{v}}(\vec{v}\,\mathbf{r}\,\vec{B}_i \to g_i\vec{v}\,\mathbf{r}\,P\vec{s}_i) \to w_iu_if_ig_i\,\mathbf{r}\,P\vec{t}_i). \tag{1}$$

We prove the step formulas of the form

$$\forall_{\vec{x}_i,u_i,f_i}(u_i\,\mathbf{r}\,A_i \to \forall_{\vec{y},\vec{v}}(\vec{v}\,\mathbf{r}\,\vec{B}_i \to I^{\mathbf{r}}(f_i\vec{v},\vec{s}_i)) \to$$
$$\forall_{\vec{y},\vec{v}}(\vec{v}\,\mathbf{r}\,\vec{B}_i \to Q(f_i\vec{v},\vec{s}_{ij})) \to Q(C_iu_if_i,\vec{t}_i)).$$

Assume the variables $\vec{x}_i,u_i,f_i$ and premises, then due to the definition of Q, it suffices to prove $\mathcal{R}_{\iota_I}^\tau(C_iu_if_i)\vec{w}\,\mathbf{r}\,P\vec{t}_i$, that is $w_iu_if_i(\lambda_{\vec{v}}\mathcal{R}_{\iota_I}^\tau(f_i\vec{v})\vec{w})\,\mathbf{r}\,P\vec{t}_i$.

To the corresponding assumption (1) we feed the following terms. Firstly $\vec{x}_i$, u_i, and f_i, then $\lambda_{\vec{v}}(\mathcal{R}^\tau_{\iota_I}(f_i\vec{v})\vec{w})$ for g_i. The first and second premises are trivial. The third one also follows, because $(\lambda_{\vec{v}}(\mathcal{R}^\tau_{\iota_I}(f_i\vec{v})\vec{w}))\vec{v}\,\mathbf{r}\,P\vec{s}_i$ is same as $Q(f_i\vec{v}, \vec{s}_i)$ due to the definition of Q. In case $K_i(I, P)$ involves a c.r. premise followed by $\to^{\mathrm{nc}}$, Inv^- should be used as necessary. If P is n.c., what we prove is same as I^-. Assume I is n.c. In case I is a one-clause non-computational inductive predicate and its competitor predicate P is c.r., there is nothing to do for I_i^+ and the identity function $\lambda_{x^\tau}x$ provably realizes the formula of I^-. Otherwise there is nothing to do because I_i^+ and I^- are all n.c. Case $\mathrm{Inv}_A^{+/-}$ Easy to show that the identity function of type $\tau(A) \to \tau(A)$ realizes the both of Inv_A^+ and Inv_A^-. $\qquad\square$

3. Case studies

We show case studies of program extraction. Our case studies are formalized and available as running examples in Minlog, which is downloadable at `http://minlog-system.de/`. You find the first case study in `http://www.mathematik.uni-muenchen.de/~miyamoto/evod.scm`, and the second and the third in the Minlog package `examples/parsing/parens.scm` and `lib/real.scm`.

3.1. *Even and odd numbers*

The natural numbers we deal with are objects of the algebra $\mathbf{N}$ which is defined to be $\mu_\xi(0^\xi, S^{\xi\to\xi})$. The predicates stating natural, even, and odd numbers are defined to be $T_\mathbf{N}$, EV, and OD as follows, respectively. Let object variables n, m range over $\mathbf{N}$ so that we can safely omit type annotation.

$$T_\mathbf{N} := \mu_X(X0, \forall_n^{\mathrm{nc}}(Xn \to^c X(Sn))),$$
$$EV := \mu_X(X0, \forall_n^{\mathrm{nc}}(Xn \to^c X(S(Sn)))), \quad OD := \{n \mid EV(Sn)\}.$$

The following axioms are available due to the above predicate definitions.

$$T_\mathbf{N}0, \quad \forall_n^{\mathrm{nc}}(T_\mathbf{N}n \to^c T_\mathbf{N}(Sn))),$$
$$\forall_n^{\mathrm{nc}}(T_\mathbf{N}n \to^c P0 \to^c \forall_m^{\mathrm{nc}}(T_\mathbf{N}m \to^c Pm \to^c P(Sm)) \to^c Pn),$$
$$EV0, \quad \forall_n^{\mathrm{nc}}(EVn \to^c EV(S(Sn))),$$
$$\forall_n^{\mathrm{nc}}(EVn \to^c P0 \to^c \forall_m^{\mathrm{nc}}(EVm \to^c Pm \to^c P(S(Sm))) \to^c Pn).$$

Proposition 1. $\forall_{n\in T}^c(EVn \vee ODn)$.

Proof. By induction on n. It suffices to prove the step formulas of T_N^-. The first one is $EV0 \vee OD0$ is trivial due to $\vee_0^+$ and EV_0^+. The second one is $\forall_m^{nc}(T_N m \to^c EVm \vee ODm \to^c EV(Sm) \vee OD(Sm))$. Assume m, $T_N m$, and $EVm \vee ODm$. Apply $\vee^-$ for $EVm \vee ODm$ to prove $EV(Sm) \vee OD(Sm)$. It is sufficient to prove $EV(Sm) \vee OD(Sm)$ from EVm and also from ODm. The first case is done by $\vee_1^+$, EV_1^+ and the assumption. The second case is by $\vee_0^+$ and the assumption. $\qquad\square$

We obtain from the derivation the extracted term, whose normal form is $\lambda_{x_u^N}(\mathcal{R}_N^B x_u \, \mathsf{T} \, \lambda_{x_v^N, x_w^B}(\mathcal{C}_B^B x_w \, \mathsf{F} \, \mathsf{T}))$. This program is implicitly expressed by f as follows, where not flips the boolean.

$$f0 = \mathsf{T}, \qquad\qquad f(Sn) = \mathrm{not}(fn),$$
$$\mathrm{not}\,\mathsf{T} = \mathsf{F}, \qquad\qquad \mathrm{not}\,\mathsf{F} = \mathsf{T}.$$

Normalizing the terms $f0$ and $f(S(S(S0)))$, they go to T and F.

3.2. *Parsing*

Consider two characters "(" and ")", and strings consisting of them. We extract a parser which determines whether each open parenthesis in a given string has the corresponding close parenthesis [5].

Our string is a list of characters, where the algebra of character is just the boolean algebra. Here we use the names L and R instead of "(" and ")". We define an inductive predicate U corresponding to the grammar of strings. What we are going to prove is the decidability of U, from a proof of which we extract a parser. The parser takes a string, and first it determines whether the string is parsable or not, and if it is, the parse tree is output as well.

Definition 20 (Parenthesis and grammar). Define the algebras of parentheses **Par** to be $\mu_\xi(\mathsf{L}^\xi, \mathsf{R}^\xi)$. Assume $\mathbf{L}_\alpha$ is the list algebra, and let p, q range over **Par** and x, y, z over $\mathbf{L_{Par}}$. We define the grammar $s, t := \mathsf{Nil} \mid s\mathsf{L}t\mathsf{R}$ as an inductive predicate $U := \mu_X^c(X\mathsf{Nil}, \forall_{x,y}^{nc}(Xx \to^c Xy \to^c X(x\mathsf{L}y\mathsf{R})))$.

When there is no confusion, list operations as prepending and appending are omitted. We can easily prove that U is same as a rather intuitive grammar defined as follows: $s, t := \mathsf{Nil} \mid st \mid \mathsf{L}s\mathsf{R}$. The associated algebra of U is given by the binary tree $\mathbf{T} := \mu_\xi(\mathsf{I}^\xi, \mathsf{C}^{\xi \to \xi \to \xi})$. We define a program constant LP of arity $(\mathbf{N}, \mathbf{L_{Par}})$, such as $\mathsf{LP}\, n\, x$ means that $U(\mathsf{L}^n y)$.

Definition 21 (Constant LP).

$$\text{LP } 0 \text{ Nil} = \text{T}, \qquad \text{LP } (\text{S}n) \text{ Nil} = \text{F}, \qquad \text{LP } n \text{ L::}x = \text{LP } (\text{S}n) \, x,$$
$$\text{LP } 0 \text{ R::}x = \text{F}, \qquad \text{LP } (\text{S}n) \text{ R::}x = \text{LP } n \, x.$$

In other words, the term $\text{LP } n \, x$ expresses that the number of superfluous R in x is n.

Lemma 1 (Totality of LP). $\forall^c_{x \in T_{\text{L}_{\text{Par}}}, n \in T_{\text{N}}} T_{\text{B}}(\text{LP } n \, x).$

Proof. Straightforward by means of induction on string, induction on natural numbers, and the definition of LP. □

Lemma 2 (LP-concatenation). $\forall^c_{x \in T_{\text{L}_{\text{Par}}}, y \in T_{\text{L}_{\text{Par}}}, n \in T_{\text{N}}, m \in T_{\text{N}}} (\text{LP } n \, x \;\rightarrow\; \text{LP } m \, y \rightarrow \text{LP}(n + m)(xy)).$

Proof. Straightforward by induction on strings and natural numbers, and the definition of LP. □

We can prove the following soundness formula.

Lemma 3 (Soundness). $\forall^{\text{nc}}_x (U x \rightarrow \text{LP } 0 \, x).$

Proof. Assume x and $U x$. Use U^- to prove $\text{LP } 0 \, x$. $\text{LP } 0 \text{ Nil}$ goes to truth by conversion. It suffices to prove the step formula $\forall^{\text{nc}}_{x,y}(U x \rightarrow^c \text{LP } 0 \, x \rightarrow^c U y \rightarrow^c \text{LP } 0 \, y \rightarrow^c \text{LP } 0 \, (x \text{L} y \text{R}))$. Due to Lemma 2 it suffices to prove $\text{LP } (\text{S}0) \, y \text{R}$, which follows by Lemma 2 and the assumption $\text{LP } 0 \, y$. □

In order to prove the completeness formula, namely, $\text{LP } 0$ is a subset of U, we use the definition of RP, such as $\text{RP } n \, x$ means $U(x\text{R}^n)$, and the following lemma.

Definition 22 (Predicate RP$_P$).

$$\text{RP} := \mu^c_X (X \, 0 \text{ Nil}, \forall^{\text{nc}}_{n,x,z}(P z \rightarrow^c X \, n \, x \rightarrow^c X \, \text{S}n \, (xz\text{L}))).$$

In other words, the formula $\text{RP } n \, x$ expresses that the number of superfluous L in x is n. It is easy to see that strings x and y with superfluous L and R, respectively, can be concatenated in order to cancel such superfluous parentheses. We state this property as follows.

Lemma 4 (Closure). $\forall^c_{y \in T} \forall^{\text{nc}}_{n,x,z}(\text{RP}_U \, n \, x \rightarrow^c U z \rightarrow^c \text{LP } n \, y \rightarrow U(xzy)).$

Proof. By induction on y. For the first case we assume n, x, z, and $\mathrm{RP}_U\, n\, x$, then use RP^-. The first subcase is trivial because there is a premise Uz which is same as our goal. As a premise of the second subcase is F, we use Theorem 1. For the second case, the goal formula is $\forall^c_{p,x}(\forall^{nc}_{n,y,z}(\mathrm{RP}_U\, n\, y \to^c Uz \to^c \mathrm{LP}\, n\, x \to U(yzx)) \to^c \forall^{nc}_{n,y,z}(\mathrm{RP}_U\, n\, y \to^c Uz \to^c \mathrm{LP}\, n\, (p{::}x) \to U(yz(p{::}x))))$. Do case distinction on p. Case L. Assume variables and premises, and prove $U(yz(\mathsf{L}{::}x))$ by means of the induction hypothesis with $\mathsf{S}n$, xzL, and Nil. The premises are canceled by RP^+_1 with the assumption Uz and $\mathrm{RP}\, n\, x$, U^+_0, and the assumption $\mathrm{LP}\,(\mathsf{S}n)\, y$. Case R. Assume x and the induction hypothesis, and also the variables n, y, and z, and an assumption $\mathrm{RP}_U\, n\, y$. Use RP^-. The first subcase is trivial by Theorem 1 because there is F in the premises. The second subcase is to prove $\forall^{nc}_{n,x,z_0}(Uz_0 \to^c \mathrm{RP}_U\, n\, x \to (Uz \to \mathrm{LP}\, n\, (\mathsf{R}{::}y) \to^c U(xz\mathsf{R}y)) \to^c Uz \to^c \mathrm{LP}(\mathsf{S}n)(\mathsf{R}{::}y) \to U(xz_0\mathsf{L}z\mathsf{R}y))$. We prove $\mathrm{LP}(\mathsf{S}n)(\mathsf{R}{::}y) \to U(xz_0\mathsf{L}z\mathsf{R}y)$ from the assumptions and the main induction hypothesis. It suffices to cancel the premises of the induction hypothesis, that are $\mathrm{RP}_U\, n\, x$ and $U(z_0\mathsf{L}z\mathsf{R})$. The first one is in the assumptions, and the second one is proven by U^+_1. $\square$

Lemma 5 (Completeness). $\forall^c_{x \in T}(\mathrm{LP}\, 0\, x \to Ux)$.

Proof. Assume x and $\mathrm{LP}\, 0\, y$, and use Lemma 4 with y, 0, Nil, and Nil. The premises are canceled by RP^+_0, U^+_0, and the assumption $\mathrm{LP}\, 0\, y$. $\square$

Proposition 2 (Decidability of U). $\forall^c_x \exists_{b\mathbf{B}}(b \to Ux \wedge \neg b \to \neg U)$.[1]

Proof. Assume x and use $\exists^+$ with $\mathrm{LP}\, 0\, x$, then assume b. For the left conjunct assume $\mathrm{LP}\, 0\, x$, then Ux follows by Lemma 5. For the right conjunct assume $\mathrm{LP}\, 0\, x \to \mathsf{F}$ and Ux to prove F. By Lemma 3 Ux implies $\mathrm{LP}\, 0\, x$. $\square$

We extract $\lambda_x\langle\mathrm{LP}\, 0\, x, g(x, \mathsf{Nil}, \mathsf{I}^\mathbf{T})\rangle$ from Theorem 2 where $g(x, as, a)$ abbreviates $\mathcal{R}^{\mathbf{L_T} \to \mathbf{T} \to \mathbf{T}}_{\mathbf{LP}}\, x M N a s a$ for $M := \lambda_{ts}\mathbf{L_T}, t\mathbf{T}(\mathcal{C}_{\mathbf{L_T}}\, ts\, t\, \lambda_{\_,\_}\mathsf{I})$ and $N := \lambda_{p,x,f,ts,t}(\mathcal{C}_{\mathbf{Par}}p(f(t{::}ts)\mathsf{I})(\mathcal{C}_{\mathbf{L_T}}\, ts\, \mathsf{I}\, \lambda_{t_0,ts_0}fts_0(\mathcal{C}t_0t)))$. We can present

---

[1] Following the existing Minlog proof script for this case study, we formulate the disjunction $A \vee B$ by $\exists_{b\mathbf{B}}(b \to A \wedge (b \to \mathsf{F}) \to B)$.

g in the following implicit form:

$$g(\mathsf{Nil}, \mathsf{Nil}, a) = a,$$
$$g(\mathsf{Nil}, a{::}as, a') = \mathsf{I},$$
$$g(\mathsf{L}{::}x, as, a) = g(x, a{::}as, \mathsf{I}),$$
$$g(\mathsf{R}{::}x, \mathsf{Nil}, a) = \mathsf{I},$$
$$g(\mathsf{R}{::}x, a{::}as, a') = g(x, as, \mathsf{C}aa').$$

Informally speaking, g takes x as a string to parse, as as a stack of parse tree generated so far, and a as the current parse tree. The first case is the successful case; nothing further to parse and the stack is empty, namely, the previous results of parsing have been used to form the current parse tree. The second and the fourth cases are unsuccessful cases. The second one happens when the parser saw an L but the corresponding R is missing. On the other hand the fourth case happens when L is missing. The third case is to read L. The parser pushes the current parse tree, corresponding to the last parse tree before the L, to the stack and continues parsing x with taking the empty tree as the current parse tree. Here we can see that the length of the stack is the number of parentheses to be closed. The fifth case is to read R and to go on. Here a' is a parse tree from the balanced string, say y, immediately before the current string $\mathsf{R}{::}x$, and $a{::}as$ is the list of parse trees obtained before y. Reading R, the parser pops a from the stack and forms a branch with the current parse tree a'.

Let t be the extracted program. We give some experiments:

$$t\mathsf{Nil} \mapsto^* \langle \mathsf{T}, \mathsf{I} \rangle,$$
$$t(\mathsf{LR}) \mapsto^* \langle \mathsf{T}, \mathsf{CII} \rangle,$$
$$t\mathsf{R} \mapsto^* \langle \mathsf{F}, \mathsf{I} \rangle,$$
$$t(\mathsf{LLRLRR}) \mapsto^* \langle \mathsf{T}, \mathsf{CI}(\mathsf{C}(\mathsf{C}(\mathsf{CII})\mathsf{I})) \rangle,$$
$$t(\mathsf{LRRLLR}) \mapsto^* \langle \mathsf{F}, \mathsf{I} \rangle.$$

3.3. *Approximate splitting property of real numbers*

In this section we introduce real numbers and study the approximate splitting property, a computational solution of which is a decision procedure to determine whether the given real number is below or above the given interval with a positive length [8]. We first define the notion of real numbers which is suitable for constructive setting, then find a computational solution of the approximate splitting property through program extraction [16].

In classical mathematics the notions of Cauchyness of a rational sequence $(a_n)_n$ is described e.g. as $\forall_k \exists_l \forall_{n,m} |a_n - a_m| \leq 2^{-k}$. One issue comes when we think about computing for a given k a modulus of convergence l. It is apparently not possible in general as one cannot read off infinite data from $(a_n)_n$ to witness the convergence. Following an idea by Bishop to overcome this problem, we develop Cauchy reals and its arithmetic based on rational arithmetic.

Definition 23 (Rational number). Define positive numbers $\mathbf{P}$, integers $\mathbf{Z}$ and rational numbers $\mathbf{Q}$ as follows.

$$\mathbf{P} := \mu_\xi(\xi, \xi \to \xi, \xi \to \xi), \quad \mathbf{Z} := \mu_\xi(\xi, \mathbf{P} \to \xi, \mathbf{P} \to \xi), \quad \mathbf{Q} := \mathbf{Z} \times \mathbf{P}.$$

Definition 24 (Cauchy real). Define $\mathbf{R} := \mu_\xi((\mathbf{N} \to \mathbf{Q}) \to (\mathbf{Z} \to \mathbf{N}) \to \xi)$, a pair of a rational sequence $(a_n)_n$ and a function $M^{\mathbf{Z} \to \mathbf{N}}$ has the Cauchyness property iff $\forall_{k,m,n}(m, n \geq Mk \to |a_n - a_m| \leq 2^{-k})$. We say that a pair $(a_n)_n$, M is a Cauchy real iff it satisfies the Cauchyness property, and then M is called a Cauchy modulus.

Definition 25 (Orders on real numbers). A real $x = \langle (a_n)_n, M \rangle$ is *non-negative* if $-\frac{1}{2^k} \leq a_{M(k)}$ for any $k \in \mathbf{Z}^+$, and written $x \in \mathbb{R}^{0+}$. x is *k-positive* if $\frac{1}{2^k} \leq a_{M(k+1)}$, and written $x \in_k \mathbb{R}^+$. We write $x \leq y$ for $y - x \in \mathbb{R}^{0+}$ and $x <_k y$ for $y - x \in_k \mathbb{R}^+$. We can omit the subscript k if there is no confusion.

While $x < y$ or $x = y$ or $x > y$ classically holds for any real numbers x, y, it is not the case in constructive mathematics. However, if we think about two Cauchy reals y, z s.t. $y <_k z$ for some k, we can constructively prove $x \leq z \lor y \leq x$ for any Cauchy real x. We determine it by approximating real numbers, so that they fall into an interval of the half length of $z - y$. Then, in order to make a decision, we can compare the rational approximation of x with the rational approximation of $\frac{y+z}{2}$, instead of comparing the reals.

Proposition 3 (Aproximate splitting property).

$$\forall^{nc}_{x,y,z,k}(\text{Real } x \to^c \text{Real } y \to^c \text{Real } z \to^c y <_k z \to^c x \leq z \lor y \leq x).$$

Proof. Let x, y, and z be $\langle (a_n)_n, M \rangle$, $\langle (b_n)_n, N \rangle$, and $\langle (c_n)_n, L \rangle$, respectively. Assume $y <_k z$, that is, $\frac{1}{2^k} \leq c_n - b_n$ for $n := \max(N(k+2), L(k+2))$. Let m be $\max(n, N(k+2))$. We make a case distinction on $a_m \leq \frac{b_n + c_n}{2}$. If it is the case, we prove $x \leq z$ as follows. For $l \geq m$, $a_l \leq c_l$ holds because $a_l \leq a_m + \frac{1}{2^{k+2}} \leq \frac{b_n + c_n}{2} + \frac{c_n - b_n}{4} = c_n - \frac{c_n - b_n}{4} \leq b_m - \frac{1}{2^{k+2}} \leq b_l$. Otherwise, namely, $a_m \not\leq \frac{b_n + c_n}{2}$, we prove $y \leq x$ as follows. For $l \geq m$, $b_l \leq a_l$ holds because $b_l \leq b_m + \frac{1}{2^{k+2}} \leq b_m + \frac{c_n - b_n}{2} \leq \frac{b_n + c_n}{2} + \frac{c_n - b_n}{4} = a_m - \frac{1}{2^{k+2}} \leq a_l$. $\square$

The term extracted from the formalized proof is the following.

$$\lambda_{x,y,z,k}(\mathcal{C}_{\mathbf{R}} x \lambda_{as,M}(\mathcal{C}_{\mathbf{R}} y \lambda_{bs,N}(\mathcal{C}_{\mathbf{R}} z \lambda_{cs,L}(r \leq q))))$$

where $r = as(max(M(k+2), N(k+2), L(k+2)))$ and $q = \frac{1}{2}(bs(max(N(k+2), L(k+2)) + cs(max(N(k+2), L(k+2)))))$.

We give a couple of experiments. Let t be the extracted term. We use the following program to compute the fast Cauchy sequence of the square root of the given rational number.

$$\text{sqrtaux}^{\mathbf{Q} \to \mathbf{N} \to \mathbf{Q}},$$
$$\text{sqrtaux}\, a^{\mathbf{Q}}\, 0 := 1,$$
$$\text{sqrtaux}\, a^{\mathbf{Q}}\, (Sn) := ((\text{sqrtaux}\, a\, n) + (a/(\text{sqrtaux}\, a\, n)))/2,$$
$$\text{sqrt}^{\mathbf{Q} \to \mathbf{N} \to \mathbf{Q}},$$
$$\text{sqrt}\, a^{\mathbf{Q}}\, n := \text{sqrtaux}\, a\, (Sn).$$

Let id embed a non-negative integer into a natural.

(1) Take an interval between 1 and $\sqrt{2}$, which satisfies $1 <_2 \sqrt{2}$, and compare $\frac{1}{\sqrt{2}}$ with it.

$$t\langle 1, \lambda_k 0\rangle \langle \text{sqrt}\, 2, \text{id}\rangle \langle \text{sqrt}\, (1/2), \text{id}\rangle 2 \mapsto^* \mathsf{T}$$

(2) Take an interval between 1 and $\sqrt{2}$, which satisfies $1 <_2 \sqrt{2}$, and compare $\frac{1}{2}$ with it.

$$t\langle 1, \lambda_k 0\rangle \langle \text{sqrt}\, 2, \text{id}\rangle \langle \lambda_n (1/2), \text{id}\rangle 2 \mapsto^* \mathsf{F}$$

(3) Take an interval between $\frac{\sqrt{2}}{2}$ and $\sqrt{2}$, which satisfies $\frac{\sqrt{2}}{2} <_1 \sqrt{2}$, and compare $\frac{\sqrt{3}}{2}$ with it.

$$t\langle \text{sqrt}\, 2, \text{id}\rangle \langle \text{sqrt}\, (3/4), \text{id}\rangle 1 \mapsto^* \mathsf{T}$$

4. Notes

4.1. *Logic for program extraction*

Some features of TCF have been omitted in this note. TCF equips general induction as well as primitive induction, and simultaneous and nested inductive/coinductive definitions as well as inductive definitions shown in this note. Coinduction is a sophisticated mean to handle infinite data and behaviors of processes [14]. An important example is the notion of bisimulation which is coinductively defined and is used for behavioral equality. The coinduction principle is realized by *corecursion* which is convenient when one deals with infinite data as streams. The notion of coinduction was studied in the context of realizability by Tatsuta, and later also by Berger and Schwichtenberg with a strong motivation towards implementation [4, 15, 17]. While only constructive proofs are dealt with in this note, TCF also supports extraction from classical proofs due to A-translation and Dialectica interpretation [13, 18]. Those features are all implemented and available in Minlog.

Decoration is a feature for fine-tuning extracted programs. By discarding computationally unimportant ingredient through extraction, TCF offers efficient extracted programs, in contrast to the paradigm of proofs-as-programs. Due to the idea of switching off non-computational ingredients in proofs, the term normalization of Minlog is sufficiently fast for a number of case studies. It is possible to automatically decorate proofs in TCF for a given decorated goal formula. An algorithm of finding the optimal decoration was studied by Ratiu and Schwichtenberg [12].

4.2. *Active research topics*

Constructive analysis is actively studied in settings of formal systems including TCF. There is a recent work in an application of coinduction for stream represented real numbers. They dealt with a specific stream called Gray-code, which suggest a logic for extracting non-deterministic programs [6]. The motivation of the work was rather in how to find a known algorithm through program extraction [19]. On the other way around, there are several attempts to find computational solutions to known theorems/axioms as Ramsey's theorem [11], open induction [1], and bar induction [3].

4.3. *Proof assistants*

We give a brief survey on proof assistants which offer program extraction or related functionalities. Coq is a proof assistant whose base logic is constructive, and supporting a feature called program extraction [9]. Differently from Minlog, the target language of their program extraction is not inside the base theory but it is generic programming languages as OCaml and Haskell. Concerning fine-tuning of extracted programs, one can switch on and off through the Set/Prop distinction, which looks similar as the computational/non-computational distinction in TCF. Isabelle has a feature for program extraction based on realizability interpretation [7]. This work by Berghofer was examined and compared with Minlog and Coq through a case study in extracting normalization algorithm [2]. Nuprl is a proof assistant based on Martin-Löf's type theory, and supports program extraction [10].

Acknowledgments

I am grateful to the organizing committee of the Autumn School Proofs and Computations 2016. I owe a lot to Prof. Helmut Schwichtenberg for his advice and encouragement. Scientific discussions with him were also helpful for me.

References

[1] U. Berger. Open induction and classical countable choice (January, 2004).

[2] U. Berger, S. Berghofer, P. Letouzey, and H. Schwichtenberg, Program extraction from normalization proofs, *Studia Logica.* **82**, 27–51 (2006).

[3] U. Berger and P. Oliva, Modified bar recursion, *Mathematical Structures in Computer Science.* **16**(2), 163–183 (2006). doi: 10.1017/S0960129506005093. URL http://dx.doi.org/10.1017/S0960129506005093.

[4] U. Berger. From coinductive proofs to exact real arithmetic. In eds. E. Grädel and R. Kahle, *Computer Science Logic*, LNCS, pp. 132–146, Springer Verlag, Berlin, Heidelberg, New York (2009).

[5] U. Berger, K. Miyamoto, H. Schwichtenberg, and M. Seisenberger. Minlog — a tool for program extraction supporting algebras and coalgebras. In eds. A. Corradini, B. Klin, and C. Cîrstea, *CALCO*, vol. 6859, *Lecture Notes in Computer Science*, pp. 393–399, Springer (2011). ISBN 978-3-642-22943-5.

[6] U. Berger. Extracting non-deterministic concurrent programs. In eds. J. Talbot and L. Regnier, *25th EACSL Annual Conference on Computer Science Logic, CSL 2016, August 29–September 1, 2016, Marseille, France*, vol. 62, *LIPIcs*, pp. 26:1–26:21, Schloss Dagstuhl - Leibniz-Zentrum fuer Informatik (2016). ISBN 978-3-95977-022-4. doi: 10.4230/LIPIcs.CSL.2016.26. URL http://dx.doi.org/10.4230/LIPIcs.CSL.2016.26.

[7] S. Berghofer. Program extraction in simply-typed higher order logic. In eds. H. Geuvers and F. Wiedijk, *TYPES*, vol. 2646, *Lecture Notes in Computer Science*, pp. 21–38, Springer (2002). ISBN 3-540-14031-X.

[8] E. Bishop and D. Bridges, *Constructive Analysis.* vol. 279, *Grundlehren der mathematischen Wissenschaften*, Springer Verlag, Berlin, Heidelberg, New York (1985).

[9] Coq http://coq.inria.fr/.

[10] Nuprl http://www.nuprl.org/.

[11] P. Oliva and T. Powell, A constructive interpretation of Ramsey's theorem via the product of selection functions, *Mathematical Structures in Computer Science.* **25** (8), 1755–1778 (2015). doi: 10.1017/S0960129513000340. URL http://dx.doi.org/10.1017/S0960129513000340.

[12] D. Ratiu and H. Schwichtenberg. Decorating proofs. In eds. S. Feferman and W. Sieg, *Proofs, Categories and Computations. Essays in honor of Grigori Mints,* pp. 171–188. College Publications (2010).

[13] D. Ratiu. *Refinement of Classical Proofs for Program Extraction.* PhD thesis, Mathematisches Institut der Universität München (2011).

[14] D. Sangiorgi, *Introduction to Bisimulation and Coinduction.* Cambridge University Press, New York, NY, USA (2011). ISBN 1107003636, 9781107003637.

[15] H. Schwichtenberg and S. S. Wainer, *Proofs and Computations.* Perspectives in Logic, Association for Symbolic Logic and Cambridge University Press (2012).

[16] H. Schwichtenberg. Constructive analysis with witnesses. Manuscript (Oct, 2016).
http://www.math.lmu.de/~schwicht/seminars/semws16/constr16.pdf.

[17] M. Tatsuta. Realizability of monotone coinductive definitions and its application to program synthesis. In ed. J. Jeuring, *MPC,* vol. 1422, *Lecture Notes in Computer Science,* pp. 338–364, Springer (1998). ISBN 3-540-64591-8.

[18] T. Trifonov. *Analysis of methods for extraction of programs from non-constructive proofs.* PhD thesis, Mathematisches Institut der Universität München (2012).

[19] H. Tsuiki, Real number computation through gray code embedding, *Theor. Comput. Sci.* **284**(2), 467–485 (2002). doi: 10.1016/S0304-3975(01)00104-9. URL http://dx.doi.org/10.1016/S0304-3975(01)00104-9.

Chapter 6

The Data Structures of the Lambda Terms

Masahiko Sato

Graduate School of Informatics
Kyoto University
Kyoto, Japan
masahiko@kuis.kyoto-u.ac.jp

The syntax of the λ-calculus is analyzed from a view point of abstract syntax. Namely, the collection of λ-terms, as well as the collection of λ-terms modulo α-equivalence, are studied as inductively generated free algebras determined by suitably given first-order context-free grammars.

Contents

1. Introduction

Starting from McCarthy's LISP programming language [5] introduced in the late 1950's, the λ-calculus has deeply influenced the development of modern programming languages. Nowadays, the λ-calculus is also used in the core part of most of the proof assistants. We think that it is therefore important to study the mathematical structure of the λ-*terms* so that we can represent λ-terms within a computer and at the same time we can mathematically analyze their structure.

We first review the existing representations of the λ-terms, and then give our approach to representing lambda terms as datatypes.

2. Church's syntax

In this section we review the traditional syntax of the λ-calculus introduced by Church, and discuss the well-known problems of the syntax mostly related to the definition of the substitution operation on lambda terms.

The λ-terms are defined by the following grammar:

$$\Lambda \ni M, N ::= x \mid \lambda_x M \mid (M\ N).$$

In the above definition, x is taken from a countably infinite set $\mathbb{X}$ of *variables*. A term $\lambda_x M$ is called an *abstraction* and $(M\ N)$ is called an *application*.

We write $[x := N]M$ for the result of *substituting* N for x in M. The intuitive idea of substitution is clear: replace each occurrence of x in M by N. But, this simplistic definition often fails to give us a correct answer.

For example, when $M = \lambda_y(x\ y)$ what is $[x := y]M$? $[x := y]\lambda_y(x\ y) = \lambda_y(y\ y)$ is not correct. The variable y, which is substituted for x in M, was a free variable before substitution, but it becomes a bound variable after substitution. The problem is solved by renaming y in M to a fresh variable z. Then, $[x := y]\lambda_z(x\ z) = \lambda_z(y\ z)$. We replaced $M = \lambda_y(x\ y)$ by $M' = \lambda_z(x\ z)$ which is obtained by renaming. Such a pair of M and M' are called *α-equivalent*.

Another example: what is $[x := y]\lambda_x y$? Clearly, $\lambda_y y$ is a wrong answer, since a correct substitution must be done only on *free* occurrences of a variable.

These examples show that defining the substitution operation on λ-terms is a subtle problem. Indeed, a correct inductive definition given by Curry [3, Page 94] is rather complex and involves a case where renaming of a bound variable in M is done to avoid an unsolicited capture of a free variable in N. Thus, for the first example above, the correct definition is:

$$[x := y]\lambda_y(x\ y) = \lambda_z[x := y][y := z](x\ y) = \lambda_z[x := y](x\ z) = \lambda_z(y\ z),$$

where z could be any variable other than x and y, but Curry had to enumerate all the variables as $e_1, e_2, \ldots$ and had to choose the first variable in the list such that it is distinct from both x and y to make the definition as a deterministic definition.

Because of the subtlety and complexity of the substitution operation, most of the modern text books (for example Barendregt [1]) on the λ-calculus do not deal with λ-terms directly but rather deal with λ-terms modulo α-equivalence.

In the rest of this section, we analyze the λ-terms by giving a new classification of λ-terms. Based on this new classification we will study the data structures of λ-terms in the following sections.

2.1. *An alternative definition of* Λ

Recall that:

$$\Lambda \ni M, N ::= x \mid \lambda_x M \mid (M\ N)$$

We note that Λ may be regarded as an initial algebra of the following two constructors

$$\lambda_x : \Lambda \to \Lambda \qquad (x \in \mathbb{X})$$

$$\mathsf{App} : \Lambda \times \Lambda \to \Lambda$$

generated by the variables $x \in \mathbb{X}$. In this setting, each λ-term is either a *variable*, a λ-*abstract* or an *application*.

But, writing $\lambda_{x_1 x_2 \cdots x_n} M$ for $\lambda_{x_1} \lambda_{x_2} \cdots \lambda_{x_n} M$ $(n \geq 0)$, any λ-term can be uniquely written in one of the following two forms.

(1) $\lambda_{x_1 x_2 \cdots x_n} y$.
(2) $\lambda_{x_1 x_2 \cdots x_n} (M\ N)$.

We will call a term of the first form a *thread* and the second form a *generalized application*.

Using the above classification, we can give an alternative definition of Λ as follows. We write $\bar{x}$ for $x_1 x_2 \cdots x_n$,

$$\frac{}{\lambda_{\bar{x}} y \in \Lambda}, \qquad \frac{\lambda_{\bar{x}} M \in \Lambda \quad \lambda_{\bar{x}} N \in \Lambda}{\lambda_{\bar{x}} (M\ N) \in \Lambda}.$$

This gives a correct definition of Λ since by these rules we can generate all the λ-terms. It is, however, inconvenient to use this as an official definition of λ-term since it does not give us an initial algebra. Nevertheless, it can be used to characterize closed λ-terms and also to define α-equivalence as we will see below.

The above classification gives us the following 3 invariants under α-equivalence. For each λ-term M, we can assign the following 2 values.

(1) If $M = \lambda_{x_1 x_2 \cdots x_n} N$ or $M = \lambda_{x_1 x_2 \cdots x_n} (N\ P)$, then the *height* of M is n.

(2) If $M = \lambda_{x_1 x_2 \cdots x_n} N$, then its *thickness* is 1. If $M = \lambda_{x_1 x_2 \cdots x_n} (N\ P)$, then its *thickness* is the sum of the thickness of $\lambda_{x_1 x_2 \cdots x_n} N$ and the thickness of $\lambda_{x_1 x_2 \cdots x_n} P$.

We also define the *shape* of a thread as follows.

(1) $\mathrm{Sh}(\lambda_{\bar{x}} y) := (n, y)$ if $y \notin \bar{x}$.

(2) $\mathrm{Sh}(\lambda_{\bar{x}} y) := (n, i)$, if $y \in \bar{x}$ and i is the largest index such that $y = x_i$.

That shape is invariant under α-equivalence is intuitively obvious. We use this intuitive idea to define the notion of α-equivalence formally by the following two rules:

$$\frac{\mathrm{Sh}(\lambda_{\bar{x}} u) = \mathrm{Sh}(\lambda_{\bar{y}} v)}{\lambda_{\bar{x}} u \equiv_\alpha \lambda_{\bar{y}} v},$$

$$\frac{\lambda_{\bar{x}} M \equiv_\alpha \lambda_{\bar{y}} M' \quad \lambda_{\bar{x}} N \equiv_\alpha \lambda_{\bar{y}} N'}{\lambda_{\bar{x}} (M\ N) \equiv_\alpha \lambda_{\bar{y}} (M'\ N')}.$$

We can check that $\lambda_{xy}(x\ y) \equiv_\alpha \lambda_{yx}(y\ x)$ as follows:

$$\frac{\dfrac{\mathrm{Sh}(\lambda_{xy} x) = \mathrm{Sh}(\lambda_{yx} y) = (2, 1)}{\lambda_{xy} x \equiv_\alpha \lambda_{yx} y} \quad \dfrac{\mathrm{Sh}(\lambda_{xy} y) = \mathrm{Sh}(\lambda_{yx} x) = (2, 2)}{\lambda_{xy} y \equiv_\alpha \lambda_{yx} x}}{\lambda_{xy}(x\ y) \equiv_\alpha \lambda_{yx}(y\ x)}.$$

2.2. The set Λ_0 of closed terms

We can inductively define the subset Λ_0 of Λ, consisting of closed λ-terms, as follows:

$$\frac{y \in \bar{x}}{\lambda_{\bar{x}} y \in \Lambda_0}, \qquad \frac{\lambda_{\bar{x}} M \in \Lambda_0 \quad \lambda_{\bar{x}} N \in \Lambda_0}{\lambda_{\bar{x}} (M\ N) \in \Lambda_0}.$$

Note that the above definition is obtained from our alternative definition of the λ-terms by simply restricting the thread construction rule by the side condition $y \in \bar{x}$. This side condition has the effect of allowing only *closed* threads may be constructed by the rule.

This construction is remarkable, since in the traditional definition of the set of closed λ-terms, one has to define the whole set of λ-terms and then obtain its subset of closed terms by using the notion of free occurrences of a variable in a term.

This definition suggests that we should be able to develop proof theory of the λ-calculus with free variables without appealing to the notion of bound variables, and of the λ-calculus of closed λ-terms without using the notion of variables.

3. From Church's syntax to de Bruin notation

In this section, we relate Church's definition of λ-terms to the notation introduced by de Bruijn [4] which eliminates variable names of the λ-binders and replaces variable names of the bound occurrences of variables by suitably chosen natural numbers called de Bruijn indices.

Although de Bruijn cites neither Quine [6] nor Bourbaki [2], from a historical point of view, we think that it is appropriate to put Quine-Bourbaki's notation in between Church's syntax and de Bruijn's notation as we will do so here.

As an illustrative example, we take up the λ-term $M = \lambda_x \lambda_y (\lambda_z (z\ x)\ (x\ y))$. We show in Figure 1 its representations in Church's syntax (left figure) and in Quine-Bourbaki notation (right figure).

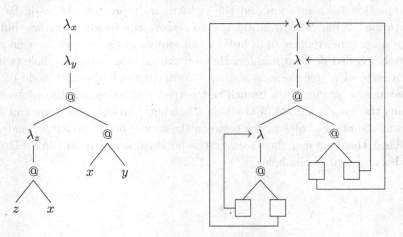

Fig. 1. Church's syntax and Quine-Bourbaki notation

The left figure represents M as a tree where each node (other than a leaf node) has one or two branches. A unary node is occupied by a λ-abstraction operator, and a binary node is occupied by the application operator designated by '@'. In Church's syntax, occurrences of variables x, y, z are categorized as either the *binding occurrences* or the *bound occurrences*. Here, each binding occurrence is prefixed by λ, and each bound occurrence occupies a leaf of the syntax tree. In [6] Quine says "they[1] serve merely to indicate cross-references to various positions of quantification.[2]" Then, Quine says "Such cross-references could be made instead by curved lines." In his example [6] Quine used curved lines and called them *bonds*. Here we used polygonal lines[3] instead. It should be clear that by using bonds we can get rid of explicit use of variables without changing the essence of the term. Here, the essence means the fact that choices of particular variables are irrelevant as long as they preserve the cross-references. This is, at the same time, the essence of the notion of α-equivalence of λ-terms.

Thus, we see that the right figure may be thought of as a representation of all the λ-terms which are α-equivalent to $\lambda_z \lambda_x (\lambda_y (y\ z)\ (z\ x))$. A theoretical drawback of the right figure is that, while the left figure is a *tree*, it is not a tree but a *graph*. That it is a graph makes it difficult to prove properties of such graphs since there is no induction principle on graphs.

Then, de Bruijn [4] introduced an ingenious way of turning a Quine-Bourbaki's graph notation into a tree notation without losing any information in the original graph. In Figure 2, the left figure represents M in Quine-Bourbaki notation and the right figure represents M in de Bruijn notation. What de Bruijn did was to remove the bondage arrows, but at the same time, replace each hole of the sources of the arrows by a natural number called *de Bruijn index*. He used the fact that from any hole (which is a leaf) of a tree there is a unique route to reach the root node of the tree always staying on a branch of the tree, and moreover, the route contains the target λ-node of the hole's bondage arrow. Then, one can just count the number of λ-nodes between the source node and target node. It is clear that one can then recover the bondage arrow from the de Bruijn index assigned to the hole.

---

[1]The variables x, y, z in the syntax-tree.

[2]Quine used the term 'quantification' since he used a universally quantified logical formula in his example.

[3]Bourbaki [2] also used polygonal lines.

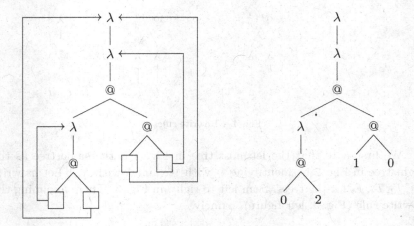

Fig. 2. Quine-Bourbaki notation and de Bruijn notation

4. de Bruijn algebra

In this section, we first extend de Bruijn notation by introducing application operators $@^n$ for each natural number n in Subsection 4.1. Then, based on this extended notation, we will define an algebra D which we call *de Bruijn algebra* in Subsection 4.2

4.1. *Extended de Bruijn notation*

As in the previous section, we first explain our idea using syntax trees shown in Fig. 3.

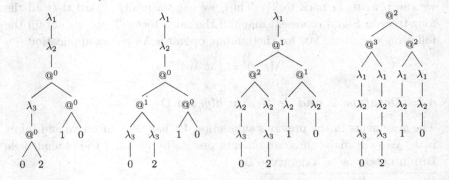

Fig. 3. Trees T_1, T_2, T_3, and T_4 in extended de Bruijn notation

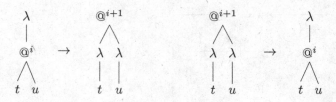

Fig. 4.   Rewrite rules

We first note that the leftmost tree in Fig. 3 is the same tree as the right tree in Fig. 2 by identifying @ with $@^0$ and λ with λ_i.[4] Let us write T_1, T_2, T_3, T_4 for the trees, from left to right, in Fig. 3. Then, assuming the rewrite rule (Fig. 4, left figure), namely,

$$\lambda(t\ u)^i \to (\lambda t\ \lambda u)^{i+1}$$

we can rewrite T_1 to T_4 (via T_2 and T_3) as follows.

$$
\begin{aligned}
T_1 = \lambda\lambda(\lambda(0\ 2)^0\ (1\ 0)^0)^0 &\to \lambda\lambda((\lambda 0\ \lambda 2)^1\ (1\ 0)^0)^0 = T_2 \\
&\to \lambda(\lambda(\lambda 0\ \lambda 2)^1\ \lambda(1\ 0)^0)^1 \\
&\to \lambda((\lambda\lambda 0\ \lambda\lambda 2)^2\ (\lambda 1\ \lambda 0)^1)^1 = T_3 \\
&\to (\lambda(\lambda\lambda 0\ \lambda\lambda 2)^2\ \lambda(\lambda 1\ \lambda 0)^1)^2 \\
&\to ((\lambda\lambda\lambda 0\ \lambda\lambda\lambda 2)^3\ (\lambda\lambda 1\ \lambda\lambda 0)^2)^2 = T_4.
\end{aligned}
$$

Also, by adding the rewrite rule (Fig. 4, right figure), namely,

$$(\lambda u\ \lambda u)^{i+1} \to \lambda(t\ u)^i$$

we can rewrite T_4 back to T_1. Thus, we can naturally regard that all the four trees in Fig. 3 represent one and the same term. Thus, we obtain the following distribute law for the binding operator λ over an application.

$$\lambda(t\ u)^i = (\lambda t\ \lambda u)^{i+1}.$$

4.2. *Datatype $\mathbb{D}$ and de Bruijn algebra* D

The arguments in the previous subsection 4.1 has been intuitive and informal. We will make these arguments precise by defining the extended de Bruijn notations as a datatype $\mathbb{D}$.

---

[4]We decorated λ with suffix i ($i = 1, 2, 3$) to identify them through the four trees in Fig. 3

Definition 1 (Datatype $\mathbb{D}$ of extended de Bruijn notation). We define the datatype $\mathbb{D}$ as follows:

$$\frac{k \in \mathbb{N}}{k \in \mathbb{D}} \text{ Idx}, \qquad \frac{D \in \mathbb{D}}{\lambda D \in \mathbb{D}} \text{ Abs}, \qquad \frac{D \in \mathbb{D} \quad E \in \mathbb{D}}{(D\ E)^i \in \mathbb{D}} \text{ App}^i.$$

□

Next, we define a binary relation $\to$ and a binary relation $\equiv$ both on $\mathbb{D}$ inductively as follows:

Definition 2 (Relations $\to$ and $\equiv$ on $\mathbb{D}$).

$$\frac{}{\lambda(D\ E)^i \to (\lambda D\ \lambda E)^{i+1}} \qquad \frac{D \to E}{\lambda D \to \lambda E}$$

$$\frac{D \to D'}{(D\ E)^i \to (D'\ E)^i} \qquad \frac{E \to E'}{(D\ E)^i \to (D\ E')^i}$$

$$\frac{}{D \to D} \qquad \frac{D \to E \quad E \to F}{D \to F} \qquad \frac{D \to F \quad E \to F}{D \equiv E}.$$

□

For example, we can derive

$$\lambda(D\ E)^i \equiv (\lambda D\ \lambda E)^{i+1}$$

as follows:

$$\frac{\lambda(D\ E)^i \to (\lambda D\ \lambda E)^{i+1} \quad (\lambda D\ \lambda E)^{i+1} \to (\lambda D\ \lambda E)^{i+1}}{\lambda(D\ E)^i \equiv (\lambda D\ \lambda E)^{i+1}}.$$

We now make $\mathbb{D}$ into an algebra D as follows. It is not difficult to show that the relation $\to$ is confluent and strongly normalizing. So, we write D^* for the normal form of D. We call D^* the *applicative normal form* of D since an application in D^* never occurs in the scope of a λ-binder. Then, we have $D \equiv E$ iff $D^* = E^*$. Hence, $\equiv$ is indeed an equivalence relation on $\mathbb{D}$. Now, we put $\mathsf{D} := \mathbb{D}/\equiv$, and write $[D]$ for the equivalence class containing D. Then,

$$[D] = [E] \in \mathsf{D} \iff D^* = E^* \in \mathbb{D}. \tag{1}$$

For example, we have

$$[\lambda(D\ E)^i] = [(\lambda D\ \lambda E)^{i+1}], \tag{2}$$

since

$$\left(\lambda(D\ E)^i\right)^* = \left((\lambda D\ \lambda E)^{i+1}\right)^* = (\lambda(D^*)\ \lambda(E^*))^{i+1}.$$

In this setting, we define the following two algebraic operations on D,

$$\lambda : \mathsf{D} \to \mathsf{D},$$
$$\mathsf{App}^i : \mathsf{D} \times \mathsf{D} \to \mathsf{D} \quad (i \in \mathbb{N}),$$

by putting[5]:

$$\lambda[D] := [\lambda D], \tag{3}$$
$$([D]\ [E])^i := [(D\ E)^i]. \tag{4}$$

Then we have:

$$\begin{aligned}
\lambda([D]\ [E])^i &= \lambda[(D\ E)^i] \\
&= [(\lambda D\ \lambda E)^{i+1}] \quad \text{by (2)} \\
&= ([\lambda D]\ [\lambda E])^{i+1} \quad \text{by (4)} \\
&= (\lambda[D]\ \lambda[E])^{i+1} \quad \text{by (3)}.
\end{aligned}$$

Thus, by allowing ourselves to use the same meta-variables (D, E etc.), we see that the de Bruijn algebra D enjoys the *distributive law*:

$$\lambda(D\ E)^i = (\lambda D\ \lambda E)^{i+1}$$

for the λ operator. It follows, from this distributive law, that the four trees T_1, T_2, T_3, T_4 in Fig. 3 are all equal to each other when interpreted as elements of the de Bruijn algebra D.

5. Extension of Church's syntax

Inspired by the extended de Bruijn notation introduced in the previous section, we extend Church's syntax and develop a new notation for λ-terms. We extend Church's syntax of λ-terms in two steps. In Subsection 5.1, we introduce the datatype $\mathbb{C}$ which extends Church's syntax by introducing, as in Subsection 4.1, a family of application operations. In Subsection 5.2, we further extend Church's syntax by introducing the nameless binder λ as well as a family of combinators I_k $(k \in \mathbb{N})$.

---

[5]The well-definedness of these equations can be verified by using (1).

5.1. *Datatype of extended Church's syntax and Church algebra*

The first step can be visualized in tree form as in Fig. 5 where the trees T_5, T_6, T_7, and T_8 (from left to right) are all intended to represent the same term $\lambda_x\lambda_y(\lambda_z(z\ x)\ (x\ y))$ as in Section 3. In fact, by comparing Fig. 3 with Fig. 5, we see that the trees T_1, T_2, T_3, T_4 are respectively isomorphic to the trees T_5, T_6, T_7, T_8.

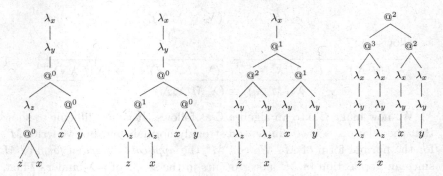

Fig. 5. Trees T_5, T_6, T_7, and T_8 in extended Church's syntax

Thus, we can formalize the idea behind Fig. 5 in a parallel manner as in Subsection 4.2. We first define the datatype $\mathbb{C}$ of extended Church's syntax. It should be clear that $\mathbb{C}$ naturally extends Church's syntax (by identifying $(M\ N)^0$ with $(M\ N)$).

Definition 3 (Datatype $\mathbb{C}$ of extended Church's syntax).
We define the datatype $\mathbb{C}$ inductively as follows. We assume that $\mathbb{X}$ is a countably infinite set of *variables* equipped with a decidable equality.

$$\frac{x \in \mathbb{X}}{x \in \mathbb{C}}\ \text{Var,} \qquad \frac{x \in \mathbb{X} \quad M \in \mathbb{C}}{\lambda_x M \in \mathbb{C}}\ \text{Abs,} \qquad \frac{M \in \mathbb{C} \quad N \in \mathbb{C}}{(M\ N)^i \in \mathbb{C}}\ \text{App}^i.$$

$\square$

Next, we define a binary relation $\to$ and a binary relation $\equiv$ both on $\mathbb{C}$ inductively as follows.

Definition 4 (Relations $\to$ and $\equiv$ on $\mathbb{C}$).

$$\frac{}{\lambda_x(M\ N)^i \to (\lambda_x M\ \lambda_x N)^{i+1}} \qquad \frac{M \to N}{\lambda_x M \to \lambda_x N}$$

$$\frac{M \to M'}{(M \ N)^i \to (M' \ N)^i} \qquad \frac{E \to E'}{(M \ N)^i \to (M \ N')^i}$$

$$\frac{}{M \to M} \qquad \frac{M \to N \quad N \to P}{M \to P} \qquad \frac{M \to P \quad N \to P}{M \equiv N}.$$

□

For example, we can derive

$$\lambda_x(M \ N)^i \equiv (\lambda_x M \ \lambda_x N)^{i+1}$$

as follows:

$$\frac{\lambda_x(M \ N)^i \to (\lambda_x M \ \lambda_x N)^{i+1} \quad (\lambda_x M \ \lambda_x N)^{i+1} \to (\lambda M \ \lambda N)^{i+1}}{\lambda_x(M \ N)^i \equiv (\lambda_x M \ \lambda_x N)^{i+1}}.$$

We now make $\mathbb{C}$ into an algebra C as follows. It is not difficult to show that the relation $\to$ is confluent and strongly normalizing. So, we write M^* for the normal form of M. We call M^* the *applicative normal form* of M since an application in M^* never occurs in the scope of a λ-binder. Then, we have $M \equiv N$ iff $M^* = N^*$. Hence, $\equiv$ is indeed an equivalence relation on $\mathbb{C}$. Now, we put $\mathsf{C} := \mathbb{C}/\equiv$, and write $[M]$ for the equivalence class containing M. Then,

$$[M] = [N] \in \mathsf{C} \iff M^* = N^* \in \mathbb{C}. \tag{5}$$

For example, we have

$$[\lambda_x(M \ N)^i] = [(\lambda_x M \ \lambda_x N)^{i+1}], \tag{6}$$

since

$$\left(\lambda_x(M \ N)^i\right)^* = \lambda_x((\lambda_x M \ \lambda N)^{i+1}) = (\lambda_x(M^*) \ \lambda_x(N^*))^{i+1}.$$

In this setting, we define the following two algebraic operations on C,

$$\lambda_x : \mathsf{C} \to \mathsf{C} \qquad (x \in \mathbb{X}),$$
$$\mathsf{App}^i : \mathsf{C} \times \mathsf{C} \to \mathsf{C} \qquad (i \in \mathbb{N}),$$

by putting[6]:

$$\lambda_x[M] := [\lambda_x M], \tag{7}$$
$$([M] \ [N])^i := [(M \ N)^i]. \tag{8}$$

---

[6]The well-definedness of these equations can be verified by using (5).

Then we have:

$$\lambda_x([M]\ [N])^i = \lambda_x[(M\ N)^i]$$
$$= [(\lambda_x M\ \lambda_x N)^{i+1}]\ \text{ by (6)}$$
$$= ([\lambda_x M]\ [\lambda_x N])^{i+1}\ \text{ by (8)}$$
$$= (\lambda_x[M]\ \lambda_x[N])^{i+1}\ \text{ by (7)}.$$

Thus, by allowing ourselves to use the same meta-variables (M, N etc.) range over elements of C, we see that the de Church algebra C enjoys the *distributive law*:

$$\lambda_x(M\ N)^i = (\lambda_x M\ \lambda_x N)^{i+1}.$$

for the λ_x operator. It follows, from this distributive law, that the four trees T_5, T_6, T_7, T_8 in Fig. 5 are all equal to each other when interpreted as elements of the Church algebra C.

5.2. *Further extension of Church's syntax*

In this subsection, we further extend Church's syntax by introducing a new binder λ, which we call *fresh binder* and a family of combinators I_k ($k \in \mathbb{N}$). We will also extend the rewrite relation $\rightarrow$. In this extended syntax and rewrite relation, we will see that we can rewrite the left tree T_8 to the right tree T_9 as shown in Fig. 6.

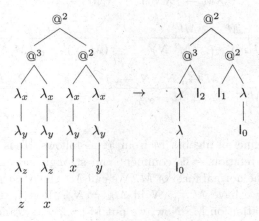

Fig. 6.   Further extended Church's syntax

Definition 5 (Datatype $\mathbb{M}$ of further extended Church's syntax).
We define the datatype $\mathbb{M}$ by the following grammar.

$$\mathbb{M} \ni M, N ::= x \mid \mathsf{I}_j \mid \lambda M \mid \lambda_x M \mid (M\ N)^i,$$

where $i, j \in \mathbb{N}$ and $x \in \mathbb{X}$.   □

We have the following inclusion relation among the datatypes Λ, $\mathbb{C}$ and $\mathbb{M}$.

$$\Lambda \subsetneq \mathbb{C} \subsetneq \mathbb{M}.$$

Next, we define the binary relation $\to$ and the binary relation $\equiv$ on $\mathbb{C}$ to those on $\mathbb{M}$ as follows.

Definition 6 (Relations $\to_\alpha$ and $\equiv$ on $\mathbb{M}$). We use the following notational convention in the following definition.

$$\lambda^0 M := M$$
$$\lambda^{i+1} M := \lambda \lambda^i M$$

$$\frac{}{\lambda_x \lambda^i \mathsf{I}_j \to_\alpha \lambda^{i+1} \mathsf{I}_j} \qquad \frac{}{\lambda_x \lambda^i x \to_\alpha \mathsf{I}_i} \qquad \frac{x \neq y}{\lambda_x \lambda^i y \to_\alpha \lambda^{i+1} y}$$

$$\frac{}{\lambda (M\ N)^i \to_\alpha (\lambda M\ \lambda N)^{i+1}} \qquad \frac{}{\lambda_x (M\ N)^i \to_\alpha (\lambda_x M\ \lambda_x N)^{i+1}}$$

$$\frac{M \to_\alpha N}{\lambda M \to_\alpha \lambda N} \qquad \frac{M \to_\alpha N}{\lambda_x M \to_\alpha \lambda_x N}$$

$$\frac{M \to_\alpha M'}{(M\ N)^i \to_\alpha (M'\ N)^i} \qquad \frac{E \to_\alpha E'}{(M\ N)^i \to_\alpha (M\ N')^i}$$

$$\frac{}{M \to_\alpha M} \qquad \frac{M \to_\alpha N \quad N \to_\alpha P}{M \to_\alpha P} \qquad \frac{M \to_\alpha P \quad N \to_\alpha P}{M \equiv_\alpha N}.$$

□

We now define an algebra M from $\mathbb{M}$ as follows. It is not difficult to show that the relation $\to$ is confluent and strongly normalizing. So, we write M_α for the normal form of M. We call M_α the α *normal form* (α-*nf*) of M. Then, we have $M \equiv_\alpha N$ iff $M_\alpha = N_\alpha$. Hence, $\equiv_\alpha$ is indeed an equivalence relation on $\mathbb{M}$. Now, we put $\mathsf{M} := \mathbb{M}/\equiv_\alpha$, and write $[M]$ for the equivalence class containing M. Then,

$$[M] = [N] \in \mathsf{M} \iff M_\alpha = N_\alpha \in \mathbb{M}.$$

In this setting, we define the following three algebraic operations on M,

$$\lambda : \mathsf{M} \to \mathsf{M},$$
$$\lambda_x : \mathsf{M} \to \mathsf{M} \qquad (x \in \mathbb{X}),$$
$$\mathsf{App}^i : \mathsf{M} \times \mathsf{M} \to \mathsf{M} \qquad (i \in \mathbb{N}),$$

by putting:

$$\lambda[M] := [\lambda M],$$
$$\lambda_x[M] := [\lambda_x M],$$
$$([M]\,[N])^i := [(M\ N)^i].$$

Then, by allowing ourselves to use the same meta-variables (M, N etc.) range over elements of M, we see that the following equations hold for elements of M.

$$\lambda_x \lambda^i \mathsf{l}_j = \lambda^{i+1} \mathsf{l}_j \tag{9}$$
$$\lambda_x \lambda^i x = \mathsf{l}_i \tag{10}$$
$$\lambda_x \lambda^i y = \lambda^{i+1} y \ \text{ if } x \neq y \tag{11}$$
$$\lambda_x (M\ N)^i = (\lambda_x M\ \lambda_x N)^{i+1} \tag{12}$$
$$\lambda (M\ N)^i = (\lambda M\ \lambda N)^{i+1}. \tag{13}$$

We note that we can characterize the structure of the algebra M syntactically as follows. We define the subset $\mathbb{M}_\alpha$ of M by the grammar:

$$\mathbb{M}_\alpha \ni M, N ::= \lambda^i x \mid \lambda^i \mathsf{l}_j \mid (M\ N)^i, \tag{14}$$

where $i, j \in \mathbb{N}$. It is easy to see that $M \in \mathbb{M}_\alpha$ iff M is an α-nf. Moreover, $\mathbb{M}_\alpha$ is closed under the operations λ and App^i ($i \in \mathbb{N}$), and we can define the operations $\lambda_x : \mathbb{M}_\alpha \to \mathbb{M}_\alpha$ ($x \in \mathbb{X}$) by putting:

$$\lambda_x \lambda^i \mathsf{l}_j := \lambda^{i+1} \mathsf{l}_j \tag{15}$$
$$\lambda_x \lambda^i x := \mathsf{l}_i \tag{16}$$
$$\lambda_x \lambda^i y := \lambda^{i+1} y \ \text{ if } x \neq y \tag{17}$$
$$\lambda_x (M\ N)^i := (\lambda_x M\ \lambda_x N)^{i+1}. \tag{18}$$

Then, $\mathbb{M}_\alpha$ becomes an algebra having the same signature as M. Here, by comparing the equations (15)–(18) with (9)–(12), we see that $\mathbb{M}_\alpha$ and M are algebraically isomorphic.[7]

---

[7]The isomorphism $\mathbb{M}_\alpha \to \mathsf{M}$ is given by $M \mapsto [M]$.

6. Datatype $\mathbb{L}$ of combinatory λ-terms

In this section, we introduce the datatype $\mathbb{L}$ of *combinatory λ-terms* and define an algebra L which will be shown to be isomorphic to the algebra M.

Definition 7 (Datatype $\mathbb{L}$ of combinatory λ-terms). We define the datatype $\mathbb{L}$ by the following grammar.
$$\mathbb{L} \ni K, L ::= x \mid \mathsf{I}_j \mid \lambda L \mid (K\ L)^i,$$
where $i, j \in \mathbb{N}$ and $x \in \mathbb{X}$. We call an element of $\mathbb{L}$ *combinatory λ-term* since it is constructed without using the variable binding operators λ_x. $\square$

Since $\mathbb{L}$ is obtained from $\mathbb{M}$ by dropping the case $\lambda_x M$, $\mathbb{L}$ can be naturally thought of as a subset of $\mathbb{M}$. So, from Definition 6 we can define relations $\to_\alpha$ and $\equiv_\alpha$ on $\mathbb{L}$ by restricting the domains of the relations to $\mathbb{L}$.

We define the subset $\mathbb{L}_\alpha$ of $\mathbb{L}$ by the grammar:
$$\mathbb{L}_\alpha \ni K, L ::= \lambda^i x \mid \lambda^i \mathsf{I}_j \mid (K\ L)^i, \tag{19}$$
where $i, j \in \mathbb{N}$. By comparing (19) with (14), we see that $\mathbb{L}_\alpha = \mathbb{M}_\alpha$. Therefore, putting $\mathsf{L} := \mathbb{L}/\equiv_\alpha$, we can make L into an algebra isomorphic to $\mathbb{L}_\alpha$ just as in Subsection 5.2. Thus, L is also isomorphic to M.

Since $\mathbb{L}$ is free from the variable binding operators λ_x, we can define the substitution operation on $\mathbb{L}$ naturally as follows.
$$[z := L]z := L,$$
$$[z := L]x := L \ \text{ if } x \neq z,$$
$$[z := L]\mathsf{I}_k := \mathsf{I}_k,$$
$$[z := L]\lambda K := \lambda[z := L]K,$$
$$[z := L](K_1\ K_2)^j := ([z := L]K_1\ [z := L]K_2)^j.$$
We see that, for fixed x and L, the substitution operation $[x := L] : \mathbb{L} \to \mathbb{L}$ is an endomorphism with respect to the constructors λ and App^j. This fact suggests that both $\mathbb{L}$ and $\mathbb{L}_\alpha$ are the data structures suitable for representing the λ-terms.

7. Conclusion

Starting from the datatype Λ introduced by Church, we first reviewed the historical process of representing Church's syntax in terms of Quine-Bourbaki notation and de Bruijn notation.

We then extended Λ to a new datatype $\mathbb{M}$ which is rich enough to contain the α normal form of each λ-term in Λ. We also identified a subdatatype $\mathbb{L}$ of $\mathbb{M}$ whose elements may be thought of as combinatory λ-terms.

Acknowledgements

We thank the Japan Society for the Promotion of Science (JSPS), Core-to-Core Program (A. Advanced Research Networks) for supporting the research.

References

[1] H. Barendregt, *The Lambda Calculus, Its Syntax and Semantics*. North-Holland (1984). revised ed.
[2] N. Bourbaki, *Éléments de Mathématique Théorie des Ensembles*. Hermann (1966). 3rd ed.
[3] H. Curry and R. Feys, *Combinatory Logic, Volume I*. North-Holland (1958).
[4] N. de Bruijn, Lambda calculus notation with nameless dummies, a tool for automatic formula manipulation, *Indagationes Mathematicae*. **34**, 381–392 (1972).
[5] J. McCarthy, Recursive Functions of Symbolic Expressions and Their Computation by Machine, Part I, *Comm. ACM*. **3** (4), 184–195 (1960).
[6] W. Quine, *Mathematical Logic*. Harvard University Press (1951). Revised ed.

Chapter 7

Provable (and Unprovable) Computability

Stanley S. Wainer

School of Mathematics
University of Leeds
Leeds, LS2 9JT, UK
s.s.wainer@leeds.ac.uk

These lecture notes form a revised and extended version of our previous [20]. The intention is to elucidate the roles played by the Fast- and Slow-growing subrecursive hierarchies in characterizing and classifying the computational content of a broad spectrum of theories lying between polynomial-time arithmetic and the considerably stronger, foundationally significant Π^1_1-CA$_0$. The principal focus lies in the notion of a *provably computable* or *provably recursive* function (Σ^0_1-definable and provably total) and on the proof-theoretic methods used to analyze and measure their complexity. The hierarchy of theories we consider is based on a "predicative" or "tiered" input/output arithmetic of elementary recursive strength, first developed by Leivant [11]. The higher levels are then generated by the addition of successively iterated inductive definitions. The ordinal analysis of these theories is where the technical proof theory enters in, and central to our approach will be the Ω-rules and collapsing methods of Buchholz [2], which demonstrate clearly the inter-relationships with subrecursive hierarchies. The final section brings out another close relationship - with combinatorial independence results, particularly Friedman's [4, 5] extended Kruskal theorem for labelled trees. For a wider survey of fast-growing functions and unprovability see Simpson [18].

Contents

1. The Computable Hierarchy

Definition 1. Let $\{e\}_s^f(x) = $ the value generated (in some nominated register) after s steps in the computation of program e on oracle f with input x. This is irrespective of whether the computation eventually terminates or not!

Definition 2. The computable, or sub-recursive jump operator is $f \mapsto f'$ where

$$f'(x) = \{x_0\}_{x_1}^f(x_2)$$

here, x_i is the i-th component coded into x under some iterated pairing function. This is just a variant of Heaton's sub-recursive jump first developed in [8].

Definition 3. The Kalmar-"elementary" functions are those definable from constants by compositions of $+$, $\dot{-}$, bounded sums $\Sigma_{i<k}$ and bounded products $\Pi_{i<k}$ (they are the same as Grzegorczyk's $\mathcal{E}^3$). The "sub-elementary" functions are those definable without bounded products (they are polynomially bounded, the same as the functions Turing machine computable in linear space, or Grzegorczyk's $\mathcal{E}^2$.)

For example: bounded search is a sub-elementary operation. The least number $m < k$ such that $g(m) = 0$ may be defined as

$$\Sigma_{i<k}(1\dot{-}\Sigma_{m\leq i}(1\dot{-}g(m))) \ .$$

Lemma 1. *For "honest" functions f, f' is elementarily inter-reducible with the iterate $n \mapsto f^n(n)$.*

Proof. Choosing e to be the program which repeatedly makes an oracle call, and letting $x = \langle e, n, n \rangle$, one sees that $f'(x) = f^n(n)$ so the iterate is clearly sub-elementarily reducible to f'. Conversely $f'(\langle e, s, x \rangle)$ may be computed elementarily as follows: run s steps of the computation, with input x, but where the oracle calls are instead supplied by a given (finite) sequence number $\sigma = \langle y_0, y_1, \ldots, y_r \rangle$. For this computation to be correct we simply need to be sure that σ is a long-enough initial segment of the function f, i.e. $y_i = f(i)$ for each $i \leq r$ and r is large enough to accommodate all oracle calls possible within s steps. The "worst" that can happen is that each step is an oracle call, so a suitable r can be taken as $f^n(n)$ where $n = \max(s, x)$, provided f is strictly increasing and positive. Then each y_i can be computed, provided we know that f itself is elementarily computable from the iterate. This is where some "honesty" is required; we say f is "honest" if it is strictly increasing and elementarily computable from any function which bounds it. □

Definition 4. For "tree ordinals" α, with specified fundamental sequences $\{\lambda_i\}$ assigned at limits λ, the Fast Growing hierarchy F_α is obtained by iterating the sub-recursive jump (analogously to the hyperarithmetic hierarchy which iterates the ordinary Turing jump).

$$F_0(n) = n + 1 \,;\ F_{\alpha+1}(n) = F_\alpha^n(n) \,;\ F_\lambda(n) = F_{\lambda_n}(n).$$

NB: (i) Each F_α will be computable, provided the tree ordinals can be coded by numerical "notations" in such a way that it is decidable whether α is zero or a successor or a limit, and in each case the predecessor $\alpha + 1 \mapsto \alpha$ or $(\lambda, i) \mapsto \lambda_i$ is computable. (ii) Unlike the hyperarithmetic hierarchy, the definition is highly sensitive to the choice of fundamental sequences. In fact by choosing the fundamental sequence appropriately, one may force even F_ω to have arbitrarily high subrecursive complexity. This motivates an emphasis on "natural" or "standard" fundamental sequences, though no convincing definition of this notion has ever emerged. Suffice it to say that some fundamental sequences are more natural than others. For example, $\{0, 1, 2, 3, \ldots\}$ is a very natural fundamental sequence to ω, whereas $g(0), g(1), g(2), g(3), \ldots\}$ is "unnatural" if g is some wildly increasing recursive function whose termination cannot be proven in your favourite theory — say Peano arithmetic.

Theorem 1. *For arithmetical theories T with "proof-theoretic ordinal" $\|T\|$, the functions provably computable in T are exactly those elementary in the F_α for $\alpha \prec \|T\|$, and hence also bounded by the F_α's. (For Peano arithmetic this goes back to Schwichtenberg and the present author independently around 1970, and then to many others since — classifying provable recursion across the broad spectrum of theories. See e.g. [15] for details and further references.)*

Definition 5. By contrast, the Slow Growing Hierarchy is given by:

$$G_0(n) = 0 \,;\ G_{\alpha+1}(n) = G_\alpha(n) + 1 \,;\ G_\lambda(n) = G_{\lambda_n}(n)$$

and this is even more sensitive to the ordinal assignments. It is a kind of rank function on tree ordinals and we will return to it later. Note however, that this definition is "pointwise-at-n" and so G is very different from F, and H below. See Weiermann [22].

1.1. *Goodstein Sequences and the Hardy Hierarchy H*

- Take any number a, and a number $n < a$ called the "base".
- For example choose $a = 16$ and $n = 2$.
- Write a in "complete base-n", thus $a = 2^{2^2}$.
- Subtract 1, so the base-2 representation is $a - 1 = 2^{2+1} + 2^2 + 2^1 + 1$.
- Increase the base by 1, to produce $a_1 = 3^{3+1} + 3^3 + 3^1 + 1 = 112$.
- Continue subtracting 1 and increasing the base, to produce the Goodstein sequence on a, n thus: $a, a_1, a_2, a_3, \ldots$.
- The Goodstein sequence on $16, 2$ begins 16, 112, 1284, 18753, 326594, ...

Theorem 2 (1. Goodstein [7], 2. Kirby & Paris [9]).
1) Every Goodstein sequence eventually terminates in 0, and this fact is formalizable in the language of Peano writhmetic PA. *2) However it is not provable in* PA. *(This was one of the earliest and simplest "arithmetical independence results".)*

Proof. (Cichon's proof [3]) Throughout any Cantor Normal Form $\alpha \prec \varepsilon_0$, replace ω by n where $n > $ all the integer coefficients in the CNF. Then we obtain a "complete base-n" representation. Subtract 1 and then put ω back for each occurrence of the base n. One thus obtains another ordinal $P_n(\alpha)$ which is *smaller* than α. E.g. With $\alpha = \omega^{\omega^\omega}$ and $n = 2$ we get $a = 2^{2^2} = 16$. Then $a - 1 = 2^{2+1} + 2^2 + 2^1 + 1$ and $P_n(\alpha) = \omega^{\omega+1} + \omega^\omega + \omega^1 + 1$.

. Repeating this process, one sees that each stage a_i of the Goodstein sequence is represented by an ordinal $\alpha_i = P_{n+i-1} \ldots P_{n+2} P_{n+1} P_n(\alpha)$ so that $\varepsilon_0 \succ \alpha \succ \alpha_1 \succ \alpha_2 \succ \ldots$. Well-foundedness then ensures that the sequence must terminate. The process is recursive, so can be formalized in the language of PA.

Now define the "Hardy hierarchy" as follows:

$$H_0(n) = n \,;\ H_{\alpha+1}(n) = H_\alpha(n+1) \,;\ H_\lambda(n) = H_{\lambda_n}(n) \,.$$

Cichon [3] noted that $H_\alpha(n) = n+$ the length of the Goodstein sequence on a, n. Therefore a proof that all G-sequences terminate says H_{ε_0} is total recursive. But, as shown below, $H_{\varepsilon_0} \simeq F_{\varepsilon_0}$ and this is not *provably* recursive in PA. Hence part 2. $\qquad\square$

The Hardy hierarchy is so-named after a paper by G.H. Hardy (1904) in which he used precisely those functions to exhibit a set of reals with cardinality $\aleph_1$. Note the contrast with the slow growing hierarchy where the successor is composed "outside" rather than "inside" (this distinction is highly significant). The Hardy functions underlie most versions of the fast growing hierarchy and seem to have a special role also in connecting to combinatorial independence results. The following relationships are easily established by appropriate transfinite inductions.

- $H_{\alpha+\beta} = H_\alpha \circ H_\beta$.
- $H_{\omega^\alpha} = F_\alpha$ since $H_{\omega^{\alpha+1}}(n) = H_{\omega^\alpha \cdot n}(n) = H_{\omega^\alpha}^n(n)$.
- Similarly if $B_\alpha = H_{2^\alpha}$ then for finite α, $B_\alpha(n) = n + 2^\alpha$ and in general,
$$B_0(n) = n + 1 ; \ B_{\alpha+1}(n) = B_\alpha(B_\alpha(n)); \ B_\lambda(n) = B_{\lambda_n}(n).$$
- Roughly, $F_\alpha \simeq B_{\omega \cdot \alpha}$ and so $\{B_\alpha : \alpha \prec \|T\|\}$ also classifies provable recursion in arithmetical theories T.

In fact the B_α version of "fast growing" seems to be particularly well-suited to the proof theory involved in such results as the above-mentioned, and this theme will be developed in the next section. However there is a simple underlying principle:

Lemma 2 (The Basic Witness-Bounding Principle). *Suppose $A(n)$ is a Σ_1^0 formula: $A(n) \equiv \exists a D(n, a)$, and suppose $A(k) \to A(n)$ is derivable by a sequence of Cuts with "height" α:*
$$\frac{\vdash^\beta A(k) \to A(m) \qquad \vdash^\beta A(m) \to A(n)}{\vdash^\alpha A(k) \to A(n)} \ (\beta \prec \alpha).$$
Then $\models \exists a \le b.D(k, a) \ \Rightarrow \models \exists a \le B_\alpha(b).D(n, a)$.

Proof. By induction on α. Since $\beta \prec \alpha$, the premises give
$$\models \exists a \le b.D(k, a) \ \Rightarrow \models \exists a \le B_\beta(b).D(m, a),$$
$$\models \exists a \le b'.D(m, a) \ \Rightarrow \models \exists a \le B_\beta(b').D(n, a).$$
Now put $b' = B_\beta(b)$ to obtain $B_\beta(b') = B_{\beta+1}(b)$. The required conclusion $\models \exists a \le b.D(k, a) \ \Rightarrow \models \exists a \le B_\alpha(b).D(n, a)$ then follows if $B_{\beta+1}(b) \le B_\alpha(b)$. $\qquad\square$

But note: this last step requires not only that $\beta + 1 \preceq \alpha$, but that $\beta + 1 \preceq_b \alpha$ where $\prec_b$ is the transitive closure of $\gamma \prec_b \gamma + 1$ and $\lambda_b \prec_b \lambda$. Thus the condition $\beta \prec \alpha$ must be strengthened to $\beta+1 \prec_b \alpha$, incorporating the initial bound b on $A(k)$. The Majorization Property

$$\beta \preceq_b \alpha \;\Rightarrow\; B_\beta(b) \leq B_\alpha(b)$$

is immediate by transfinite induction on α: If $\alpha = 0$ the result trivially holds. If α is a limit and $\beta \prec_b \alpha$ then $\beta \preceq_b \alpha_b$. Then by the induction hypothesis, $B_\beta(b) \leq B_{\alpha_b}(b) = B_\alpha(b)$ as required. If $\alpha = \gamma + 1$ and $\beta \prec_b \alpha$ then $\beta \preceq_b \gamma$ and so the induction hypothesis yields $B_\beta(b) \leq B_\gamma(b) \leq B_\gamma B_\gamma(b) = B_\alpha(b)$. The necessity of such $\prec_b$ orderings for majorization was first noted by Schmidt [14].

2. "Input-Output" Arithmetics

Definition 6. The basic I/O theory is EA(I;O) — "elementary arithmetic". It has the language of arithmetic, with quantified, "output" variables $a, b, c, \ldots$.

- In addition there are numerical constants ("inputs") $x, y, z, \ldots$.
- There are defining equations for primitive recursive functions.
- *Basic terms* are those built from the constants and variables by successive application of successor and predecessor.
- Only *basic terms* are allowed as "witnesses" in the logical rules for $\forall$ and $\exists$. E.g. $A(t) \to \exists a A(a)$ only for basic t.
- However the equality axioms give $t = a \wedge A(t) \to A(a)$, hence, writing $t{\downarrow}$ for $\exists a(t = a)$, one has $t{\downarrow} \wedge A(t) \to \exists a A(a)$ and $t{\downarrow} \wedge \forall a A(a) \to A(t)$.
- The "predicative" induction axioms, for closed basic terms t are:

$$A(0) \wedge \forall c(A(c) \to A(c+1)) \to A(t) \,.$$

Remark. The principle of numerical induction:

$$A(0) \wedge \forall_a(A(a) \to A(a+1)) \to A(n)$$

may be viewed as being "impredicative", since establishing that n has property A might entail quantification over all numbers, in particular n itself, even before that number is completely understood (see Nelson [12]). One way to unravel such impredicativities is to think of the input n as existing on a higher "tier" than the domain of output variables quantified over in A. Such tiered or predicative inductions severely restrict the computational strength of an arithmetical theory. Leivant [11] was the first to develop these ideas in a proof-theoretic context, following the Bellantoni-Cook [1] characterization of polytime, which in turn used the "normal/safe" variable separation introduced by Simmons [16].

2.1. *Working in* $\mathbf{EA(I;O)}$

Note: if t is not basic one cannot pass directly from $t = t$ to $t \downarrow$. But $a + 1$ is basic, and $t = a \rightarrow t + 1 = a + 1$ so $t \downarrow \rightarrow t + 1 \downarrow$.

Some easy examples:

- From $b + c \downarrow \rightarrow b + (c + 1) \downarrow$ one gets $b + x \downarrow$ by Σ_1-induction "up to" x. Then $\forall b(b + x \downarrow)$ and we write $I\Sigma_1(I;O) \vdash \forall b(b + x \downarrow)$.
- Then $b + x \cdot c \downarrow \rightarrow b + x \cdot (c+1) \downarrow$. Therefore, by another Σ_1-induction, $b + x \cdot x \downarrow$.
- Hence $\forall b(b + x^2 \downarrow)$, $\forall b(b + x^3 \downarrow)$ etc.
- Similarly, $I\Sigma_1(I;O) \vdash \forall b(b + p(\vec{x}) \downarrow)$ for any polynomial p. More generally, letting $IndA(a)$ stand for $A(0) \wedge \forall a(A(a) \rightarrow A(a+1))$ we have for all polynomials p, $\vdash IndA(a) \rightarrow A(p(\vec{x}))$. For, informally in $EA(I;O)$ and assuming inductively that this result already holds for polynomials p_1 and p_2 where $p = p_1 + p_2$, one notes easily that $IndA(a) \rightarrow IndA'(a)$ where $A'(a) \equiv p_1 + a \downarrow \wedge A(p_1 + a)$. Then from $IndA'(a) \rightarrow A'(p_2)$ one obtains $IndA(a) \rightarrow A(p)$. If, on the other hand, $p = p_1 \cdot p_2$ let $A''(a) \equiv p_1 \cdot a \downarrow \wedge A(p_1 \cdot a)$. Then since $A(b) \wedge IndA(a) \rightarrow IndB(a)$ where $B(a) \equiv b + a \downarrow \wedge A(b + a)$, one has $IndB(a) \rightarrow B(p_1)$ by assumption, and hence $IndA(a) \rightarrow IndA''(a)$. Therefore again by the assumption, $IndA(a) \rightarrow A''(p_2)$, and so $IndA(a) \rightarrow A(p)$ as required.
- Exponential requires a Π_2 induction on $A(a) \equiv \forall b(b + 2^a \downarrow)$, since $IndA(a)$ is easily checked using two calls on the universal quantifier. For assume $\forall b(b + 2^a \downarrow)$. Then, for arbitrary b, we have firstly: $b + 2^a \downarrow$ and then again $(b + 2^a) + 2^a \downarrow$. Therefore $\forall b(b + 2^a \downarrow) \rightarrow \forall b(b + 2^{a+1} \downarrow)$ and $\forall b(b + 2^0 \downarrow)$ because $b + 1$ is basic. Hence by a predicative Π_2 induction, $\vdash \forall b(b + 2^x \downarrow)$ and from the above, $\vdash \forall b(b + 2^{p(\vec{x})} \downarrow)$ for all polynomials p. Another quantifier jump then gives $\vdash \forall b(b + 2^{2^x} \downarrow)$ by predicative Π_3 induction, etc. These arguments are simplified numerical versions of Gentzen's proofs of transfinite induction below ε_0 in PA.

2.2. *Provably Computable Functions in* EA(I;O)

Definition 7. A provably computable/recursive function of EA(I;O) is one which is Σ_1^0-definable and provably total on inputs, i.e. $\vdash f(\vec{x}) \downarrow$.

Lemma 3. *Witnesses for Σ_1 theorems $A(n) \equiv \exists a D(n,a)$, proven in* EA(I;O) *by Σ_1-inductions up to $x := n$, are bounded by $B_{\log n}$ where $\log n = \mu m(2^m \geq n)$.*

Proof. Sketch: first, any induction up to $x := n$ can be unravelled, inside EA(I;O), to a binary tree of Cuts of height $h = \log n$. This is because, for any c, one has $\vdash A(c) \rightarrow A(c + 2^h)$ with cut-height h. The case $h = 0$ is just the given induction step, and for $h > 0$, an additional Cut gives

$$\frac{A(c) \rightarrow A(c + 2^h) \quad A(c + 2^h) \rightarrow A(c + 2^h + 2^h)}{A(c) \rightarrow A(c + 2^{h+1})}.$$

Therefore $\vdash^h A(0) \rightarrow A(n)$ with cut-height $h = \log n$. The Witness-Bounding Principle then gives $\models \exists a \leq B_h(b).D(n,a)$ where b is the witness for $A(0)$. $\qquad\square$

Theorem 3 (Leivant [11], Ostrin-W. [13]). *(i) The provably computable functions of $I\Sigma_1$(I;O) are exactly the sub-elementary functions $\mathcal{E}^2$. (ii) The provably computable functions of $I\Pi_2$(I;O) are those computable in exp-time $2^{p(n)}$. (iii) The provably computable functions of EA(I;O) are exactly the elementary functions $\mathcal{E}^3$.*

(See Leivant's Ramified Inductions [11] where such characterizations were first obtained. There, the numbers are formalised in binary notation rather than unary, as here, so that the provably computable functions of $I\Sigma_1$(I;O) are exactly the polytime functions rather than, as here, the subelementary ones, computable in polytime but with *unary* numerals. Spoors-W. [19] develops hierarchies of ramified extensions of EA(I;O) classifying primitive recursion.)

Proof. Fix $x := n$ in a given proof of $f(x) \downarrow$ say with d nested inductions. Partial cut-elimination yields a "free-cut-free" proof so, after unravelling, only cuts on the induction formulas remain. The height of the proof-tree will be (of the order of) $h = \log n \cdot d$.

(i) For $I\Sigma_1$(I;O) the Bounding Principle applies, to give polynomial-in-n complexity bounds (noting $B_\alpha(b) = b + 2^\alpha$ when α is finite):

$$B_{\log n \cdot d}(b) = b + 2^{\log n \cdot d} \leq b + (2n)^d \text{ for some given } b.$$

The function f is therefore subelementary. Conversely let $D(a, x, s)$ be a formula with only bounded quantifiers expressing that a register machine computing $f(x)$ has state a after s steps on input x. Then if f is subelementary there is a polynomial $p(x)$ such that $\exists a D(a, x, p(x))$ expresses termination $f(x) \downarrow$. But the Σ_1 formula $\exists a D(a, x, s)$ will be provably inductive on s and so by the foregoing, $\exists a D(a, x, p(x))$ hence $f(x) \downarrow$ is provable by predicative Σ_1 induction.

(ii) For Π_2 one must first reduce all cuts to Σ_1 form by Gentzen cut-reduction, which further increases the height by a base-2 exponential, so in that case the complexity bounds will be of the exponential form $2^{p(n)}$ thus:

$$B_{2^{\log n \cdot d}}(b) \leq B_{(2n)^d}(b) = b + 2^{(2n)^d}.$$

Conversely if f is computable in exp-time its termination $\exists a D(a, x, 2^{p(x)})$ will be provable by predicative Π_2 induction.

(iii) Increasing the quantifier complexity of predicative inductions corresponds (by Gentzen cut-reduction) to increasing the exponential stack-height in the complexity bounds. Thus $\vdash f(x) \downarrow$ in EA(I;O) if and only if $f(x)$ is computable in a number of steps bounded by an arbitrarily iterated base-2 exponential with $p(x)$ on the top. This is equivalent to being elementary. □

2.3. *Peano Arithmetic — by adding an Inductive Definition*

Definition 8. $ID_1(I;O)$ is obtained from EA(I;O) by adding, for each (unit-erated) inductive form $F(X, a)$ wherein the predicate variable X occurs positively, a new predicate P, and Closure and Least-Fixed-Point axioms:

$$\forall a (F(P, a) \rightarrow P(a))$$

$$\forall a (F(A, a) \rightarrow A(a)) \rightarrow \forall a (P(a) \rightarrow A(a))$$

for each formula A in the extended language.

One could add as many such inductive definitions as one wishes but we shall consider just one, that is the simple inductive form

$$F(X, a) :\equiv a = 0 \vee \exists b (X(b) \wedge a = b + 1)$$

with associated predicate N for the set of natural numbers. Thus the axioms become

$$N(0) \wedge \forall b (N(b) \rightarrow N(b + 1))$$

$$A(0) \wedge \forall b (A(b) \rightarrow A(b + 1)) \rightarrow \forall b (N(b) \rightarrow A(b)).$$

Theorem 4. *Peano Arithmetic is interpretable in* $\mathrm{ID}_1(\mathrm{I:O})$. *If* $\mathrm{PA} \vdash A(a)$ *then* $\mathrm{ID}_1(\mathrm{I;O}) \vdash N(a) \to A^N(a)$ *where* A^N *is the result of relativizing all quantifiers in* A *to* N.

Proof. The axioms of PA defining successor, addition and multiplication are included as axioms of EA(I;O), and the induction axioms occur relativized in $\mathrm{ID}_1(\mathrm{I;O})$ as above. If $\forall a A(a)$ arises from $A(b)$ by a $\forall$-rule, the premise relativizes to $N(b) \to A^N(b)$ and then the $\forall$-rule gives $\forall a(N(a) \to A^N(a))$. If $\exists a A(a)$ arises from $A(t(\vec{a}))$ by the $\exists$-rule, one needs first to show, for the term t, that $N(a_1) \wedge \ldots \wedge N(a_r) \to N(t(\vec{a}))$. But this is quite easy, by induction over the build-up of term t. Furthermore $N(t)$ now formalizes "$t =$ an input" and so, as can be checked, $N(t) \to t \downarrow$. Therefore the $\exists$-rule of EA(I;O) can be used to derive $N(a_1) \wedge \ldots \wedge N(a_r) \to \exists a(N(a) \wedge A(a))$. The other rules relativize to N straightforwardly. $\square$

Note that from the closure axiom $N(0) \wedge \forall a(N(a) \to N(a+1))$ a predicative induction immediately yields $\mathrm{ID}_1(\mathrm{I;O}) \vdash N(x)$ for any input x. Hence if f is provably recursive in PA then, by the embedding,

$$\mathrm{ID}_1(\mathrm{I;O}) \vdash \forall a(N(a) \to \exists b(N(b) \wedge f(a) = b))$$

and therefore,

$$\mathrm{ID}_1(\mathrm{I;O}) \vdash f(x) \downarrow \wedge N(f(x)).$$

2.4. *Unravelling the LFP-Axiom by Buchholz' Ω-Rule*

Still working in the I/O context, we can fix an input $\vec{x} := \vec{n}$ and unravel inductions into iterated Cuts as before. However the resulting $\mathrm{ID}_1(\mathrm{I;O})$-derivations will be further complicated by the presence of Least-Fixed-Point axioms. These must be "unravelled" as well in order to read off computational content, and Buchholz' Ω-Rule [2] does this job. (In the I/O context this was worked out in W.-Williams [21] and generalized to higher levels of finitely-iterated inductive definitions by Williams in his thesis [23].)

Definition 9. The infinitary system $\mathrm{ID}_1(\mathrm{I;O})^\infty$ has Tait-style sequents $n{:}N \vdash^\alpha \Gamma$ where n bounds the declared numerical inputs $\vec{n}$, α is a tree ordinal bounding the derivation-height, and Γ is an arbitrary set of closed formulas in the language of $\mathrm{ID}_1(\mathrm{I;O})$, written in negation normal form. $\Gamma = \{A_1, A_2, \ldots, A_r\}$ is to be interpreted disjunctively, as A_1 or A_2 or $\ldots$ or A_r. There may be other numerical parameters $\vec{m}$ as well, instantiating any output variables, and where necessary these are displayed by writing $\Gamma = \Gamma(\bar{n}, \bar{m})$. The n and m are of course numerals like $\bar{m} = SSS \ldots S0$ but the context should make this clear.

The axioms are $n{:}N \vdash^{\alpha} \Gamma$ where Γ contains either a true atom (equation or inequation between primitive recursive terms) or $N(m)$ for any $m \leq n$, or $\neg N(m), N(m)$ for any numeral m. Also included as axioms are $t = m$ where t is any "basic" term and m is its numeral value.

The rules are as follows, with ordinal heights β and α assigned to their premises and conclusions according to the rule $\beta + 1 \preceq_n \alpha$ (this combination of input with ordinal height-assignment is important; recall that $\prec_n$ is the transitive closure of $\gamma \prec_n \gamma + 1$ and, at limits, $\lambda_n \prec_n \lambda$).

$$(\exists)\ \frac{n : N \vdash^{\beta} \Gamma, A(t)}{n : N \vdash^{\alpha} \Gamma, \exists a A(a)}\ (t\ \text{``basic''}), \qquad (\forall)\ \frac{n : N \vdash^{\beta} \Gamma, A(m)\ \text{for all}\ m}{n : N \vdash^{\alpha} \Gamma, \forall a A(a)},$$

plus ($\vee$) and ($\wedge$) rules, and

$$(Cut)\ \frac{n : N \vdash^{\beta} \Gamma, C \qquad n : N \vdash^{\beta} \Gamma, \neg C}{n : N \vdash^{\alpha} \Gamma}.$$

The (N) rule is obvious:

$$(N)\ \frac{n : N \vdash^{\beta} N(m), \Gamma}{n : N \vdash^{\alpha} N(m + 1), \Gamma}$$

and the (Ω) rule embodies a cut on the N predicate, with β a limit ordinal:

$$(\Omega)\ \frac{n : N \vdash^{\beta_0} N(m), \Gamma_0 \qquad n : N \vdash^{h}_{0} N(m), \Delta\ \Rightarrow\ n : N \vdash^{\beta_h} \Gamma_1, \Delta}{n : N \vdash^{\alpha} \Gamma_0, \Gamma_1}.$$

Here $\vdash^{h}_{0}$ denotes a cut-free derivation of finite height h, and Δ is an arbitrary set of "positive-in-N" formulas.

Lemma 4 (Weakening). $n{:}N \vdash^{\alpha} \Gamma$ *and* $\alpha \prec_n \alpha'$ *and* $\Gamma \subset \Gamma' \Rightarrow n{:}N \vdash^{\alpha'} \Gamma'$. *Furthermore if* $n{:}N \vdash^{\alpha} \Gamma$ *then for any* γ, $n{:}N \vdash^{\gamma + \alpha} \Gamma$.

Lemma 5 (Embedding). *If* $\mathrm{ID}_1(\mathrm{I{:}O}) \vdash A(x; a)$ *then there is a* k *such that, for any* n *and* m, *one may derive* $n : N \vdash^{k + 2 \cdot \omega + k} A(n; m)$ *in* ID_1 $(\mathrm{I{:}O})^{\infty}$.

Proof. The primitive recursive defining equations of $\mathrm{EA}(\mathrm{I{:}O})$ easily embed into $\mathrm{ID}_1(\mathrm{I{:}O})^{\infty}$, as do the ($\vee$), ($\wedge$) and (Cut) rules.

If $\forall a A(x; a)$ is a consequence of the $\forall$-rule in $\mathrm{ID}_1(\mathrm{I{:}O})$ then the premise is $A(x; b)$ and inductively one may assume $n{:}N \vdash^{k + 2 \cdot \omega + k} A(n; m)$ in $\mathrm{ID}_1(\mathrm{I{:}O})^{\infty}$, for all n and m. Then by the ($\forall$) rule of $\mathrm{ID}_1(\mathrm{I{:}O})^{\infty}$, $n{:}N \vdash^{k + 2 \cdot \omega + (k+1)} \forall a A(n; a)$, and the first k can be increased to $k + 1$ by weakening.

If $\exists a A(x;a)$ is a consequence of the existential rule in $\mathrm{ID}_1(\mathrm{I};\mathrm{O})$ then the premise is $A(x;t)$ where the term t is "basic" (recall that this was one of the original restrictions on the I/O logic, meaning t is built up from an input or an output variable by applying only predecessors or successors). Inductively therefore, one may assume $n : N \vdash^{k+2\cdot\omega+k} A(n;t)$ in $\mathrm{ID}_1(\mathrm{I};\mathrm{O})^\infty$, for any n and m, where the numerical value of t is $n+$ a constant or $m+$ a constant. The $(\exists)$-rule of $\mathrm{ID}_1(\mathrm{I};\mathrm{O})^\infty$ then gives, after another weakening, $n : N \vdash^{(k+1)+2\cdot\omega+(k+1)} \exists a A(n;a)$ as required.

The closure axiom is derived immediately from the axioms $\vdash^0 N(0)$ and $\vdash^0 \neg N(m), N(m)$ by an application of the (N)-rule, giving $\vdash^1 \neg N(m), N(m+1)$ and then $\vdash^3 \forall b(\neg N(m) \vee N(m+1))$. Hence by $(\wedge)$ and appropriate weakening, $\vdash^{k+2\cdot\omega+k} N(0) \wedge \forall b(N(b) \to N(b+1))$.

For the least fixed point axiom apply the (Ω)-rule. As left-hand premise choose $\vdash^0 N(m), \neg N(m)$. (We can drop the $n{:}N$ declaration as it plays only a subsidiary role here.) For the right-hand premise, we need to transform a (direct) cut-free derivation $\vdash_0^h N(m), \Delta$ with finite height h, into a derivation of $\vdash^{\beta h} \neg IndA, A(m), \Delta$ where $IndA \equiv A(0) \wedge \forall a(A(a) \to A(a+1))$. This is done by induction on h. If the last rule applied is the (N) rule with premise $\vdash_0^{h-1} N(m-1), \Delta$ then inductively we may assume $\vdash^{k+\beta h-1} \neg IndA, A(m-1), \Delta$ where k is fixed such that $\vdash^k \neg A(m), A(m)$ for all m. Combining by "weakening" and a $(\wedge)$ rule, one obtains $\vdash^{k+\beta h-1+1} \neg IndA, (A(m-1) \wedge \neg A(m)), A(m), \Delta$ and then by $(\exists)$, $\vdash^{k+\beta h-1+2} \neg IndA, A(m), \Delta$. All other rules in deriving $\vdash_0^h N(m), \Delta$ (either the axiom $N(0)$ or rules "inside" Δ) are transformed straightforwardly. Therefore by choosing $\beta = 2 \cdot \omega$ we obtain by (Ω), $\vdash^{k+2\cdot\omega+1} \neg IndA, \neg N(m), A(m)$ and this holds for every m. By applying an $(\vee)$-rule followed by a $(\forall)$-rule and another $(\vee)$, one derives the least fixed point axiom $\vdash IndA \to \forall a(N(a) \to A(a))$ with ordinal bound $k + 2\cdot\omega + k$. $\qquad\square$

2.5. *Cut Elimination and Collapsing in* $\mathrm{ID}_1(\mathrm{I};\mathrm{O})^\infty$

As usual, Gentzen-style cut-reduction raises derivation-height exponentially. It cannot be done directly in PA because of the induction axioms, but now they are transformed into infinitary rules which support cut-reduction. The "cut-rank" of a derivation is the maximum "size" of cut-formulas appearing in it (assuming, of course, that there is a finite bound on the sizes of cut-formulas in a given derivation — there may not be, but the derivations we consider will all have such a bound). The size of a cut formula C is denoted $|C|$ and is always positive, so "cut-rank zero" means there are no cuts of any kind.

Lemma 6 (Cut-reduction). *(i) If $n{:}N \vdash^\gamma \Gamma, \neg C$ and $n{:}N \vdash^\alpha \Gamma, C$ both with cut-rank $\leq r \preceq_n \gamma$, where C is either existential, disjunctive, or a negated atom, and $|C| = r + 1$, then $n{:}N \vdash^{\gamma + \alpha} \Gamma$ with cut-rank $\leq r$.*
(ii) Hence if $n{:}N \vdash^\delta \Gamma$ with cut-rank $r + 1$ then $n{:}N \vdash^{2^{r+\delta}} \Gamma$ with cut-rank $\leq r$.

Proof. (i) By induction on α. We omit the $n{:}N$ declaration. If $C = \exists a A(a)$ and $\vdash^\alpha \Gamma, C$ comes from $\vdash^\beta \Gamma, \exists a A(a), A(t)$ by an $(\exists)$ rule then $\vdash^{\gamma + \beta} \Gamma, A(t)$ by the induction hypothesis. Inverting the first premise gives $\vdash^\gamma \Gamma, \neg A(m)$ where m is the numerical value of the basic term t. Since $t = m$ is an axiom, it is not difficult to check that $A(m), \neg A(t)$ is derivable with height $|A| = r \preceq_n \gamma$, so by a cut on $A(m)$ we obtain $\vdash^{\gamma + 1} \Gamma, \neg A(t)$. Now another cut, this time on $A(t)$, gives $\vdash^{\gamma + \alpha} \Gamma$, still with cut-rank r. If $\vdash^\alpha \Gamma, C$ comes about by any other rule then the induction hypothesis may be applied to the premises and the rule then re-applied to give the required conclusion. Note that if $\beta \prec_n \alpha$ then $\gamma + \beta \prec_n \gamma + \alpha$, so the assignment of ordinal bounds remains valid.

If $C = A_0 \vee A_1$ then a similar argument applies. If C is a negated atom $\neg P$ (both P and $\neg P$ are defined to have size 1, so $r = 0$) then it is easy to see, using weakening where necessary, that as one runs through the cut-free derivation of $\Gamma, \neg P$ the other given derivation of Γ, P allows one to delete $\neg P$, at each step changing the ordinal bound from β to $\gamma + \beta$. In the end one obtains a cut-free derivation $\vdash^{\gamma + \alpha} \Gamma$ as required.

(ii) By induction on δ: at a cut on C of size $r + 1$ apply the induction hypothesis to both premises. Then apply (i) with $\gamma = \alpha = 2^{r + \alpha'}$ where $\alpha' \prec_n \delta$. Then $\gamma + \alpha = 2^{r + \alpha' + 1} \preceq_n 2^{r + \delta}$ as required. $\qquad\square$

Lemma 7 (Buchholz Collapsing). *Suppose, for fixed input $x := n > 1$, we have a cut-free derivation $n{:}N \vdash_0^\alpha \Gamma$ with Γ positive in N. Then there is a derivation of finite height $n{:}N \vdash_0^h \Gamma$ where $h \leq B_\alpha(n)$. This derivation no longer contains any (Ω)-rules because each (Ω)-rule entails an infinite ordinal height.*

Proof. Again, this is done by induction on α. Checking through the rules, one sees that the only point at which the given derivation needs to take on infinite height is at the application of Ω rules. At an application of any other rule, one simply applies the induction hypothesis to the premises and then re-applies that rule with the finite height-bound increased from $h' = B_\beta(n)$ to $h = B_\alpha(n)$.

At an (Ω)-rule, assume the result holds for the premises, choosing $\Delta = \Gamma_0$:

$$n{:}N \vdash_0^{\beta_0} N(m), \Gamma_0 \quad \text{and} \quad n{:}N \vdash_0^h N(m), \Gamma_0 \Rightarrow n{:}N \vdash_0^{\beta_h} \Gamma$$

where $\beta + 1 \preceq_n \alpha$ and β is a limit. Then for the left premise, $n{:}N \vdash_0^h N(m), \Gamma_0$ where $h \leq B_{\beta_0}(n) < B_\beta(n)$. Now the right premise can be applied to this, yielding $n{:}N \vdash_0^k \Gamma$ where $k \leq B_{\beta_h}(n)$. Hence $k \leq B_\beta(\max(n, h)) \leq B_\beta B_\beta(n) = B_{\beta+1}(n) \leq B_\alpha(n)$. (That $B_{\beta_h}(n) \leq B_\beta(\max(n, h))$ is a basic property of "standard" fundamental sequences.) The (Ω) rule has thus been eliminated. $\square$

An immediate consequence is "another" proof of an Old Theorem:

Theorem 5. *Every Σ_1^0 theorem of* PA *has witnesses bounded by B_α for some $\alpha \prec \varepsilon_0$. Therefore the provably recursive functions of* PA *are those computable in B_α-bounded resource for $\alpha \prec \varepsilon_0$.*

Proof. Embed the PA proof as $\mathrm{ID}_1(\mathrm{I;O}) \vdash \exists a(N(a) \wedge A(n, a))$ with $x := n$ input. Translate this into $\mathrm{ID}_1(\mathrm{I;O})^\infty$ with proof-height $k + 2 \cdot \omega + k$ and cut-rank r. Choose $k \geq r$ and then eliminate cuts to obtain proof-height $\alpha = 2_r(k + 2 \cdot \omega + k)$ where $2_r(\delta) = \delta$ if $r = 0$ and $2^{2_{r-1}(\delta)}$ if $r > 0$. Now collapse to obtain $n{:}N \vdash_0^h \exists a(N(a) \wedge A(n, a))$ with finite height $h \leq B_\alpha(n)$. Since the derivation is cut-free, the $(\exists)$-rule may be inverted, as can the $(\wedge)$. Therefore the derivation of height h contains witnessing terms t whose values are computed from the input n by at most h applications of the N-rule, and amongst them must be a correct witness. Hence the size of the collapsed proof gives a bound $n + B_\alpha(n) \leq B_{\alpha+1}(n)$ on the existential quantifier. Note that the tree ordinal α computed here is not, strictly speaking, a sub-tree ordinal of the usual ε_0, although in set-theoretical terms it is certainly less than ε_0. However there are various ways in which it could be manipulated into an alternative form $\prec \varepsilon_0$; for example the embedding would allow $k + 2 \cdot \omega$ to be replaced by $\omega \cdot (k + 2)$. $\square$

2.6. $\mathrm{ID}_2(\mathrm{I;O})$ *and* $\mathrm{ID}_2(\mathrm{I;O})^\infty$

The next theory in the hierarchy of inductive definitions is $\mathrm{ID}_2(\mathrm{I;O})$ which axiomatizes "twice-iterated" inductive definitions. The classic example, and the one we shall concentrate on, is the set $\mathcal{O}$ of Kleene-style recursive tree ordinals, generated by the inductive operator

$$F(X, a) :\equiv a = 0 \vee \exists b(X(b) \wedge a = b \oplus 1) \vee \forall i(N(i) \to X(a_i))$$

where a_i denotes the i-th value of the (primitive) recursive function coded by a. This is a twice-iterated definition because although X only occurs positively, the N predicate may now occur freely, negatively as well as positively. The closure and least-fixed-point axioms are thus:

$$F(\mathcal{O}, a) \to \mathcal{O}(a))$$

$$\forall a(F(A, a) \to A(a)) \to \forall b(\mathcal{O}(b) \to A(b)).$$

These are now added to $\mathrm{ID}_1(\mathrm{I;O})$ to form $\mathrm{ID}_2(\mathrm{I;O})$.

Note. The classical theory ID_1 also axiomatizes $\mathcal{O}$, but regards it as an uniterated definition because Peano arithmetic is there taken for granted as a basis. This in contrast to our predicative I/O theory where N must first be introduced by its own inductive definition. Thus the classical ID_i theories occur as $\mathrm{ID}_{i+1}(\mathrm{I;O})$ in the hierarchy described here. In particular, as has already been shown, just as PA is interpretable in $\mathrm{ID}_1(\mathrm{I;O})$, so ID_1 is interpretable in $\mathrm{ID}_2(\mathrm{I;O})$.

To illustrate the computational power of $\mathrm{ID}_2(\mathrm{I;O})$, let $A(a)$ be a Π_2^0 formula expressing $\forall m(N(m) \to N(B_a(m)))$ or, as shorthand, $B_a{:}N \to N$. Then $\forall a(F(A, a) \to A(a))$ is easily proven, and so by the least-fixed-point axiom, $\forall b(\mathcal{O}(b) \to A(b))$. Therefore we can prove $B_a{:}N \to N$ whenever we can prove $\mathcal{O}(a)$. Suppose, for example, that a is defined by a recursion $a = \varphi_\eta(b)$ of the following form, where η denotes an exponential term, or Cantor normal form, with base ω_1:

$$\varphi_0(b) = b \oplus 1 \; ; \; \varphi_{\eta+1}(b) = \varphi_\eta(\varphi_\eta(b)) \; ; \; \varphi_\eta(b) = \langle \varphi_{\eta_i}(b) \rangle_{i \in N} \; ; \; \varphi_\eta(b) = \varphi_{\eta_b}(b)$$

with the last two cases depending on whether η is a limit of the "small" kind (ω-sequence) or the "large" kind ($\mathcal{O}$-sequence) (see below). Then inductively over normal forms ξ, and using the definition of φ together with the closure and least-fixed-point axioms for $\mathcal{O}$, one may prove, in $\mathrm{ID}_2(\mathrm{I;O})$,

$$\forall \delta((\varphi_\delta{:}\mathcal{O} \to \mathcal{O}) \to (\varphi_{\delta+\omega_1^\xi}{:}\mathcal{O} \to \mathcal{O})).$$

This requires Gentzen's argument [6] for lifting transfinite induction to higher levels of exponent, using N- and $\mathcal{O}$-induction on formulas of higher quantifier complexity. Since any such exponential form η may be constructed by an appropriate finite iteration of the function $(\delta; \xi) \mapsto \delta + \omega_1^\xi$, it follows that $\mathrm{ID}_2(\mathrm{I};\mathrm{O}) \vdash \mathcal{O}(\varphi_\eta(\omega))$ for every $\eta \prec \varepsilon_{\omega_1+1}$, and hence that the function B_a is provably computable in $\mathrm{ID}_2(\mathrm{I};\mathrm{O})$ for each $a \prec$ the Bachmann-Howard ordinal $\varphi_{\varepsilon_{\omega_1+1}}(\omega)$. The converse, that every provably computable function of $\mathrm{ID}_2(\mathrm{I};\mathrm{O})$ is bounded by such a B_a, is proven in what follows, by embedding, cut elimination and Buchholz collapsing in the "next" infinitary system $\mathrm{ID}_2(\mathrm{I};\mathrm{O})^\infty$.

Definition 10. The infinitary system $\mathrm{ID}_2(\mathrm{I};\mathrm{O})^\infty$ requires two kinds of declared input $n{:}N$ and $a{:}\mathcal{O}$. Sequents are now of the form $n{:}N, a{:}\mathcal{O} \vdash^\alpha \Gamma$ where Γ is a finite set of closed formulas in the language of $\mathrm{ID}_2(\mathrm{I};\mathrm{O})$, and the height-bound α is now a tree-ordinal of the *third* number class, that is either zero, or a successor, or a "small" limit $\alpha = \langle \alpha_n \rangle_{n \in N}$ or a "large" limit $\alpha = \langle \alpha_a \rangle_{a \in \mathcal{O}}$. The control over ordinal assignments β to the premises, and α to the conclusion of each rule, is exercised as follows: $\beta + 1 \preceq_{n,a} \alpha$ where n, a are the declared inputs. The ordering $\prec_{n,a}$ is the transitive closure of $\gamma \prec_{n,a} \gamma + 1$, $\lambda_n \prec_{n,a} \lambda$ if λ is a small limit, and $\lambda_a \prec_{n,a} \lambda$ if λ is a large limit.

The axioms and rules are those of $\mathrm{ID}_1(\mathrm{I};\mathrm{O})^\infty$ plus the new axioms $n{:}N, a{:}\mathcal{O} \vdash^\alpha \Gamma$ when Γ contains $\mathcal{O}(b)$ for $b \prec_n a$ or $\neg\mathcal{O}(b), \mathcal{O}(b)$ for any b. The new rules are

$$(\mathcal{O}) \quad \frac{n : N, a : \mathcal{O} \vdash^\beta F(\mathcal{O}, m), \Gamma}{n : N, a : \mathcal{O} \vdash^\alpha \mathcal{O}(m), \Gamma}$$

and, with β a large limit:

$$(\Omega_2) \frac{n : N, a : \mathcal{O} \vdash^{\beta_0} \mathcal{O}(m), \Gamma_0 \quad n : N \vdash^\xi_0 \mathcal{O}(m), \Delta \Rightarrow n : N, a : \mathcal{O} \vdash^{\beta_\xi} \Gamma_1, \Delta}{n : N, a : \mathcal{O} \vdash^\alpha \Gamma_0, \Gamma_1}$$

where ξ is a countable tree-ordinal and Δ is an arbitrary set of "positive-in-$\mathcal{O}$" formulas.

Lemma 8 (Embedding). $\mathrm{ID}_2(\mathrm{I};\mathrm{O})$ *embeds into* $\mathrm{ID}_2(\mathrm{I};\mathrm{O})^\infty$ *with ordinal height bounds* $k \cdot \omega_1 + k$ *for sufficiently large* k, *and where* $\omega_1 = \langle \xi + 1 \rangle_{\xi \in \mathcal{O}}$.

Proof. This goes essentially along the same lines as before, the main task being to embed the new closure and least-fixed-point axioms. The closure axiom is easy — from $\vdash^k \neg F(\mathcal{O}, m), F(\mathcal{O}, m)$ for a fixed k depending on F, one obtains by the $(\mathcal{O})$-rule $\vdash^{k+1} \neg F(\mathcal{O}, m), \mathcal{O}(m)$. This is for every m and so by the $(\vee)$ and $(\forall)$ rules, $\vdash^{k'} \forall m(F(\mathcal{O}, m) \to \mathcal{O}(m))$ for a sufficiently large k'. A weakening would then increase the height bound to $k' \cdot \omega_1 + k'$.

To embed the least-fixed-point axiom scheme apply the (Ω_2)-rule with $\Gamma_0 = \neg\mathcal{O}(m)$, $\Gamma_1 = \exists a(F(A, a) \wedge \neg A(a)), A(m)$ and $\beta = k \cdot \omega_1$ where k is at least the size of A and $\omega_1 = \langle \xi + 1 \rangle_{\xi \in \mathcal{O}}$. Then β has fundamental sequence $\langle k \cdot (\xi + 1) \rangle_{\xi \in \mathcal{O}}$. Choose, as left-hand premise, the axiom $\mathcal{O}(m), \neg \mathcal{O}(m)$. For the right-hand premise one needs to check by induction on ξ that, for any set Δ of positive-in-$\mathcal{O}$ formulas,

$$n : N \vdash_0^\xi \mathcal{O}(m), \Delta \;\Rightarrow\; n : N, a : \mathcal{O} \vdash^{\beta_\xi} \exists a(F(A, a) \wedge \neg A(a)), A(m), \Delta \,.$$

This goes as follows, omitting the declarations $n : N, a : \mathcal{O}$: an ($\mathcal{O}$)-rule used to derive $\mathcal{O}(m), \Delta$ (cut-free) may be inverted to yield $m = 0, \Delta$ or $\mathcal{O}(m'), \Delta$ where $m = m' \oplus 1$, or $\mathcal{O}(m_i), \neg N(i), \Delta$ for all i. Each of these has ordinal height η where $\eta + 1 \preceq \xi$ so the induction hypothesis gives, in each case, $\vdash^{\beta_\eta} \exists a(F(A, a) \wedge \neg A(a)), A(m'), \Delta$ (noting that, in the limit case $m' = m_i$, the $\mathcal{O}$-positive $\neg N(i)$ has been temporarily absorbed into Δ). Now $\beta_\eta = k \cdot (\eta + 1) \preceq k \cdot \xi$ so a weakening produces $\vdash^{k \cdot \xi} \exists a(F(A, a) \wedge \neg A(a)), A(m'), \Delta$. Then by a fixed, finite number of rules, one obtains $\vdash^{k \cdot \xi + \ell} \exists a(F(A, a) \wedge \neg A(a)), F(A, m), \Delta$. Conjunction with $\vdash^k \neg A(m), A(m)$ gives $\vdash^{k \cdot \xi + \ell + 1} \exists a(F(A, a) \wedge \neg A(a)), F(A, m) \wedge \neg A(m), A(m), \Delta$ and by an ($\exists$)-rule, $\vdash^{k \cdot \xi + \ell + 2} \exists a(F(A, a) \wedge \neg A(a)), A(m), \Delta$. Choosing $k \geq \ell + 2$ we have $\vdash^{\beta_\xi} \exists a(F(A, a) \wedge \neg A(a)), A(m), \Delta$ as required.

Now the (Ω_2)-rule can be applied:

$$n : N, a : \mathcal{O} \vdash^{\beta + 1} \exists a(F(A, a) \wedge \neg A(a)), \neg \mathcal{O}(m), A(m) \,.$$

This holds for all m and so by ($\vee$) and ($\forall$) rules one obtains

$$n : N, a : \mathcal{O} \vdash^{\beta + 4} \forall a(F(A, a) \to A(a)) \to \forall b(\mathcal{O}(b) \to A(b)) \,.$$

With $k \geq 4$, the ordinal bound weakens to $\beta + k = k \cdot \omega_1 + k$. $\qquad\square$

Cut reduction, at iterated exponential cost in derivation-height, works as well in $\mathrm{ID}_2(\mathrm{I};\mathrm{O})^\infty$ as it does in $\mathrm{ID}_1(\mathrm{I};\mathrm{O})^\infty$, though of course the ordinal bounds are now in the third (rather than second) number-class. Buchholz collapsing works too, but requires new ordinal functions to measure the cost of collapsing. These are nothing other than a "lifting" of the fast growing B-functions to the next number class.

Definition 11. Define $\varphi : \Omega_2 \times \Omega_1 \to \Omega_1$ by:

$$\varphi_\gamma(\alpha) = \begin{cases} \alpha + 1 & \text{if } \gamma = 0 \\ \varphi_{\gamma'} \circ \varphi_{\gamma'}(\alpha) & \text{if } \gamma = \gamma' + 1 \\ \varphi_{\gamma_\alpha}(\alpha) & \text{if } \gamma = \langle \gamma_\xi \rangle_{\xi \in \Omega} \\ \langle \varphi_{\gamma_i}(\alpha) \rangle_{i \in N} & \text{if } \gamma = \langle \gamma_i \rangle_{i \in N} . \end{cases}$$

Lemma 9 (Buchholz collapsing at the next level).
Suppose, for fixed inputs $n{:}N, a{:}\mathcal{O}$ with $n > 1$, we have a cut-free derivation $n{:}N, a{:}\mathcal{O} \vdash_0^\gamma \Gamma$ with Γ positive in $\mathcal{O}$. Then there is a derivation of countable ordinal height $n{:}N \vdash_0^\beta \Gamma$ where $\beta = \varphi_\gamma(a)$. This derivation no longer contains any (Ω_2)-rules because any (Ω_2)-rule entails an uncountable ordinal height.

Proof. This is done by induction on γ. Note that at an application of any rule other than (Ω_2) it is only necessary to apply the induction hypothesis to the premises and then re-apply that same rule with the height-bound increased from $\beta' = \varphi_{\gamma'}(a)$ to $\beta = \varphi_\gamma(a)$.

At an (Ω_2)-rule, assume the result holds for the premises, choosing $\Delta = \Gamma_0$:

$$n{:}N, a{:}\mathcal{O} \vdash_0^{\delta_0} \mathcal{O}(m), \Gamma_0 \quad \text{and} \quad n{:}N \vdash_0^\xi \mathcal{O}(m), \Gamma_0 \Rightarrow n{:}N, a{:}\mathcal{O} \vdash_0^{\delta_\xi} \Gamma$$

where $\delta + 1 \preceq_{n,a} \gamma$. Weaken the left premise to $n{:}N, a{:}\mathcal{O} \vdash_0^{\delta_a} \mathcal{O}(m), \Gamma_0$ and then collapse to give $n{:}N \vdash_0^\xi \mathcal{O}(m), \Gamma_0$ where $\xi = \varphi_{\delta_a}(a) = \varphi_\delta(a)$. Now the right premise can be applied to this ξ, yielding $n{:}N, a{:}\mathcal{O} \vdash_0^{\delta_\xi} \Gamma$, and the declared $a{:}\mathcal{O}$ may be increased to ξ. Collapsing this gives $n{:}N \vdash_0^{\beta'} \Gamma$ where $\beta' = \varphi_{\delta_\xi}(\xi) = \varphi_\delta(\xi) = \varphi_{\delta+1}(a)$. Since $\delta + 1 \preceq_{n,a} \gamma$ we thus have $\beta' \prec_n \beta$ where $\beta = \varphi_\gamma(a)$ and $n{:}N \vdash_0^\beta \Gamma$ as required. $\qquad\square$

Theorem 6. *Every Σ_1^0 theorem of ID_1 has witnesses bounded by B_β for some $\beta \prec \varphi_{\varepsilon_{\omega_1+1}}(0)$, the Bachmann-Howard ordinal. Therefore the provably recursive functions of ID_1 are those computable in B_β-bounded resource for $\beta \prec$ Bachmann-Howard.*

Proof. Embed the ID_1 proof as $\mathrm{ID}_2(\mathrm{I};\mathcal{O}) \vdash \exists a(N(a) \wedge A(n,a))$ with input $x := n$. Translate this into $\mathrm{ID}_2(\mathrm{I};\mathcal{O})^\infty$ with proof-height $k \cdot \omega_1 + k$ and cut-rank r. The declared $a{:}\mathcal{O}$ will be $a \preceq \omega$. Choose $k \geq r$ and then eliminate cuts to obtain uncountable proof-height $\gamma = 2_r(k{\cdot}\omega_1+k) \prec \varepsilon_{\omega_1+1}$. Now collapse once to obtain $n{:}N \vdash_0^\beta \exists a(N(a) \wedge A(n,a))$ in $\mathrm{ID}_1(\mathrm{I};\mathcal{O})^\infty$ with height $\beta = \varphi_\gamma(0) \prec \varphi_{\varepsilon_{\omega_1+1}}(\omega)$. We can then collapse again to obtain a finite-height derivation $n{:}N \vdash_0^h \exists a(N(a) \wedge A(n,a))$ where $h \leq B_\beta(n)$. As before, since the derivation is cut-free, the $(\exists)$-rule may be inverted, as can the $(\wedge)$. Therefore the derivation of height h contains witnessing terms t whose values are computed from the input n by at most h applications of the N-rule, and amongst them must be a correct witness. Hence the size of the doubly-collapsed proof gives a bound $n + B_\beta(n)$ on the existential quantifier. $\qquad\square$

2.7. Generalizing to $\text{ID}_{<\omega}$

- Williams' thesis (Leeds 2004) generalizes the foregoing to theories of finitely iterated inductive definitions $\text{ID}_i(\text{I};\text{O})$.
- Higher-level Ω-rules are then needed, and they require ordinals in successively higher number-classes $\Omega_1, \Omega_2, \ldots, \Omega_i$.
- Collapsing (and Bounding) from one level $i+1$ down to the one below is then computed in terms of higher-level extensions of the B_β hierarchy: $\varphi_\gamma^{(i)}(\beta)$ for $\gamma \in \Omega_{i+1}, \beta \in \Omega_i$. Thus in particular, $B = \varphi^{(0)}$.
- As shown, the ordinal bound of $\text{ID}_2(\text{I};\text{O})$ is then the Bachmann-Howard:

$$\tau_3 = \varphi_{\varepsilon_{\omega_1+1}}^{(1)}(\omega) = \varphi_{\varphi_{\varphi_\omega^{(3)}(\omega_2)}^{(2)}(\omega_1)}^{(1)}(\omega)$$

where, for each i, $\omega_i = \langle \xi \rangle_{\xi \in \Omega_i} \in \Omega_{i+1}$.

Definition 12. Define $\varphi^{(k)} : \Omega_{k+1} \times \Omega_k \to \Omega_k$ by:

$$\varphi_\alpha^{(k)}(\beta) = \begin{cases} \beta + 1 & \text{if } \alpha = 0 \\ \varphi_\gamma^{(k)} \circ \varphi_\gamma^{(k)}(\beta) & \text{if } \alpha = \gamma + 1 \\ \varphi_{\alpha_\beta}^{(k)}(\beta) & \text{if } \alpha = \langle \alpha_\xi \rangle_{\xi \in \Omega_k} \\ \langle \varphi_{\alpha_\xi}^{(k)}(\beta) \rangle_{\xi \in \Omega_{<k}} & \text{if } \alpha = \langle \alpha_\xi \rangle_{\xi \in \Omega_{<k}}. \end{cases}$$

Define $\tau = \langle \tau_i \rangle_{i \in \mathbb{N}}$ where $\tau_0 = \omega$ and

$$\tau_1 = \varphi_\omega^{(1)}(\omega) \,, \quad \tau_2 = \varphi_{\varphi_\omega^{(2)}(\omega_1)}^{(1)}(\omega) \,, \quad \tau_3 = \varphi_{\varphi_{\varphi_\omega^{(3)}(\omega_2)}^{(2)}(\omega_1)}^{(1)}(\omega) \,, \quad \ldots.$$

Theorem 7. *The proof-theoretic ordinal of* $\text{ID}_i \equiv \text{ID}_{i+1}(\text{I};\text{O})$ *is* τ_{i+2}. *The provably computable functions of* $\text{ID}_{<\omega} = \bigcup_i \text{ID}_i$, *and hence of* $\Pi_1^1\text{-CA}_0$, *are those computable within resource-bounds* $\{B_\beta\}_{\beta \prec \tau}$.

3. Friedman's Independence Result for $\Pi_1^1 - \text{CA}_0$: Kruskal's Theorem for Labelled Trees

Kruskal's Theorem asserts that every infinite sequence $\{T_j\}$ of (rooted) finite trees must have a pair $j_1 < j_2$ such that T_{j_1} is embeddable in T_{j_2}. Friedman [5] generalized this to labelled trees (whose nodes are assigned labels from a fixed finite set). He also derived from it (by König's lemma) the miniaturized form below, which is a Π_2^0 statement and therefore asserts the existence of a certain (extremely fast growing) recursive function K. As indicated below, this function dominates B_τ. Therefore neither the miniaturized version nor the Kruskal theorem for labelled trees can be proven in Π_1^1–CA_0, though individual instances of it are.

Theorem 8 (Friedman's Miniaturized Version). *For each constant c there is a number $K(c)$ so large that in every sequence $\{T_j\}_{j < K(c)}$ of finite trees with labels from a given finite set $\{0, 1, 2, \ldots, n\}$, and satisfying the size-bound $|T_j| \leq c \cdot 2^j$, there are $j_1 < j_2$ where $T_{j_1} \hookrightarrow T_{j_2}$. This embedding $f : T_{j_1} \hookrightarrow T_{j_2}$ must preserve infs, labels, and satisfy a certain "gap condition": if node x comes immediately below node y in T_{j_1} and if node z lies between their images $f(x)$ and $f(y)$ in T_{j_2}, then the label of z must be $\geq$ the label of $f(y)$.*

Lemma 10. *The (natural) computation sequence for $B_{\tau_i}(n)$ satisfies the size-bound above, and is a "bad" sequence, i.e. it has no gap-embeddings.*

For a proof see Chapter 5 of [15], and for an excellent overview of this topic see Simpson [17].

Corollary 1. *For a simple c_n we have $B_\tau(n) = B_{\tau_n}(n) < K(c_n)$ for all n. Therefore K is not provably recursive in $ID_{<\omega}$, nor in Π_1^1-CA_0. Therefore Kruskal's theorem is not provable in Π_1^1-CA_0.*

3.1. *The Computation Sequence for τ_n*

By reducing/rewriting τ_n according to the defining equations of the φ-functions, one passes through all the tree-ordinals $\prec_n \tau$. Each term $\varphi_\alpha^{(i)}(\beta)$ is a binary tree with root-label i, and with immediate sub-trees α and β. Each one-step reduction at most doubles the size of the tree. For example with $n = 2$, and with some steps omitted, the sequence begins:

$$\tau_2 \to \varphi_{\varphi_2^{(2)}(\omega_1)}^{(1)}(\omega) \to \varphi_{\varphi_1^{(2)}\varphi_1^{(2)}(\omega_1)}^{(1)}(\omega) \to \varphi_{\varphi_0^{(2)}\varphi_0^{(2)}\varphi_1^{(2)}(\omega_1)}^{(1)}(\omega) \to$$

$$\varphi_{\varphi_0^{(2)}\varphi_1^{(2)}(\omega_1)}^{(1)}(\varphi_{\varphi_0^{(2)}\varphi_1^{(2)}(\omega_1)}^{(1)}(\omega)) \to \varphi_{\varphi_1^{(2)}(\omega_1)}^{(1)}(\varphi_{\varphi_1^{(2)}(\omega_1)}^{(1)}(\varphi_{\varphi_0^{(2)}\varphi_1^{(2)}(\omega_1)}^{(1)}(\omega))) \to$$

$$\varphi_{\omega_1}^{(1)}\varphi_{\omega_1}^{(1)}\varphi_{\varphi_0^{(2)}(\omega_1)}^{(1)}\varphi_{\varphi_1^{(2)}(\omega_1)}^{(1)}\varphi_{\varphi_0^{(2)}\varphi_1^{(2)}(\omega_1)}^{(1)}(\omega) \to \varphi_{\varphi_{\omega_1}^{(1)}(-)}^{(1)}(\varphi_{\omega_1}^{(1)}(-)) \to \cdots.$$

Since the slow growing $G_n(\tau)$ counts up the number of successors in the descending sequence $\prec_n \tau$, and since G_n collapses $\varphi^{(1)}$ onto the function B (see the final subsection below), it follows that the length of the entire sequence (down to zero) is therefore $\geq G_n(\tau) = G_n(\tau_n) = B_{\tau_{n-1}}(n)$.

Furthermore, the sequence is bad — no term is gap-embeddable in any follower (see [15] for details). Hence the length of the computation sequence is $< K(c_n)$ where c_n is a simple elementary function measuring the size of the initial term τ_n. Thus $B_\tau(n-1) < B_{\tau_{n-1}}(n) < K(c_n)$ and K cannot therefore be provably computable in $\mathrm{ID}_{<\omega}$.

3.2. *Recall the Slow Growing Hierarchy* $G_\alpha(n)$

Definition 13. The "slow growing hierarchy" $G_\alpha(n)$ is defined "pointwise" for each number n. With n fixed it is often written $G_n(\alpha)$ instead of $G_\alpha(n)$, thus

$$G_n(0) = 0\,;\; G_n(\alpha + 1) = G_n(\alpha) + 1\,;\; G_n(\lambda) = G_n(\lambda_n).$$

For each n, G_n homomorphically collapses the arithmetic of countable tree ordinals onto ordinary arithmetic. In particular it collapses $\varphi^{(1)}$ onto B.

Theorem 9. *Let* $\varphi = \varphi^{(1)}$. *Then for "well behaved"* $\alpha \in \Omega_2, \beta \in \Omega_1$,

$$G_n(\varphi_\alpha(\beta)) = B_{G_n(\alpha)}(G_n(\beta))\,.$$

Proof. By induction on α:

- If $\alpha = 0$,
  $G_n(\varphi_0(\beta)) = G_n(\beta + 1) = G_n(\beta) + 1 = B_0(G_n(\beta))$.
- For $\alpha \mapsto \alpha + 1$,
  $G_n(\varphi_{\alpha+1}(\beta)) = G_n(\varphi_\alpha \varphi_\alpha(\beta)) = B_{G_n(\alpha)} B_{G_n(\alpha)}(G_n(\beta))$
  $= B_{G_n(\alpha)+1}(G_n(\beta)) = B_{G_n(\alpha+1)}(G_n(\beta))$.
- If $\alpha = \langle \alpha_i \rangle_{i \in N}$,
  $G_n(\varphi_\alpha(\beta)) = G_n(\langle \varphi_{\alpha_i}(\beta) \rangle_{i \in N}) = G_n(\varphi_{\alpha_n}(\beta))$
  $= B_{G_n(\alpha_n)}(G_n(\beta)) = B_{G_n(\alpha)}(G_n(\beta))$.
- If $\alpha = \langle \alpha_\xi \rangle_{\xi \in \Omega_1}$,
  $G_n(\varphi_\alpha(\beta)) = G_n(\varphi_{\alpha_\beta}(\beta)) = B_{G_n(\alpha_\beta)}(G_n(\beta))$
  $=^* B_{G_n(\alpha)_{G_n(\beta)}}(G_n(\beta)) = B_{G_n(\alpha)}(G_n(\beta))$.

The $=^*$ indicates that α is required here to be "well-behaved", meaning $G_n(\alpha_\beta) = G_n(\alpha)_{G_n(\beta)}$. For further analysis of this see [15]. $\square$

Examples: (i) For $\tau_1 = \varphi_{\omega_1}^{(1)}(\omega)$ we have $G_n(\tau_1) = B_\omega(n) = B_n(n) = n + 2^n$. (ii) $\tau_2 = \varphi_{\varphi_\omega^{(2)}(\omega_1)}^{(1)}(\omega)$ is a version of ε_0. One has $G_n(\tau_2) = B_{\varphi^{(1)}(\omega)}(n) = B_{\tau_1}(n)$ where $\tau_1 = \omega + 2^\omega$. (iii) Similarly τ_3 is a version of the Bachmann-Howard ordinal and $G_n(\tau_3) = B_{\tau_2}$. In general, $G_n(\tau_{i+1}) = B_{\tau_i}(n)$.

(iv) Another example is $\varphi_{\omega_1^2}^{(1)}(\omega)$ which is a tree ordinal version of Γ_0. Its collapse under G_n is $B_{\omega^2}(n)$, which is a version of the Ackermann function.

In each of these cases, the inverse of the collapse is a direct limit. Thus B_{τ_i} is a functor on the category N and its direct limit is the well-ordering given by τ_{i+1}. The Ackermann $B_{\omega^2}(n)$ is a functor with direct limit Γ_0. (See Chapter 5 of [15] for much further detail.)

References

[1] S. Bellantoni, S. Cook, *A new recursion theoretic characterization of the polytime functions*, Computational Complexity 2 (1992), 97–110.

[2] W. Buchholz, *An independence result for Π_1^1-$CA+BI$*, Annals of Pure and Applied Logic 33 (1987), 131–155.

[3] E.A. Cichon, *A short proof of two recently discovered independence results using recursion theoretic methods*, Proc. American Math. Society 87 (1983) 704–706.

[4] H. Friedman, *Independence results in finite graph theory I-VII*, unpublished manuscripts, Ohio State University (1981) 76 pages.

[5] H. Friedman, *Beyond Kruskal's theorem I-III*, unpublished manuscripts, Ohio State University (1982) 48 pages.

[6] G. Gentzen, *Beweisbarkeit und Unbeweisbarkeit von Anfangsfällen der transfiniten Induktion in der reinen Zahlentheorie*, Math. Annalen 119 (1943) 287–311.

[7] R.L. Goodstein, *On the restricted ordinal theorem*, Journal of Symbolic Logic 9 (1944) 33–41.

[8] A.J. Heaton, *A jump operator for subrecursion theories*, Journal of Symbolic Logic 64 (1999), 460–468.

[9] L.A. Kirby, J.B. Paris, *Accessible indepencence results for Peano arithmetic*, Bulletin London Math. Society 14 (1982) 285–293.

[10] J. Kruskal, *Well-quasi-ordering, the tree theorem and Varsonyi's conjecture*, Trans. American Math. Society 95 (1960) 210–225.

[11] D. Leivant, *Intrinsic theories and computational complexity*, in D. Leivant (Ed.) Logic and Computational Complexity, Springer Lecture Notes in Computer Science 960, Springer-Verlag (1995), 177–194.

[12] E. Nelson, *Predicative Arithmetic*, Princeton University Press (1986).

[13] G.E. Ostrin, S.S. Wainer, *Elementary arithmetic*, Annals of Pure and Applied Logic 133 (2005), 275–292.

[14] D. Schmidt, *Built-up systems of fundamental sequences and hierarchies of number-theoretic functions*, Archiv für Mathematische Logik und Grundlagenforschung 18 (1976), 47–53.

[15] H. Schwichtenberg, S.S. Wainer, *Proofs and Computations*, ASL Perspectives in Logic, Cambridge University Press (2012).

[16] H. Simmons, *The realm of primitive recursion*, Archive for Mathematical Logic 27 (1988), 177–188.

[17] S.G. Simpson, *Non-provability of certain combinatorial properties of finite trees*, in L. Harrington, M. Morley, A. Scedrov, S. Simpson (Eds.) "Harvey Friedman's Research on the Foundations of Mathematics", North-Holland (1985) 87–117.

[18] S.G. Simpson, *Unprovable theorems and fast-growing functions*, in S.G. Simpson (Ed.) "Logic and Combinatorics", Contemporary Math. 65, AMS (1987) 359–394.

[19] E.J. Spoors, S.S. Wainer, *A hierarchy of ramified theories below PRA*, in U. Berger, H. Diener, P. Schuster, M. Seisenberger (Eds.) Logic, Construction, Computation, Ontos Mathemaical Logic Vol. 3, Ontos Verlag (2012), 475–499.

[20] S.S. Wainer, *Computing bounds from arithmetical proofs*, in R. Schindler (Ed.) Ways of Proof Theory, Ontos Verlag (2010), 469–486.

[21] S.S. Wainer, R.S. Williams, *Inductive definitions over a predicative arithmetic*, Annals of Pure and Applied Logic 136 (2005) 175–188.

[22] A. Weiermann, *What makes a (pointwise) subrecursive hierarchy slow growing?*, in S.B. Cooper, J.K. Truss (Eds.) Sets and Proofs, Logic Colloquium '97, LMS Lecture Notes Vol. 258, Cambridge University Press (1999), 403–423.

[23] R.S. Williams, *Finitely iterated inductive definitions over a predicative arithmetic*, Ph.D. dissertation, University of Leeds (2004).

Chapter 8

Introduction to Minlog

Franziskus Wiesnet*

Department of Mathematics
University of Trento
Via Sommarive, 14, 38123 Trento, Italy
franziskus.wiesnet@unitn.it

In this chapter we would like to get the handling of the proof assistent Minlog across to the reader. Minlog is based on the Theory of Computable Functionals, which is introduced in Chapter 5 by Kenji Miyamoto and it is based on the second chapter of my master's thesis [4]. On the website of the Minlog system [2] there are instructions for the download and installation of Minlog. We will use the dev branch. How Minlog can be changed to the dev branch is also described on this page. In the Minlog file there is the folder `doc`, which contains the tutorial `tutor.pdf` and the reference manual `ref.pdf`. Some examples in this chapter are taken from the tutorial. In the reference manual many more commands for Minlog can be found.

Contents

*Marie Skłodowska-Curie fellow of the Istituto Nazionale di Alta Matematica

1. Fundamental Commands in Minlog

1.1. *Declaration of Predicate Variables*

To deal with variables of any kind it is useful to declare them first. In the next example we need propositional variables, which are nullary predicate variables, to prove a theorem. A predicate variable can be declared with the command `add-pvar-name`. This command has the form

```
(add-pvar-name NAME (make-arity TYPES)).
```

Instead of `NAME` a list of strings is expected. For `TYPES` a list of types of the arguments should be inserted. In our case we would like to declare nullary predicate variables, which we denote as `A`, `B`, and `C`. The list of types are therefore empty and we enter

```
(add-pvar-name "A" "B" "C" (make-arity)).
```

As the output of Minlog we get

```
ok, predicate variable A: (arity) added
ok, predicate variable B: (arity) added
ok, predicate variable C: (arity) added
```

For demonstrations we declare a predicate variable P with one argument of type N and one argument of type $N \to N$:

```
(add-pvar-name "P" (make-arity (py "nat") (py "nat=>nat")))
```

As one might expect, **nat** is the name of the algebras of natural numbers. This algebra is considered more deeply in later sections. The arrow between types is denoted by **=>** and the command **(py STRING)** indicates that **STRING** should be interpreted as a type.

We also use often the commands **(pt STRING)**, **(pv STRING)**, and **(pf STRING)**, which analogously indicates that **STRING** shoudl be interpreted as term, variable, or formula.

1.2. *First Proof*

After the implementaion of the propositional variables **A**, **B**, and **C** we use them to prove the theorem $(A \to B \to C) \to ((C \to A) \to B) \to A \to C$. A formal derivation of this short formula can be done easily. Writing a proof in Minlog does not mean giving a derivation tree of the formula. This would be too complex. Minlog processes the entered commands internally to a derivation tree, which can be displayed, as we will see.

At the beginning of each proof we enter the formula which we want to prove. We do this with the command

```
(set-goal FORMULA).
```

Instead of **FORMULA** we insert the formula in quotes. Therefore we write

```
(set-goal "(A -> B -> C) -> ((C-> A) -> B) -> A -> C").
```

As one sees, **->** stands for the implication arrow $\to$. As output we get

```
-----------------------------------------------------------------
?_1:(A -> B -> C) -> ((C -> A) -> B) -> A -> C
```

Below the dashed line there is the current goal formula. The assumptions which are given for the proof are located above the dashed line. We call this domain context. In this state of the proof there are no assumptions in the context but, since the goal formula is an implication with three premises, we move these premises as assumptions into the context and we prove the conclusion from these assumptions. The command to do this is

`(assume NAMES).`

Instead of `NAMES` we insert a list of names for the assumptions. Our list consists of three names. For example we write

`(assume "assumption1" "assumption2" "assumption3")`

and the output of the program is

```
ok, we now have the new goal

  assumption1:A -> B -> C
  assumption2:(C -> A) -> B
  assumption3:A
-----------------------------------------------------------------
?_2:C
```

The three assumptions are in the context. If we had written two names, we would only have the first two premises as assumptions in the context, and if we had written more then three names, we would get an error message. We now have to prove `C` by using the assumptions above the line. Since `assumption1` has `C` as conclusion, we ought to use this assumption. The general command for this is

`(use ASSUMPTION),`

whereby for `ASSUMPTION` we put in the name of the assumption which we would like to use. This assumption must have the goal formula as conclusion, otherwise we would get an error message. Therefore we enter

`(use "assumption1").`

The assumption with name `assumption1` has two premises. So Minlog requires a proof for each of these premises and we get two new goal formulas:

```
ok, ?_2 can be obtained from

  assumption1:A -> B -> C
  assumption2:(C -> A) -> B
  assumption3:A
-----------------------------------------------------------------
?_4:B
```

```
assumption1:A -> B -> C
assumption2:(C -> A) -> B
assumption3:A
```

```
?_3:A
```

First we have to prove the lower formula, which is exactly `assumption3`. Therefore we write

`(use "assumption3")`.

The computer informs us that the lower formal is proven and we now have to give a prove of the upper formula:

`ok, ?_3 is proved.  The active goal now is`

```
assumption1:A -> B -> C
assumption2:(C -> A) -> B
assumption3:A
```

```
?_4:B
```

To prove this goal we use `assumption2` by writing (use "assumption2"). Then we have to prove $A \rightarrow B$, which we do by moving A into the context with (assume "assumption4") and using `assumption3` with (use "assumption3"). This finishes the proof and the output is

`ok, ?_6 is proved.  Proof finished.`

1.3. *Representation of Proofs*

After proving a formula Minlog is able to display this proof in different ways. With the command

`(display-proof)`

we get a representation which is similar to the derivation tree. An expansion of this command is

`(cdp)`,

which is an abbreviation for (check-and-display-proof). In addition to display-proof the proof is checked for correctness. If the proof is correct, the output is the same. In our case we get

```
.....A -> B -> C by assumption assumption1295
.....A by assumption assumption3297
....B -> C by imp elim
.....(C -> A) -> B by assumption assumption2296
......A by assumption assumption3297
.....C -> A by imp intro assumption4301
....B by imp elim
...C by imp elim
..A -> C by imp intro assumption3297
.((C -> A) -> B) -> A -> C by imp intro assumption2296
(A -> B -> C) -> ((C -> A) -> B) -> A -> C
 by imp intro assumption1295
```

It ought to be considered as a derivation tree. The number of dots at the
beginning of each line indicates in which level of the derivation tree the
subsequent formula is. After the formula there is the name of the applied
derivation rule. A representation as derivation term is also possible with
the commands

```
(proof-to-expr)
```

and

```
(proof-to-expr-with-formulas).
```

Both commands provide a derivation term of the proof. With the last
command also the type of each variable is displayed. In our case the output
of (proof-to-expr-with-formulas) is

```
assumption1: A -> B -> C
assumption2: (C -> A) -> B
assumption3: A
assumption4: C

(lambda (assumption1)
  (lambda (assumption2)
    (lambda (assumption3)
      ((assumption1 assumption3)
        (assumption2 (lambda (assumption4) assumption3))))))
```

which can easily be understood as derivation term.

1.4. *Saving Theorems*

Usually when a theorem is proven, one would like to use this theorem later again, maybe one reduces it on a special case or uses it as a lemma to prove a larger theorem. With the command

```
(save NAME)
```

the currently proven statement will be saved with the name `NAME`. In the previous section we proved the formula

```
(A -> B -> C) -> ((C -> A) -> B) -> A -> C
```

and now we can save it with

```
(save "Theorem1")
```

in Minlog. The output is

```
ok, Theorem1 has been added as a new theorem.
ok, program constant cTheoremOne: (alpha148=>alpha149=>alpha147)=>
((alpha147=>alpha148)=>alpha149)=>alpha148=>alpha147
of t-degree 1 and arity 0 added
```

When a theorem is saved we can use it in subsequent proofs, for example with `(use "Theorem1")`.
With the command

```
(display-theorems NAME)
```

Minlog shows us the theorem with name `NAME`. One can also insert a list of names instead of `NAME`, then all theorems with these names are shown. If we enter for example `(display-theorems "Theorem1")`, Minlog returns

```
Theorem1 (A -> B -> C) -> ((C -> A) -> B) -> A -> C
```

The `pritty-print` command, which has the form

```
(pp NAME),
```

displays also the formula with name `NAME`. But here the output is without the name

```
(A -> B -> C) -> ((C -> A) -> B) -> A -> C
```

and it takes only one argument. In contrast to `display-theorems` with `pp` it is possible to display any saved formula in the system and also terms can be shown with it. Therefore we often use `pp` later on.

1.5. *Display Settings*

Before proving new theorems in Minlog we would like to mention some useful commands, which help to get a clear representation of the output. The first statement, which we announce here, is

```
(set! COMMENT-FLAG BOOL).
```

If one inserts #f instead of BOOL, Minlog does not show comments anymore. Only irregularities are shown such as error messages and some warnings. Eliminating the comments is useful to enter many commands on a row, because the computer does not have to give an output and is therefore faster. If we place #t instead of BOOL, the comments are shown again.
To remove some assumptions from the context one can use the command

```
(drop STRINGS).
```

With this command Minlog does not display anymore the assumptions whose names are written instead of STRINGS. As an example in the last proof we had the output

```
ok, we now have the new goal

  assumption1:A -> B -> C
  assumption2:(C -> A) -> B
  assumption3:A
-------------------------------------------------------------
?_2:C
```

after promoting the premises in the context. If we enter

```
(drop "assumption1" "assumption2" "assumption3")
```

we get back

```
ok, we now have the new goal

-------------------------------------------------------------
?_3:C
```

but we can continue the proof analogously as above. Dropping assumptions can be helpful to make goals more readable, by removing useless assumptions. However, these are still present, they are just hidden.
With the instruction

```
(display STRING)
```

the text STRING is shown, even if (set! COMMENT-FLAG #f) was entered. In this way one can point out important declarations, definitions or theorems. In the example above it may be useful to enter

```
(display "A, B and C are now propositional variable")
```

to point out that A, B and C are now taken. With

```
(newline)
```

one can make a line break.
As a last resort the command

```
(undo)
```

is to mention. With it one can revert the last step in a proof. Then Minlog shows the state before the last command.

1.6. *Loading External Files*

The folder lib of the Minlog file contains very helpful data files. One of them is nat.scm. There the natural numbers are introduced, some functions on the natural numbers like $+$, $*$ and the Boolean ones $\leq$ and $<$ are defined and many simple properties of these are proven. For ones own proofs one would like to use this preparatory work. With entering

```
(libload NAME)
```

the file NAME from folder lib is loaded into the system. It is recommended to use (set! COMMENT-FLAG #f) before loading some files. For loading nat.scm we write

```
(set! COMMENT-FLAG #f)
(libload "nat.scm")
(set! COMMENT-FLAG #t)
```

A general version of libload is the command load. For example

```
(load "lib/nat.scm")
```

is equivalent to (libload "nat.scm").

1.7. *Proofs in Predicate Logic*

Now we would like to prove formulas with universal quantifiers. As first example we take the formula $\forall_n(Pn \to Qn) \to \forall_n Pn \to \forall_n Qn$ where n has type $\mathbb{N}$ and P, Q are unary predicate variables. First we load `nat.scm` as in the section above. In this file `nat` is set as the name of the algebra of natural numbers and `n`, `m` are already declared as variables of type `nat`. We define P and Q via the command

```
(add-pvar-name "P" "Q" (make-arity (py "nat")).
```

and set goal formula with

```
(set-goal "all n(P n -> Q n) -> all n P n -> all n Q n").
```

As one can see, the universal quantifier is denoted by `all` and requests two arguments. The first argument is the variable to be quantified in the formula given by the second argument. Minlog returns

```
--------------------------------------------------------------
?_1:all n(P n -> Q n) -> all n P n -> all n Q n
```

With

```
(assume "assumption1" "assumption2")
```

we move the premises into the context. Then the goal formula is `all n Q n`. To prove it we put a new variable into the context and prove the formula specialized to this variable. Therefore we enter

```
(assume "m")
```

and the output is

```
 ok, we now have the new goal

  assumption1:all n(P n -> Q n)
  assumption2:all n P n
  m
--------------------------------------------------------------
?_3:Q m
```

Of course, we could also write (`assume "n"`) to get the goal formula `Q n`. But it has always to be a variable of type `nat`. Therefore (`assume "a"`) would lead to an error message.

To prove `Q m` we use `assumption1` via (use `"assumption1"`), since it is an abstraction of `Q m`. As output we get

```
ok, ?_3 can be obtained from

  assumption1:all n(P n -> Q n)
  assumption2:all n P n
  m
-------------------------------------------------------------
?_4:P m
```

and with (use `"assumption2"`) the proof is already finished.

1.8. *use-with*

Sometimes Minlog does not recognize how a formula in the `use`-command has to be specialized. In such cases one has to tell directly how to use a formula. To do this there is the instruction

(use-with NAME LIST).

For `NAME` one inserts the name of the formula which should be specialised. Instead of `LIST` we need a list of formulas and terms. For each universal quantifier the corresponding term `TERM` have to be stated with (pt TERM). For each premise one has to provide the name of an assumption of this formula. Optionally one can write `"?"` for a premise. Then Minlog requires a proof of this premise afterwards.

As an example we consider the proof of the last section. After the instructions

```
(set-goal "all n(P n -> Q n) -> all n P n -> all n Q n")
(assume "assumption1" "assumption2")
(assume "m")
```

we have the output

```
ok, we now have the new goal

  assumption1:all n(P n -> Q n)
  assumption2:all n P n
  m
-------------------------------------------------------------
?_3:Q m
```

Here the goal formula is a specialization of `assumption1`. So instead of
(use "assumption1") we can also write

```
(use-with "assumption1" (pt "m") "?")
```

and the output again is

```
 ok, ?_3 can be obtained from

  assumption1:all n(P n -> Q n)
  assumption2:all n P n
  m
-----------------------------------------------------------------
?_4:P m
```

Note that for each universal quantification and each implication in front
of the goal formula there has to be exactly one argument in the `use-with`
command. The goal formula `P m` can be proven with (use "assumption2")
or with (use-with "assumption2" (pt "m")). Here one sees that the
`use` instruction is a short form of `use-with` and actually one only needs
`use-with`, if Minlog does not recognize how to use `use`.

1.9. *inst-with*

The instruction `inst-with` is similar to the command `use-with`. One also
specializes a formula. But the specialized formula does not have to be the
goal formula but it is added to the context. The command has the form

```
(inst-with NAME LIST).
```

Analogously to `use-with`, `NAME` has to be the name of the formula which
will be specialized and `LIST` is a list of formula names and terms. As
example we take the initial situation of the last section:

```
 ok, we now have the new goal

  assumption1:all n(P n -> Q n)
  assumption2:all n P n
  m
-----------------------------------------------------------------
?_3:Q m
```

Now we specialize `assumption2` to m with the command

```
(inst-with "assumption2" (pt "m"))
```

and get

```
ok, ?_3 can be obtained from

  assumption1:all n(P n -> Q n)
  assumption2:all n P n
  m  3:P m
-----------------------------------------------------------------
?_4:Q m
```

As we see, the specialized formula appears in the context and has the name 3. If we use this formula later on, we must not put 3 in quotes.
If one would like to give a name to the new formula one can use the command

```
(inst-with-to NAME LIST NAME1).
```

The arguments `NAME` and `LIST` are analog to the corresponding ones in `inst-with` but `NAME1` is the name of the new formula in the context. In our example we can write

```
(inst-with-to "assumption1" (pt "m") 3 "goal")
```

to get the output

```
ok, ?_4 can be obtained from

  assumption1:all n(P n -> Q n)
  assumption2:all n P n
  m  3:P m
  goal:Q m
-----------------------------------------------------------------
?_6:Q m
```

and by entering `(use "goal")` the proof is finished.

1.10. *assert and cut*

Many proofs, especially long proofs, one would like to split up into different parts such that in each part another formula is proven. In informal proofs

this is done by expressions like "Claim:..." or "It is sufficient to prove ...".
To do this in Minlog there are the two commands

```
(assert FORMULA)
```

and

```
(cut FORMULA).
```

The proof of the goal formula `GOAL` is divided by both commands into
a proof of `FORMULA` and a proof of `FORMULA->GOAL`. The difference between these two commands are the order of the new parts. By using
`assert` Minlog firstly requires a proof of `FORMULA` and afterwards a proof
of `FORMULA->GOAL`. By using `cut` it is reversed.
In the example of the previous sections after the output

```
ok, we now have the new goal

  assumption1:all n(P n -> Q n)
  assumption2:all n P n
  m
-----------------------------------------------------------------
?_3:Q m
```

we can enter (`assert "P m"`) and get

```
ok, ?_3 can be obtained from

  assumption1:all n(P n -> Q n)
  assumption2:all n P n
  m
-----------------------------------------------------------------
?_4:P m -> Q m


  assumption1:all n(P n -> Q n)
  assumption2:all n P n
  m
-----------------------------------------------------------------
?_5:P m
```

back. The new goal can be proven with (`use "assumption2"`) followed by
(`use "assumption1"`). The output after using (`cut "P m"`) is

```
ok, ?_3 can be obtained from

  assumption1:all n(P n -> Q n)
  assumption2:all n P n
  m
---------------------------------------------------------------
?_5:P m
```

```
  assumption1:all n(P n -> Q n)
  assumption2:all n P n
  m
---------------------------------------------------------------
?_4:P m -> Q m
```

These are the same goals in reversed order.

1.11. *Proof Search*

Minlog is able to search for proofs of formulas by itself. Of course, this is only successful for short proofs. The command to search for the current goal is

```
(search).
```

In the last sections we have seen many unnecessarily long proofs of `all n(P n -> Q n) -> all n P n -> all n Q n`. Now we see the shortest proof: after entering

```
(set-goal "all n(P n -> Q n) -> all n P n -> all n Q n")
(search)
```

we get the output

```
ok, ?_1 is proved by minimal quantifier logic. Proof finished.
```

and the proof is done. The probability that a proof is found by `search` is indirect proportional to the length of the proof. Especially the probability is very low if Minlog has to find a witness for an existential quantifier or if it has to use not only the context formulas but some other theorems.

To use the `search` command several times in a row one can use the instruction

```
(auto).
```

This command enters the command (`search`) until the proof is done or (`search`) does not find a proof anymore.

1.12. *Cheating in Minlog*

It often occurs that it takes very long to prove formulas by Minlog although they are obviously true for humans. In such cases it could be reasonable to declare these formulas as globals assumptions first and prove them later. By the instruction

```
(add-global-assumption NAME FORMULA)
```

the formula `FORMULA` is saved under the name `NAME`. After adding this formula as a global assumption it can be used in each proof, for example by the `use` command. The instruction

```
(display-global-assumption NAME)
```

shows the global assumption with name `NAME`. Instead of `NAME` one can also put in a list of names. Then all global assumptions in this list are shown. If this list is empty, all global assumptions are shown. If we enter (`display-global-assumption`) directly after starting Minlog, we get the output

```
Stab ((Pvar -> F) -> F) -> Pvar
Efq F -> Pvar
StabLog ((Pvar -> bot) -> bot) -> Pvar
EfqLog bot -> Pvar
```

We see, that the ex-falso-quodlibet and the stability are global assumptions by default.
The command

```
(remove-global-assumption LIST)
```

removes the global assumptions in list `LIST`. Therefore, if we want to use Minlog without any global assumptions, we accomplish this with the command

```
(remove-global-assumption "Stab" "Efq" "StabLog" "EfqLog").
```

While proving a formula it sometimes occurs that one would like to set a subgoal as a global assumption. In such cases one can use the command

`(admit)`.

By entering `(admit)` the current goal formula becomes a global assumption and is considered to be proven. In this way one can actually prove each formula, which justifies the title of this section. Therefore it is recommended that a finished proof contains `admit` or a global assumption as rarely as possible.

1.13. *Searching in Minlog*

While doing a proof about a certain issue, for example about the addition on naturals numbers, one often wants to know which statements about this issue are already given. A list of the given statements can be displayed by the command

`(search-about LIST)`.

Instead of `LIST` Minlog requires a list of strings and the output are all theorems and global assumptions whose names contain each of the strings in the list. Therefore the name of a theorem should always be as meaningful as possible.
Now if `nat.scm` is loaded, we can search for theorems about the addition on naturals numbers by the command

`(search-about "Nat" "Plus")`.

As ouput we get

```
Theorems with name containing Nat and Plus
NatPlusDouble
all n,m NatDouble n+NatDouble m=NatDouble(n+m)
.

.

.

NatPlusComm
all n,m n+m=m+n
No global assumptions with name containing Nat and Plus
```

which is a long list of theorems and an empty list of global assumptions. Of course, instead of the dots there are many more theorems.

2. Algebras and Inducively Defined Predicates

In this section we introduce algebras and inductively defined predicates in Minlog.

2.1. *Algebras*

The general command to define algebras in Minlog is given by

```
(add-algs NAME LISTS).
```

Here NAME is the name of the defined algebra in quotes and LISTS is a List of pais

```
'(NAME TYPE)
```

with a name for the constructor of type TYPE. Here every constructor gets his name during the definition of the algebra. We will consider some examples:
in nat.scm the algebra of natural numbers is given by

```
(add-algs "nat" '("Zero" "nat") '("Succ" "nat=>nat")).
```

As we see, zero is denoted by Zero and the successor is denoted by Succ. By entering the instruction

```
(display-alg NAME)
```

Minlog shows us a list of the constructors of the algebra NAME. In our case (display-alg "nat") leads to the output

```
nat
Zero: nat
Succ: nat=>nat
```

The boolean algebra is implemented in Minlog by default. Its two constructors have the names True and False. We can see this see by entering (display-alg "boole"):

```
boole
True: boole
False: boole
```

In Minlog we can also introduce algebras with parameters. The strings `alpha`, `beta`, `gamma`, `alpha0`, `beta0`, `gamma0` and so on stand for type variables.

In the library file `list.scm` the algebra of lists is defined by

```
(add-algs "list" '("list" "Nil") '("alpha=>list=>list" "Cons")).
```

The list type has the paramenter `alpha`. By using the list algebra this parameter has to be written explicitly. We see this by displaying the algebra with `(display-alg "list")`:

```
list
Nil: list alpha
Cons: alpha=>list alpha=>list alpha
```

2.2. *Declaration of Term Variables*

Before using variables Minlog has to know which variables have which type. For each type with name `TYPE` the string `TYPE` is by default also the name of a variable with the same type. Therefore a command like

```
(set-goal "all nat nat=nat")
```

can be understood by Minlog.

If `v` is a variable of type `TYPE`, then also `v0`, `v1` and so on are variables of type `TYPE`. Hence

```
(set-goal "all nat1000 nat1000=nat1000")
```

is also a valid instruction.

Of course, it is also possible to add new names for variables. The command to do this is

```
(add-var-name NAMES TYPE).
```

For `NAMES` we give a list of the variables names which should have type `TYPE`. In `nat.scm` the letters `n`, `m` and `l` are introduced as variables names for natural numbers by the instruction

```
(add-var-name "n" "m" "l" (py "nat")).
```

2.3. *Inductively Defined Predicates*

Inductively defined predicates are defined by the command

```
(add-ids (list (list NAME (make-arity TYPES) ALGNAME)) LIST).
```

NAME stands for the name of the defined predicate. The types of the arguments of the predicate are inserted for TYPES. ALGNAME is the name of the algebra of the inductively defined predicate and the introduction rules of the predicate are written instead of LIST in the form

```
'(FORMULA NAME),
```

whereby FORMULA is the introduction axiom and NAME its name.

As an example, we define the unary inductively defined predicate EvenI, which says that a natural number is even. To define the property that a natural numer is even, we define that 0 is even and, if n is even, then also n+2 is even. In Minlog we do this as follows:

```
(add-ids
  (list (list "EvenI" (make-arity (py "nat")) "algEvenI"))
  '("EvenI 0" "InitEvenI")
  '("all n(EvenI n -> EvenI(Succ( Succ n)))" "GenEvenI"))
```

With the command (display-alg "algEvenI") we take a look at the algebra of EvenI:

```
algEvenI
CInitEvenI: algEvenI
CGenEvenI: nat=>algEvenI=>algEvenI
```

We see that the constructor which corresponds to the introduction axiom Axiom has the name CAxiom. This holds in general.

One also sees that algEvenI and nat have the same constructor types. Therefore one could also write nat instead of algEvenI in the definition of EvenI. In general if one writes an already defined algebra instead of ALGNAME, Minlog checks whether the constructor types are the same. If this is the case this algebra is used as the type of the predicate.

By the command

```
(display-idpc NAME)
```

the introduction axioms of the inductively defined predicate NAME and its algebra are shown.

For the predicate EvenI we enter (display-idpc "EvenI") and get

```
EvenI with content of type algEvenI
InitEvenI: EvenI 0
```

```
GenEvenI: all n(EvenI n -> EvenI(Succ(Succ n)))
```

back.

The introduction axioms are saved as theorems in the system and can be used for example with the **use** command. There is also the possibility to use the instruction

```
(intro N).
```

Then Minlog uses the N-th introduction axiom of the predicate in the goal formula. The numbering starts with 0.

2.4. *Proofs with Inductively Defined Predicates*

As a counterpart to EvenI we define the predicate OddI with says that a natural number is odd:

```
(add-ids
 (list (list "OddI" (make-arity (py "nat")) "algOddI"))
 '("OddI 1" "InitOddI")
 '("all n(OddI n -> OddI(Succ( Succ n)))" "GenOddI"))
```

We prove the theorem that the successor of an even number is odd. Therefore we enter the instruction

```
(set-goal "all n(EvenI n -> OddI(Succ n))").
```

By using (assume "n" "EvenIn") we put the variable n and the premise EvenI n into the context and the output is

```
ok, we now have the new goal

  n  EvenIn:EvenI n
-----------------------------------------------------------
?_2:OddI(Succ n)
```

To prove the goal formula we would like to apply the elimination axiom of EvenI to the predicate $\{n|$OddI(Succ $n)\}$. In general, if NAME is the name of a formula which is an inductively defined predicate, we can enter the command

```
(elim NAME)
```

to use the elimination axiom of this predicate. The goal formula becomes the parameter in the elimination axiom and exactly the free variables in the formula `NAME` are abstracted. The elimination axiom of `EvenI` with parameter P is given by

$$\forall_n(\texttt{EvenI }n \to Pn \to \forall_n(\texttt{EvenI }n \to Pn \to P(SSn)) \to Pn).$$

Since `EvenI` n is already given, Minlog requires a proof of $P0$ and $\forall_n(\texttt{EvenI }n \to Pn \to P(SSn))$. Here $Pn := \texttt{OddI }Sn$ and indeed after `(elim "EvenIn")` we get the output:

```
ok, ?_2 can be obtained from

  n   EvenIn:EvenI n
----------------------------------------------------------------
?_4:all n(EvenI n -> OddI(Succ n) -> OddI(Succ(Succ(Succ n))))



  n   EvenIn:EvenI n
----------------------------------------------------------------
?_3:OddI 1
```

For the goal `OddI 1` we just write `(use "InitOddI")`. To prove the other goal we move with

```
(assume "m" "EvenIm" "OddISm")
```

the variable and the two premises into the context. The new goal is given by:

```
ok, we now have the new goal

  n   EvenIn:EvenI n
  m   EvenIm:EvenI m
  OddISm:OddI(Succ m)
----------------------------------------------------------------
?_5:OddI(Succ(Succ(Succ m)))
```

We prove it by the second introduction axiom of `OddI`, which is given by

```
GenOddI: all n(OddI n -> OddI(Succ(Succ n))).
```

So we enter `(use "GenOddI")` followed by `(use "OddISm")` and the proof is finished. With `(cdp)` we take a look at the proof tree:

```
......allnc n8136(EvenI n8136 -> OddI 1 ->
      all n(EvenI n -> OddI(Succ n) -> OddI(Succ(Succ(Succ n))))
      -> OddI(Succ n8136)) by axiom Elim
......n
....EvenI n -> OddI 1 -> all n(EvenI n -> OddI(Succ n) ->
    OddI(Succ(Succ(Succ n)))) -> OddI(Succ n) by allnc elim
.....EvenI n by assumption EvenIn3363
....OddI 1 -> all n(EvenI n -> OddI(Succ n) ->
    OddI(Succ(Succ(Succ n)))) -> OddI(Succ n) by imp elim
....OddI 1 by axiom Intro
...all n(EvenI n -> OddI(Succ n) -> OddI(Succ(Succ(Succ n)))) ->
   OddI(Succ n) by imp elim
........all n(OddI n -> OddI(Succ(Succ n))) by axiom Intro
........Succ m
.......OddI(Succ m) -> OddI(Succ(Succ(Succ m))) by all elim
.......OddI(Succ m) by assumption OddISm3367
......OddI(Succ(Succ(Succ m))) by imp elim
.....OddI(Succ m) -> OddI(Succ(Succ(Succ m)))
     by imp intro OddISm3367
....EvenI m -> OddI(Succ m) -> OddI(Succ(Succ(Succ m)))
     by imp intro EvenIm3366
...all n(EvenI n -> OddI(Succ n) -> OddI(Succ(Succ(Succ n))))
   by all intro
..OddI(Succ n) by imp elim
.EvenI n -> OddI(Succ n) by imp intro EvenIn3363
all n(EvenI n -> OddI(Succ n)) by all intro
```

Although the output is complex, we see that the first formula is derived by the elimination axiom and the formulas `all n(OddI n ->`
`OddI(Succ(Succ n)))` and `OddI 1` are derived by the introduction axioms.
This goes well, because we have used the elimination axiom exactly once
and the introduction axioms exactly twice.

3. Decorations

Minlog allows usage of decorated connectives $\to^{nc}$ and $\forall^{nc}$ to indicate that
the premise or the quantified variable are not used computationally. Their
introduction and elimination rules are are like the ones for $\to$ and $\forall$, except
that in the introduction rules $\to^{nc+}$ and $\forall^{nc+}$ the premise or the quantified
variable are not used computationally. Since these are rather intuitive
concepts, we do not develop the theory here but only explain their usage
in Minlog.

Especially when dealing with decorated logical symbols computer support is very useful. As just explained, we have to calculate the free computational variables, if we want to use the rules $\to^{nc+}$ and $\forall^{nc+}$. With computer support this is trivial but to do it with pen and paper takes some time.

3.1. *Non-computational Universal Quantifier*

We first take a look at the non-computational universal quantifier. In Minlog it is denoted by `allnc` and has the same syntactical rules as the normal universal quantifier. In Section 1.7 we have proven the formula $\forall_n(Pn \to Qn) \to \forall_n Pn \to \forall_n Qn$ and now we would like to prove the decorated form $\forall_n^{nc}(Pn \to Qn) \to \forall_n^{nc} Pn \to \forall_n^{nc} Qn$. In order to do this we load the library file `nat.scm` and declare the predicates P and Q as we have done in Section 1.7. Then we enter the instruction

```
(set-goal "allnc n(P n -> Q n) -> allnc n P n -> allnc n Q n").
```

The output is as expected:

```
-------------------------------------------------------------
?_1:allnc n(P n -> Q n) -> allnc n P n -> allnc n Q n
```

By the command (assume "assumption1" "assumption2" "m") we move the two premises and the variable m into the context. The output is similar to the one we have had in Section 1.7:

```
ok, we now have the new goal

  assumption1:allnc n(P n -> Q n)
  assumption2:allnc n P n
  {m}
-------------------------------------------------------------
?_2:Q m
```

The difference is that the variable m is in curly brackets. This tells us that m should not be used computationally. In our case both quantifiers in the assumptions are non-computational. Therefore we can enter (use "assumption1") followed by (use "assumption2") without any problems and the proof is done. The output after (cdp) is

```
.....allnc n(P n -> Q n) by assumption assumption12189
.....m
....P m -> Q m by allnc elim
.....allnc n P n by assumption assumption22190
.....m
....P m by allnc elim
...Q m by imp elim
..allnc n Q n by allnc intro
.allnc n P n -> allnc n Q n by imp intro assumption22190
allnc n(P n -> Q n) -> allnc n P n -> allnc n Q n
by imp intro assumption12189
```

We see that the rules `allnc elim` and `allnc intro` occur. These are the rules for the non-computational universal quantifier.

While dealing with the non-computational universal quantifier one should note that Minlog just returns a warning if one applies a non-computational variable to a computational universal quantifier. The proof can be continued normally. As an example we try to prove the formula above where the first universal quantifier is changed to a non-computational one:

```
(set-goal "all n(P n -> Q n) -> allnc n P n -> allnc n Q n")
```

Here we also start with `(assume "assumption1" "assumption2" "m")`. If we now enter `(use "assumption1")`, we just get a warning:

```
Warning: nc-intro with cvar(s)
m
ok, ?_2 can be obtained from

  assumption1:all n(P n -> Q n)
  assumption2:allnc n P n
  {m}
---------------------------------------------------------------
?_3:P m
```

Minlog does not stop the proof with an error message, because at this point it is not clear whether usage of the rule was really incorrect. If we finish the "proof" with `(use "assumption2")` and enter `(cdp)`, we will be informed that the "proof" is incorrect:

```
warning: allnc-intro with cvarm
```

```
.....all n(P n -> Q n) by assumption assumption12193
.....m
....P m -> Q m by all elim
.....allnc n P n by assumption assumption22194
.....m
....P m by allnc elim
...Q m by imp elim
..allnc n Q n by allnc intro
.allnc n P n -> allnc n Q n by imp intro assumption22194
all n(P n -> Q n) -> allnc n P n -> allnc n Q n
by imp intro assumption12193
Incorrect proof: nc-intro with computational variable(s)
m
```

Here we also see a difference between (cdp) and (dp). The command (dp) would not inform the user that the proof is incorrect.

3.2. *Non-computational Implication*

The non-computational implication is denoted by --> and has the same syntactical rules as ->. As an application example we show that the non-computational implication is transitive. Therefore we introduce three propositional variables by (add-pvar-name "A" "B" "C" (make-arity)) and set the goal formula:

```
(set-goal "(A-->B)-->(B-->C)->A-->C")
```

As first step we move the three assumptions into the context with the command (assume "assumption1" "assumption2" "assumption3"). The output of Minlog is

```
ok, we now have the new goal

  {assumption1}:A --> B
  assumption2:B --> C
  {assumption3}:A
-------------------------------------------------------------
?_2:C
```

Similar to the non-computational variables we see that the non-computational assumptions are in curly brackets. These assumptions can be only used in non-computational parts of the proof. Whether we are in a non-computational or not is not shown by Minlog and must be checked by the user.

In our example we first use `assumption2`, which is not in curly brackets, with (use "assumption2"). According to `assumption2` B implies C non-computationally. Therefore while proving B we are in a non-computational part of the proof, hence we can enter (use "assumption1") followed by (use "assumption3") which finishes the proof. We see that the proof is correct, if we enter (cdp):

```
....B --> C by assumption assumption22198
.....A --> B by assumption assumption12197
.....A by assumption assumption32199
....B by impnc elim
...C by impnc elim
..A --> C by impnc intro assumption32199
.(B --> C) -> A --> C by imp intro assumption22198
(A --> B) --> (B --> C) -> A --> C by impnc intro assumption12197
```

The proof tree uses the new rules `impnc intro` and `impnc elim`.

It should be noted that one implication in the proven formula is computational otherwise we could not prove the formula, because C could have computational content and if every implication were non-computational, this content would come from nothing.

Analogously to the non-computational universal quantifier Minlog only returns a warning if one uses a non-computational assumption in a computational part.

3.3. *Decorated Predicates*

The non-computational universal quantifier and implication can be used for the definition on an inductively defined predicate. In Section 2.3 we have defined the predicate `EvenI` with the second introduction axiom

```
GenEvenI: all n(EvenI n -> EvenI(Succ (Succ n)))
```

This axiom can be changed to

```
GenEvenI: allnc n(EvenI n -> EvenI(Succ (Succ n)))
```

since heuristically one can say that the information on n is already given in `EvenI n`. Compared to inductively defined predicate without decoration there is not an essential change.

Also the definition of a non-computational inductively defined predicate is quite similar to the definition of a computational one. The only change is that one does not declare a name for the algebra of the predicate.

To give a good example we use the algebra of lists, which are introduced in the library file `list.scm`. We would like to define the predicate `RevI` which takes two lists as arguments and says that the first list reversed is the second list. Therefore we load `nat.scm` and `list.scm` form the library. The list type is defined with a type variable `alpha` as we have seen in section 2.1. We first declare two variables `xs` and `ys` of type `list alpha` and one variable `x` of type `alpha`. In `list.scm` also the infix notation `::` for the constructor `Cons alpha` is declared. For `x ::(Nil alpha)` we have the abbreviation `x:` and `++` denotes the concatenation of two lists. The predicate `revI` is therefore defined by the command:

```
(add-ids (list (list "RevI" (make-arity (py "list alpha")
                                        (py "list alpha")))))
 '("RevI(Nil alpha)(Nil alpha)" "InitRevI")
 '("all xs,ys,x(RevI xs ys -> RevI(xs++x:)(x::ys))" "GenRevI"))
```

As an application we prove that `RevI` is symmetric. Hence we enter

```
(set-goal "all xs,ys(RevI xs ys -> RevI ys xs)").
```

After the command (assume "xs" "ys" "RevIxsys") the output is

```
ok, we now have the new goal

  xs  ys  RevIxsys:RevI xs ys
-----------------------------------------------------------------
?_2:RevI ys xs
```

Here we use the elimination rule for `RevI xs ys` to the goal formula by the instruction (elim "RevIxsys"). This is possible since `RevI` is non-computational. Minlog gives

```
ok, ?_2 can be obtained from

  xs  ys  RevIxsys:RevI xs ys
-----------------------------------------------------------------
?_4:all xs,ys,x(RevI xs ys -> RevI ys xs
                -> RevI(x::ys)(xs++x:))
```

```
   xs   ys   RevIxsys:RevI xs ys
-------------------------------------------------------------------
?_3:RevI(Nil alpha)(Nil alpha)
```

back. The goal formula is easily proven by (use "InitRevI"). To prove
the second formula we move via

(assume "xs1" "ys1" "x" "RevIxs1ys1" "RevIys1xs1")

everything into the context and again use the elimination axiom for the
assumption RevIys1xs1 with (elim "RevIys1xs1"). The output is

ok, ?_5 can be obtained from

```
   xs   ys   RevIxsys:RevI xs ys
   xs1  ys1  x  RevIxs1ys1:
     RevI xs1 ys1
   RevIys1xs1:RevI ys1 xs1
-------------------------------------------------------------------
?_7:all xs,ys,x0(
     RevI xs ys -> RevI(x::xs)(ys++x:)
     -> RevI(x::xs++x0:)((x0::ys)++x:))


   xs   ys   RevIxsys:ŘevI xs ys
   xs1  ys1  x  RevIxs1ys1:
     RevI xs1 ys1`
   RevIys1xs1:RevI ys1 xs1
-------------------------------------------------------------------
?_6:RevI(x:)((Nil alpha)++x:)
```

We know that :x and (Nil alpha)++x: are equal terms and Minlog also
knows it. Form InitRevI we get RevI(Nil alpha)(Nil alpha) and with
GenRevI this leads to RevI :x :x. Since Minlog does not recognize how
to apply the use command we enter

(use-with "GenRevI" (pt "(Nil alpha)") (pt "(Nil alpha)")
                                    (pt "x") "InitRevI").

After this we get

ok, ?_6 is proved. The active goal now is

```
xs  ys  RevIxsys:RevI xs ys
xs1 ys1 x  RevIxs1ys1:
  RevI xs1 ys1
RevIys1xs1:RevI ys1 xs1
```

```
?_7:all xs,ys,x0(
    RevI xs ys -> RevI(x::xs)(ys++x:)
    -> RevI(x::xs++x0:)((x0::ys)++x:))
```

back. `RevI(x::xs)(ys++x:)->RevI(x::xs++x0:)((x0::ys)++x:))` has the form of the axiom `GenRevI`. Therefore we move with (`assume "xs2" "ys2" "x0" "assumption1"`) everything but the last premise into the context and finish the proof by

`(use-with "GenRevI" (pt "x::xs2") (pt "ys2++x:") (pt "x0"))`.

Finally we take a look at the predicate `RevI` by (`display-idpc "RevI"`) and we see that `RevI` is non-computational:

```
RevI non-computational
InitRevI: RevI(Nil alpha)(Nil alpha)
GenRevI: all xs,ys,x(RevI xs ys -> RevI(xs++x:)(x::ys))
```

3.4. *Leibniz Equality and Simplification*

An important example of a non-computational inductively defined predicate is the Leibniz equality. It is saved in Minlog by default and denoted by `EqD`. We see its introduction axiom if we enter (`display-idpc "EqD"`):

```
EqD non-computational
InitEqD: allnc alpha^ alpha^ eqd alpha^
```

The system also uses the infix notation `T1 eqd T2` instead of `EqD T1 T2`. Since the type of the computational version of the Leibniz equality would be the unit type each predicate, not only the non-computational ones, can be used in the elimination axiom.

The characteristic property of the equality is $\forall_{x,y}(\text{Eq}xy \to A(x) \to A(y))$. This property can be used conveniently in Minlog by the `simp` command. If we have a formula `FORMULA` of the form `T1 eqd T2`, which is a Leibniz equality or an abstraction of it, with the command

`(simp FORMULA)`

each term T1 in the current goal formula is replaced by T2. If FORMULA has premises, Minlog requires also a proof of these premises. Formally the theorem $\forall_{x,y}(\text{Eq}xy \to A(y) \to A(x))$, which is obviously equivalent to the characteristic property of the equality above, is used. The first premise $\text{Eq}xy$ is already given and Minlog requires a proof of $A(y)$.

As an example we prove for a natural number n and a function $f : \mathbb{N} \to \mathbb{N}$ that $f(n) = n \to f(n) = f(f(n))$ holds. Hence we enter

```
(set-goal "all f,n(f n eqd n -> f (f n) eqd f n)").
```

With (assume "f" "n" "fn=n") we move the variables and the premise into the context and the new goal is:

```
ok, we now have the new goal

  f   n   fn=n:f n eqd n
----------------------------------------------------------------------
?_2:f(f n)eqd f n
```

Here we directly use (simp "fn=n"), which leads to the output

```
ok, ?_2 can be obtained from

  f   n   fn=n:f n eqd n
----------------------------------------------------------------------
?_3:f n eqd n
```

and the proof is finished by (use "fn=n").

It is also possible to use the other direction of the equality, i.e., the theorem $\forall_{x,y}(\text{Eq}xy \to A(x) \to A(y))$. One can do this by the command

```
(simp "<-" FORMULA).
```

In our proof above we could also use this command, even though it is not expedient. Nevertheless, for demonstration purposes we show the output after entering (simp "<-" "fn=n") instead of (simp "fn=n"):

```
ok, ?_2 can be obtained from

  f   n   fn=n:f n eqd n
----------------------------------------------------------------------
?_3:f(f(f n))eqd f(f n)
```

Each occurrence of n was replaced by f n.

It is not only possible to insert the name of an already known formula for FORMULA. One can also write (pf EQUALITY) instead of FORMULA, where EQUALITY has to be an abstraction of T1 eqd T2. In this case Minlog also requires a proof of EQUALITY.

3.5. *Examples of Inductively Defined Predicates*

In this section we discuss disjunction, conjunction and the existential quantifier in Minlog. We start with conjunction. In Minlog there are four versions of it:

```
AndD with content of type yprod
InitAndD: Pvar1 -> Pvar2 -> Pvar1 andd Pvar2
AndL with content of type identity
InitAndL: Pvar1 -> Pvar2 --> Pvar1 andl Pvar2
AndR with content of type identity
InitAndR: Pvar1 --> Pvar2 -> Pvar1 andr Pvar2
AndNc non-computational
InitAndNc: Pvar1 --> Pvar2 --> Pvar1 andnc Pvar2
```

There is no AndU since the type of it would be the unit type and therefore AndU and AndNc are equivalent. To use the introduction axiom of each conjuction one also has the command (split). For the predicates above (split) and (intro 0) are synonyms.

Similar to conjunction the existential quantifier also has four versions:

```
ExD with content of type yprod
InitExD: all alpha^((Pvar alpha)alpha^ ->
                   exd alpha^0 (Pvar alpha)alpha^0)
ExL with content of type identity
InitExL: all alpha^((Pvar alpha)alpha^ -->
                 exl alpha^0 (Pvar alpha)alpha^0)
ExR with content of type identity
InitExR: allnc alpha^((Pvar alpha)alpha^ ->
               exr alpha^0 (Pvar alpha)alpha^0)
ExNc non-computational
InitExNc: allnc alpha^((Pvar alpha)alpha^ -->
                  exnc alpha^0 (Pvar alpha)alpha^0)
```

Since usage of the elimination axiom of the existential quantifier is complex, there is the command

```
(by-assume ASSUMPTION VAR NAME).
```

With this instruction the new variable `VAR`, which fulfils the existential statement `ASSUMPTION` $=: \exists_x A(x)$, is introduced and the formula $A(\text{VAR})$ is saved in the context with name `NAME`.

We demonstrate the usage of `by-assume` by proving $\forall_n (P(n) \to Q(n)) \to \exists_n P(n) \to \exists_n Q(n)$. Hence we enter

```
(set-goal "all n(P n -> Q n) -> exd n P n -> exd n Q n")
(assume "assumption1" "assumption2")
```

and the output is

```
 ok, we now have the new goal

  assumption1:all n(P n -> Q n)
  assumption2:exd n P n
-----------------------------------------------------------
?_2:exd n Q n
```

Here we use the `by-assume` command with

```
(by-assume "assumption2" "n" "assumption2Inst").
```

Minlog returns

```
ok, we now have the new goal

  assumption1:all n(P n -> Q n)
  n  assumption2Inst:P n
-----------------------------------------------------------
?_5:exd n Q n
```

We see that `assumption2` has been dropped. But it still exists and one could use it. To prove the formula `exd n Q n` we use the introduction axiom of EqD by writing (`intro 0 (pt "n")`) and Minlog shows

```
ok, ?_5 can be obtained from

  assumption1:all n(P n -> Q n)
  n  assumption2Inst:P n
-----------------------------------------------------------
?_6:Q n
```

With (use "assumption1") and (use "assumption2Inst") we finishes the proof.

In contrast to Ex and And the disjunction has five variations, because the type of OrU is not the unit type but the Boolean algebra.

```
 OrD with content of type ysum
InlOrD: Pvar1 -> Pvar1 ord Pvar2
InrOrD: Pvar2 -> Pvar1 ord Pvar2
OrR with content of type uysum
InlOrR: Pvar1 --> Pvar1 orr Pvar2
InrOrR: Pvar2 -> Pvar1 orr Pvar2
OrL with content of type ysumu
InlOrL: Pvar1 -> Pvar1 orl Pvar2
InrOrL: Pvar2 --> Pvar1 orl Pvar2
OrU with content of type boole
InlOrU: Pvar1 --> Pvar1 oru Pvar2
InrOrU: Pvar2 --> Pvar1 oru Pvar2
OrNc non-computational
InlOrNc: Pvar1 -> Pvar1 ornc Pvar2
InrOrNc: Pvar2 -> Pvar1 ornc Pvar2
```

The axioms of the disjuctions can be applied straightforwardly so there are no special commands for the disjuction.

For each of these inductively defined predicates there are also interactive versions andi, exi, and ori. These are not new inductively defined predicates but the system takes the suitable version.

For example, if we want to prove the formula

$$(A \to B \wedge C) \to (A \to B) \wedge (A \to C),$$

we can do this by entering

```
(set-goal "(A -> B andi C) -> (A -> B) andi (A -> C)").
```

Minlog recognizes that B and C could be computational and therefore it replaces andi by andd:

```
----------------------------------------------------------------
?_1:(A -> B andd C) -> (A -> B) andd (A -> C)
```

The proof of this formula in Minlog is left as an exercise.

4. Terms

4.1. *define Command*

Expressions can be abbreviated by the `define` command. These expressions do not have to be terms. With `define` one can also save formulas, types or just strings. The instruction has the form

`(define STRING EXPR)`.

Instead of `STRING` one writes an arbitrary string and thus ensures that a later usage of `STRING` will be replaced by `EXPR`.

We will use the `define` command to abbreviate terms. For example, if we want to define the natural number 4, we can do this by

`(define four (pt "Succ(Succ(Succ(Succ 0)))"))`

but after entering `(pp four)` Minlog displays 4 and not `Succ(Succ(Succ(Succ 0)))`, since the decimal representation of natural numbers is already implemented. The extracted term of a proof will become long. Therefore the `define` command is very useful.

But if one would like to use some formulas or types several times, it can be reasonable to set them with the `define` command. Examples for this are

`(define Goal (pf "(A -> B -> C) -> ((C -> A) -> B) -> A -> C"))`

or

`(define sequences (py "nat=>alpha"))`.

It should be noted that one can overwrite defined strings without any problems and Minlog not even returns a warning.

4.2. *Program Constants*

Program constants or programmable constants are an important part of terms in Minlog. There are two steps to introduce program constants in Minlog. First one declares the name `NAME` and the type `TYPE` of the program constant by the command

`(add-program-constant NAME TYPE)`.

We take the addition on the natural numbers as example. In the file `nat.scm` it is declared by

`(add-program-constant "NatPlus" (py "nat=>nat=>nat"))`.

In the second step one adds the computation rules of the program constant by the command

`(add-computation-rule TERM1 TERM2)`.

A computation rule has the form

$$D\vec{P} = M$$

where $\vec{P}$ is a list of patterns built from distinct variables and constants, and M is a term with no more free variables than those in $\vec{P}$. In the instruction above TERM1 should be replaced by $D\vec{P}$ and TERM2 by M. In the case of the addition on the natural numbers we write

```
(add-computation-rule (pt "NatPlus n Zero") (pt "n"))
(add-computation-rule (pt "NatPlus n (Succ m)")
                      (pt "Succ (NatPlus n m)")).
```

There is also an abbreviation by the command

```
(add-computation-rules
 "NatPlus n Zero" "n"
 "NatPlus n (Succ m)" "Succ(NatPlus n m)").
```

This command does the same as the two commands above but it is shorter, since one does not have to write `pt` and `add-computation-rule` for each computation rule.

If we want to see the computation rules of a program constant, we have the command

`(display-pconst LIST)`

which shows us the rules for the program constants in the List LIST. In our case entering (`display-pconst "NatPlus"`) after loading `nat.scm` leads to the output

```
NatPlus
  comprules
0 n+0 n
1 n+Succ m Succ(n+m)
  rewrules
0 0+n n
1 Succ n+m Succ(n+m)
2 n+(m+1) n+m+1
```

We also see the rewrite rules of the program constant, which we consider later.

To remove the program constant `NAME` we can use the instruction

`(remove-program-constant NAME).`

Minlog removes the name of the program constant and deletes each computation rule of it. But one should be careful with a deleted program constant. The theorems about the deleted program constant are not removed. Therefore, if a new program constant with the same name is introduced, Minlog assumes that the theorems of the old program constant also holds for the new one. But obviously this does not have to be true.

4.3. *Examples of Program Constants*

The boolean operators `andb`, `orb`, and `impb` are program constants which often occur. They have the names `AndConst`, `OrConst`, and `ImpConst` and also the infix notation is declared in Minlog. The computation rules are the following:

```
AndConst
  comprules
0 True andb boole^ boole^
1 boole^ andb True boole^
2 False andb boole^ False
3 boole^ andb False False
OrConst
  comprules
0 True orb boole^ True
1 boole^ orb True True
2 False orb boole^ boole^
3 boole^ orb False boole^
ImpConst
  comprules
0 False impb boole^ True
1 True impb boole^ boole^
2 boole^ impb True True
```

Another important program constant is the recursion operator. In Minlog it is denoted by

`(Rec ALG=>TYPE).`

Strictly speaking, this is the recursion operator from the algebra `ALG` into the typ `TYPE`. In the next section we see a term with the recursion operator. At this point we would like to consider the type of the recursion operator. The command

```
(term-to-type TERM)
```

gives the type of an general term `TERM` back. To see the type of the recursion operator form the natural numbers into a type `alpha` we write

```
(pp (term-to-type (pt "(Rec nat=>alpha)")))
```

and we get the expected output:

```
nat=>alpha=>(nat=>alpha=>alpha)=>alpha
```

The last example in this section is the decidable equality for objects of a finitary algebra. The decidable equality for a finitary algebra is automatically implemented when the algebra is defined. It is denoted by = and the computation rules are so deeply implemented in Minlog that we can not see them explicitly.

4.4. *Abstraction and Application*

In Minlog the λ abstraction of a variable is denoted is denoted by putting these variables in square brackets and separated by commas in front of the term. The application of a term `N` to a term `M` is just denoted by `M N`. To show this in detail we define the addition of the natural numbers as a term of type $\mathbb{N} \to \mathbb{N} \to \mathbb{N}$ by using the recursion operator:

```
(define Plus (pt "[n,m](Rec nat=>nat) n m ([n0,n1]Succ n1)"))
```

If we would like to apply an already defined term `TERM1` to another term `TERM2`, we can use the command

```
(make-term-in-app-form TERM2 TERM1).
```

For example the application of 1 to the term `Plus` can be entered by

```
(make-term-in-app-form Plus (pt "1")).
```

We also can normalize this term with `nt`, which is discussed two sections later, by entering

```
(pp (nt (make-term-in-app-form Plus (pt "1"))))
```

and get just `Succ` as output. This is consistent with the fact that addition with 1 gives the successor.

The abstraction of a variable `VAR` from a defined term `TERM` is implemented by the command

`(make-term-in-abst-form VAR TERM)`.

So, if we enter something like the instruction `(pp (make-term-in-abst-form (pv "n") (pt "n1")))`, the output is `[n]n+1`.

4.5. *Boolean Terms as Formulas*

In Minlog a boolean term `b` can be also identified with the formula `b eqd True`. So Minlog easily understands commands like

`(set-goal "all boole1,boole2(boole1 andb boole2 -> boole1)")`.

The identification of a boolean term with a formula is done by the theorems `EqDTrueToAtom` and `AtomToEqDTrue`, which are saved in Minlog by default:

```
AtomToEqDTrue all boole^(boole^ -> boole^ eqd True)
EqDTrueToAtom all boole^(boole^ eqd True -> boole^)
```

While working with boolean variables as formulas one often needs the theorem

`Truth T`

which says `T eqd T`. With this theorem one easily can prove boolean terms which normalizes to `T`.

4.6. *Normalisation*

In Minlog two terms are considered equal if they have a common reduct w.r.t. the computation and rewrite rules. This equality is also implemented in Minlog and in this chapter we take a look at the conversion rules in Minlog.

Program constants are defined by their computation rules. In the system the two sides of the rules are automatically identified. The computation rules are also saved as theorems. The X-th computation rule of the program constant `NAME` has the name `NAMEXCompRule`. This is one reason why the rules are numbered in the output of `display-pconst`. The computation rules do not appear in the output of search-about. But if one would like to see these rules, too, one can expand search-about by the string 'all. So we can explicitly search for the computation rules of boolean operators by the command

```
(search-about 'all "CompRule" "Const").
```

The output is the following list:

```
Theorems with name containing CompRule and Const
NegConst1CompRule
negb False eqd True
NegConst0CompRule
negb True eqd False
OrConst3CompRule
all boole^ (boole^ orb False)eqd boole^
OrConst2CompRule
all boole^ (False orb boole^)eqd boole^
OrConst1CompRule
all boole^ (boole^ orb True)eqd True
OrConst0CompRule
all boole^ (True orb boole^)eqd True
ImpConst2CompRule
all boole^ (boole^ impb True)eqd True
ImpConst1CompRule
all boole^ (True impb boole^)eqd boole^
ImpConst0CompRule
all boole^ (False impb boole^)eqd True
AndConst3CompRule
all boole^ (boole^ andb False)eqd False
AndConst2CompRule
all boole^ (False andb boole^)eqd False
AndConst1CompRule
all boole^ (boole^ andb True)eqd boole^
AndConst0CompRule
all boole^ (True andb boole^)eqd boole^
No global assumptions with name containing CompRule and Const
```

These theorems can be used with the simp command.

Often one would like to normalize a term TERM completely. In Minlog one does this with the instruction

(nt TERM).

Of course, this is not always possible and therefore this instruction could lead to an infinite loop, which had to be stopped manually by the user.
In Section 4.4 we defined the term PLUS and now we apply this term to two natural numbers and normalize it. So we enter

```
(define 7+2 (pt "([n,m](Rec nat=>nat) n m ([n0,n1]Succ n1))7 2"))
(pp (nt 7+2))
```

and indeed the output is 9.
Normalization in a proof is also possible with the command

(ng NAME).

Each term in the assumption with name NAME is normalized by this instruction. If NAME is replaced by #t, each term in the goal formula is normalized and if one just writes (ng), each term in the goal and in the context is normalized.
Since the normalisation algorithm does not need to terminate, it is sometimes better just to do β and η conversion. With

(term-to-beta-eta-nf TERM)

the term TERM is normalized referring to β and η conversion. As example the output of (pp (term-to-beta-eta-nf (pt "([n]n+2)1"))) is 1+2.
It is also possible to add new rewrite rules, which are used by the normalisation with **nt** or **ng**. By entering the command

(add-rewrite-rule TERM1 TERM2)

Minlog adds the global assumption that TERM1 and TERM2 are equal. If one uses add-rewrite-rule after the proof that TERM1 is equal to TERM2 this rewrite rule is added as a theorem and is not added as an assumption. In both cases this rule is always used by the commands **nt** and **ng** and appears in the output of display-pconst. We have already seen this in the output of (display-pconst "NatPlus") after loading nat.scm:

```
NatPlus
  comprules
0 n+0 n
1 n+Succ m Succ(n+m)
```

```
rewrules
0 0+n n
1 Succ n+m Succ(n+m)
2 n+(m+1) n+m+1
```

Similar to the computation rules the X-th rewrite rule is saved with the name NAMEXRewRule.

4.7. *The Extracted Term*

After proving a formula in Minlog one can access the computational content of the proof by the command

```
(proof-to-extracted-term).
```

As an example we prove that for every even natural number n there exists a natural number m such that $m+m = n$ holds. Here we use the decorated version of the predicate EvenI, which is implemented by

```
(add-ids
 (list (list "EvenI" (make-arity (py "nat")) "algEvenI"))
 '("EvenI 0" "InitEvenI")
 '("allnc n(EvenI n -> EvenI(Succ (Succ n)))" "GenEvenI")).
```

To prove the statement above we set

```
(set-goal "allnc n(EvenI n -> exl m m+m=n)").
```

Here we use exl, since the equality m+m=n does not have any computational content, and we will use n only non-computationally, therefore it is bounded with an nc quantifier.
To prove the goal formula we first move with (assume "n" "EvenIn") the variable n and the premise into the context and we enter (elim "EvenIn") to use the elimination axiom of EvenI. For each introduction axiom of EvenI we get two new goal formulas:

```
ok, ?_2 can be obtained from

  {n}  EvenIn:EvenI n
----------------------------------------------------------------
?_4:allnc n(EvenI n -> exl m m+m=n -> exl m m+m=Succ(Succ n))
```

```
{n}  EvenIn:EvenI n
```

```
?_3:exl m m+m=0
```

We prove the goal formula `exl m m+m=0` by using the introduction axiom of ExL with the term 0. So we enter `(intro 0 (pt "0"))` and prove the equality 0+0=0 with `(use "Truth")`. For proving the goal formula `?_4` we first move with `(assume "n1" "EvenIn1" "IH")` everything into the context. From the induction hypotheses IH we get m with `m+m=n1` by the command `(by-assume "IH" "m" "IHInst")`. For proving the goal formula `exl m m+m=Succ(Succ n)` we first enter `(intro 0 (pt "m+1"))`, which leads to the output

```
ok, ?_9 can be obtained from
```

```
{n}   EvenIn:EvenI n
{n1}  EvenIn1:EvenI n1
m  IHInst:m+m=n1
```

```
?_10:m+1+(m+1)=Succ(Succ n1)
```

After normalization by `(ng)` the new goal is exactly `IHInst`. So we finish the proof with `(use "IHInst")`.
Now we get the extracted term by the command

```
(define eterm (proof-to-extracted-term)).
```

The output after `(pp eterm)` is hardly readable:

```
[algEvenI3985]
 (Rec algEvenI=>nat)algEvenI3985(([n^3986]n^3986)0)
 ([algEvenI3991,n3989]
   ([n3988,(nat=>nat)_3987](nat=>nat)_3987 n3988)n3989
   ([m]([n^3990]n^3990)(m+1)))
```

But the normalized form of it is more readable. Therefore we enter `(pp (nt eterm))` and get

```
[algEvenI0](Rec algEvenI=>nat)algEvenI0 0([algEvenI1]Succ)
```

back. It is always recommended to normalize an extracted term. We see that there is no variable for the algebra `algEvenI` declared, which expands the extracted term. So we define f as a variable of type algEvenI. The new normalized term is

```
[f0](Rec algEvenI=>nat)f0 0([f1]Succ)
```

The occurrence of the recursions operator matches with the fact that we used the elimination axiom of EvenI exactly once. The type of the extracted term algEvenI=>nat is also plausible, since for a witness that n is even we get a natural number m with m+m=n.

5. Totality

5.1. *Implementation of Totality*

To each algebra ALG, which is defined in Minlog, one can add the totality predicate by the instruction

```
(add-totality ALG).
```

The newly added predicate has the name TotalAlg, where the first letter of ALG is a capital letter. In list.scm the totality of lists are introduced. With (display-idpc "TotalList") we take a look at the introduction axiom:

```
TotalList with content of type list
TotalListNil: TotalList(Nil alpha)
TotalListCons: allnc alpha^(
 Total alpha^ ->
 allnc (list alpha)^0(
  TotalList(list alpha)^0 -> TotalList(alpha^ ::(list alpha)^0)))
```

We see that the premise of TotalListCons is the totality of alpha. This is the absolute totality. A generalisation of absolute totality is relative totality. In Minlog it can be introduced by

```
(add-rtotality ALG).
```

It is denoted by RTotalALG. We take a look at relative totality of list by the instruction (display-idpc "RTotalList") and get the output

```
RTotalList with content of type list
RTotalListNil: (RTotalList [...])(Nil alpha)
RTotalListCons: allnc alpha^(
 (Pvar alpha)_356 alpha^ ->
 allnc (list alpha)^0(
```

```
(RTotalList [...])(list alpha)^0 ->
(RTotalList [...])
(alpha^ ::(list alpha)^0)))
```

For better readability

```
(RTotalList (cterm (alpha^1) (Pvar alpha)_356 alpha^1))
```

is replaced by (RTotalList [...]). As we see, relative totality has a type variable alpha and also a predicate variable (Pvar alpha)_356. Compared to absolute totality, relative totality has this predicate variable instead of totality of alpha in the premise of TotalListCons. Relative totality is a generalization of absolute totality. Each totality of another type is replaced by a predicate variable. If one inserts for each predicate variable in the relative totality predicate the predicate $\mu_X(\forall_x X x)$, one gets structural totality, which is also a variation of totality.

5.2. *Implicit Representation of Totality*

The reader may have seen in previous sections that some variable names in the output of Minlog are equipped with ^. Now we see why. For each quantification over a variable without ^ Minlog adds implicitly the assumption that this variable is total. Therefore expressions like $\forall_x A$ and $\forall_x^{nc} A$ are abbreviations for $\forall_{\hat{x}}^{nc}(\text{Total } \hat{x} \to A)$ and $\forall_{\hat{x}}^{nc}(\text{Total } \hat{x} \to^{nc} A)$. The character ^ after the variable name indicates that the variable does not have to be total. In Minlog this abbreviation is introduced by two theorems

```
AllncTotalIntro
allnc alpha^(Total alpha^ --> (Pvar alpha)alpha^)
 -> allnc alpha(Pvar alpha)alpha
AllTotalIntro
allnc alpha^(Total alpha^ -> (Pvar alpha)alpha^)
 -> all alpha(Pvar alpha)alpha
```

and it is eliminated by

```
AllncTotalElim allnc alpha(Pvar alpha)alpha ->
 allnc alpha^(Total alpha^ --> (Pvar alpha)alpha^)
AllTotalElim all alpha(Pvar alpha)alpha ->
 allnc alpha^(Total alpha^ -> (Pvar alpha)alpha^)
```

For each version of the existential quantifier there are the introduction axioms for the totality abbreviation

```
ExDTotalIntro
 exr alpha^(Total alpha^ andd (Pvar alpha)alpha^) ->
 exd alpha(Pvar alpha)alpha
ExLTotalIntro
 exr alpha^(Total alpha^ andl (Pvar alpha)^' alpha^) ->
 exl alpha(Pvar alpha)^' alpha
ExRTotalIntro
 exr alpha^(TotalNc alpha^ andr (Pvar alpha)alpha^) ->
 exr alpha(Pvar alpha)alpha
ExNcTotalIntro
exnc alpha^(Total alpha^ andnc (Pvar alpha)alpha^) ->
exnc alpha(Pvar alpha)alpha
```

and also the elimination axioms of the totality abbreviation.

```
ExNcTotalElim
 exnc alpha (Pvar alpha)alpha ->
 exnc alpha^(Total alpha^ andnc  (Pvar alpha)alpha^)
ExRTotalElim
 exr alpha (Pvar alpha)alpha ->
 exr alpha^(TotalMR alpha^ alpha^ andr (Pvar alpha)alpha^)
ExLTotalElim
 exl alpha (Pvar alpha)^ alpha ->
 exr alpha^(Total alpha^ andl (Pvar alpha)^ alpha^)
ExDTotalElim
 exd alpha (Pvar alpha)alpha ->
 exr alpha^(Total alpha^ andd (Pvar alpha)alpha^)
```

Of course, these theorems only can be used if the corresponding (absolute) totality predicate is defined. But even if the totality of the corresponding type is not defined, one can use variables with and without ^. As long as the totality of this type is not defined, there is no semantical difference between x and x^. Nevertheless for example, the system never allows a specialisation of a formula $\forall_x A$ to a variable $\hat{x}$.

5.3. *Totality of Program Constants*

For proving that a program constant PCONST is total there is the command

```
(set-totality-goal PCONST).
```

The computer generates as goal formula the totality of this program constant.

In the file nat.scm the totality of the addition on the natural numbers is proven. This is done by

```
(set-totality-goal "NatPlus").
```

Minlog gives the detailed goal

```
?_1:allnc n^(TotalNat n^ ->
             allnc n^0(TotalNat n^0 -> TotalNat(n^ +n^0)))
```

back. To prove this we move with (assume "n^" "Tn" "m^" "Tm") everything into the context and prove TotalNat(n^ +m^) by the elimination axiom for Tm. Therefore we enter (elim "Tm") and get the output

```
ok, ?_2 can be obtained from

  {n^}  Tn:TotalNat n^
  {m^}  Tm:TotalNat m^
```

```
?_4:allnc n^0(TotalNat n^0 ->
              TotalNat(n^ +n^0) -> TotalNat(n^ +Succ n^0))



  {n^}  Tn:TotalNat n^
  {m^}  Tm:TotalNat m^
```

```
?_3:TotalNat(n^ +0)
```

After normalisation with (ng #t) the first goal is exactly the assumption Tn. So we prove it with (use "Tn"). For the second goal we enter first the instruction (assume "l^" "Tl" "IH") to move the premises and the variables into the context. This leads to the output:

```
ok, we are back to goal

  {n^}  Tn:TotalNat n^
  {m^}  Tm:TotalNat m^
```

```
{1^}   Tl:TotalNat 1^
IH:TotalNat(n^ +1^)
```

```
?_5:TotalNat(n^ +Succ 1^)
```

When we normalise the goal formula with (ng #t), we get the new goal
TotalNat(Succ(n^ +1^)). Here we use the totality of the successor function, which is exactly the second introduction axiom of TotalNat. So we
enter (use "TotalNatSucc"). Minlog requires a proof that the argument
of Succ is total, i.e. TotalNat(n^ +1^). This is done by (use "IH") and
the proof is finished.
After proving the totality of a program constant PCONST one can save the
totality by

```
(save-totality).
```

This instruction saves the totality of PCONST as a theorem with name
PCONSTTotal.

5.4. *Totality of Boolean Terms*

For total boolean terms the logical connectives →, ∧, and ∨ are equivalent
to the boolean once impb, andb, and orb. Especially for andb this fact is
also implemented in Minlog. Therefore one can use the command (split)
also to prove a formula of the form a andb b, where a and b are total
boolean terms. In the other direction, one can prove a or b directly by
using the assumption a andb b. For example the goal

```
all boole1,boole2(boole1 andb boole2 -> boole1)
```

is easily proven by

```
(assume "boole1" "boole2" "assumption")
(use "assumption").
```

To demonstrate the usage of (split) we prove the goal

```
all boole1,boole2(boole1 andb boole2 -> boole2 andb boole1).
```

In Minlog this is done by entering

```
(assume "boole1" "boole2" "assumption")
(split)
```

```
(use "assumption")
(use "assumption").
```

Here it is important that both variables are total. The statement is also true if only one of both variables is total. But then (split) and (use "assumption") does not lead to a correct proof and one has to prove it in another way.

For total variables of a finitary algebra the decidable equality is equivalent to the Leibniz equality. In Minlog this is not automatically implemented such that one has to prove it for each algebra individually. In the library file this is done for some algebras. In nat.scm the theorem NatEqToEqD and in list.scm the theorems ListBooleEqToEqD, ListNatEqToEqD and ListListNatEqToEqD are proven.

If for an finitary algebra ALG the theorem

```
all a,b(a = b -> a eqd b)
```

is saved with the name ALGEqToEqD, then we also can use formulas of the form T1 = T2 instead of T1 eqd T2 as arguments of simp, if T1 and T2 are total terms with type ALG. If such a theorem is not saved, the theorem ALGEqToEqD becomes a global assumption by the usage of simp as above.

5.5. *Induction*

To use the abbreviation of totality efficiently there is the command .

```
(ind).
```

This command can be used if the goal formula has the form $\forall_x A(x)$. Minlog applies the elimination axiom of the totality of x to the predicate $\{x|A(x)\}$ and the premises of this axiom become the new goal formulas. An extended command of ind is

```
(ind TERM)
```

for a total term TERM. This instruction can be applied to each goal formula $A(\text{TERM})$. In this case the elimination axiom of the totality of TERM is applied to $\{x|A(x)\}$.

As an example we prove, that each total natural number is even or odd. Therefore we load the file nat.scm and define EvenI and OddI as in Section 2.3 and 2.4. To set the goal we enter

```
(set-goal "all n (EvenI n ord OddI n)").
```

Of course, here we use the computational disjunction OrD since EvenI and
OddI have computational content. After setting the goal formula we directly
use induction on n. Therefore we enter (ind) and get

```
ok, ?_1 can be obtained from

  n3987
-----------------------------------------------------------------
?_3:all n(EvenI n ord OddI n
          -> EvenI(Succ n) ord OddI(Succ n))


  n3987
-----------------------------------------------------------------
?_2:EvenI 0 ord OddI 0
```

back. The goal ?_2 holdes because of EvenI 0. So we enter (intro
0) twice and this goal is proven. For the goal ?_3 we first move with
(assume "n" "IH") the variable and the premise into the context and then
use the elimination axiom of OrD by the command (elim "IH"). Then
Minlog requires twice a proof of EvenI(Succ n) ord OddI(Succ n) once
with the assumption EvenI n and once with the assumption OddI n.

```
ok, ?_5 can be obtained from

  n3990  n   IH:EvenI n ord OddI n
-----------------------------------------------------------------
?_7:OddI n -> EvenI(Succ n) ord OddI(Succ n)


  n3990  n   IH:EvenI n ord OddI n
-----------------------------------------------------------------
?_6:EvenI n -> EvenI(Succ n) ord OddI(Succ n)
```

If EvenI n holds, we prove Odd(Succ n). Therefore we enter
(assume "Evenn") followed by (intro 1). The formula OddI(Succ n)
can be proven by using the elimination axiom of (EvenI n) by
(elim "Evenn"). The output of Minlog is

```
ok, ?_9 can be obtained from
```

```
 n3996   n   IH:EvenI n ord OddI n
 Evenn:EvenI n
-----------------------------------------------------------------
?_11:all n(EvenI n -> OddI(Succ n)
              -> OddI(Succ(Succ(Succ n))))

 n3996   n   IH:EvenI n ord OddI n
 Evenn:EvenI n
-----------------------------------------------------------------
?_10:OddI 1
```

OddI 1 is the first introduction axiom of OddI and so we prove it with
(intro 1). The goal ?_11 follows from the second introduction axiom of
OddI. We just enter (assume "n1" "Evenn1") and (use "GenOddI") and
it is proven. So the case EvenI n is done. If OddI n holds, the proof is
done analogously:

```
(assume "Oddn")
(intro 0)
(elim "Oddn")
(use "GenEvenI")
(intro 0)
(assume "n1" "Oddn1")
(use "GenEvenI")
```

As normalized extracted term we get

```
[n0]
 (Rec nat=>algEvenI ysum algOddI)n0
     ((InL algEvenI algOddI)CInitEvenI)
 ([n1,(algEvenI ysum algOddI)_2]
   [if (algEvenI ysum algOddI)_2
     ([algEvenI3]
      (InR algOddI algEvenI)
      ((Rec algEvenI=>algOddI)algEvenI3 CInitOddI
       ([n4,algEvenI5]CGenOddI(Succ n4))))
     ([algOddI3]
      (InL algEvenI algOddI)
      ((Rec algOddI=>algEvenI)algOddI3(CGenEvenI 0 CInitEvenI)
       ([n4,algOddI5]CGenEvenI(Succ n4))))])
```

The first recursion operator corresponds to the elimination axiom of the totality, which we used with (ind). The other two recursion operators come from the elimination of EvenI n and OddI n. The elimination of EvenI n ord OddI n with the command (elim "IH") provides a recursion operator on the sum of types. In the normalized extraced term this occurs as if. In Minlog if denotes the case operator. For a sum of types the case operator and the recursion operator are program constants with the same computational rules and Minlog uses the case operator, since it is simpler. In the next section we take a deeper look on the case operator.

5.6. *Case Distinction*

In a proof by induction it often occurs that one does not need the induction hypothesis or there is not even an induction hypothesis. For example, we have the last case for the boolean totality. The elimination axiom of the boolean totality is $\forall_b(\mathbf{T}_\mathbb{B}b \to Ptt \to Pff \to Pb)$. As we see, this is just case distinction by the two cases $b = tt$ and $b = ff$. But also for total natural numbers there are formulas which can be proven just by distinction between the zero case and the successor case. As example we take the (modified) predecessor function on the natural numbers, which is a program constant Pred given by the computation rules Pred $0 := 0$ and Pred $Sn := n$. One way of proving that this program constant is total, i.e. $\forall_n(\mathbf{T}_\mathbb{N}n \to \mathbf{T}_\mathbb{N}(\text{Pred } n))$, is by using the elimination axiom of totality. In Minlog this proof is done as follows:

```
(set-totality-goal "Pred")
(use "AllTotalElim")
(ind)
(intro 0)
(use "AllTotalIntro")
(assume "n^" "Tn^" "Spam")
(use "Tn^")
```

After entering (use "AllTotalIntro") we get the output

```
 ok, ?_4 can be obtained from

 n4149
----------------------------------------------------------------
?_5:allnc n^(TotalNat n^ -> TotalNat(Pred n^)
            -> TotalNat(Pred(Succ n^)))
```

but the premise `TotalNat(Pred n^)` is not used in the proof. As extracted term we have

```
[n0][if n0 0 ([n1]n1)]
```

which do not have a recursion operator, although we have used the command `(ind)`. Therefore we rather expect an extracted term like

```
[n0](Rec nat=>nat) n0 0 ([n1,n2]n1).
```

But, since we have not used the induction hypothesis, the bounded variable n2 does not appear anywhere else. In such cases the recursion operator can be replaced by the simpler case operator. In Minlog this is done automatically by the normalisation of such a term. The case operator is denoted by `[if ...]`, where ... stands for the arguments of the case operator.
If one just would like to do case distinction on a total variable, there is the instruction

```
(cases),
```

which are analogously used as `(ind)`. We have also the expansion

```
(cases TERM)
```

of the `cases` command to do case distinction on a total term TERM. So we can also prove the totality of `Pred` with `cases`:

```
(set-totality-goal "Pred")
(use "AllTotalElim")
(cases)
(intro 0)
(use "AllTotalIntro")
(assume "n^" "Tn^")
(use "Tn^")
```

The proof is almost similar but after `(use "AllTotalIntro")` we have the output

```
ok, ?_4 can be obtained from

  n4312
-----------------------------------------------------------------
?_5:allnc n^(TotalNat n^ -> TotalNat(Pred(Succ n^)))
```

Here the induction hypothesis `TotalNat(Pred n^)` does not occur as premise.

One often uses `cases` for a variable whose algebra just have constructor types of the form $\kappa(\xi) = \vec{\sigma} \to \xi$. This is the case for the boolean algebra and the sum algebra. But also for the type product the `case` command is useful, even though there is only one case. As an educational example we define the program constant `sort`, which rearranges a pair of natural numbers such that the smaller number is on the left-hand side. In Minlog we do this by the instruction

```
(add-program-constant "sort"
    (py "(nat yprod nat)=>(nat yprod nat)"))
(add-computation-rules
  "sort (n pair m)" "[if (m<n) (m pair n) (n pair m)]")
```

We declare x as a variable of type **nat yprod nat** and then we enter the goal formula

```
(set-goal "all x lft(sort x) <= rht(sort x)")
```

Here we directly use the command (`cases`) and Minlog shows all possible cases how x can be built. There is only the case x = n pair m for total natural numbers n and m. Therefore the output is

```
 ok, ?_1 can be obtained from

    x4406
  ---------------┬------------------------------------------------
?_2:all n,n0 lft(sort(n pair n0))<=rht(sort(n pair n0))
```

and we have reached that x is decomposed. Now we enter (`assume "n" "m"`) and normalize the goal formula by (`ng`). Since `sort` is defined by case distinction on m<n, we also use case distinction on m<n for the proof and write therefore (`cases (pt "m<n")`). So we get two new goals:

```
 ok, ?_4 can be obtained from

    x4414  n   m
  --------------------------------------------------------------
?_6:(m<n -> F) ->
    lft[if False (m pair n) (n pair m)]
```

```
   <=rht[if False (m pair n) (n pair m)]


 x4414   n   m
-----------------------------------------------------------
?_5:m<n -> lft[if True (m pair n) (n pair m)]
         <=rht[if True (m pair n) (n pair m)]
```

In each case the term **m<n** was already replaced. After (assume "case1") and **ng** we get for the first case:

```
ok, the normalized goal is

 x4430   n   m   case1:m<n
-----------------------------------------------------------
?_8:m<=n
```

We prove this with the theorem NatLtToLe: all n,m(n<m -> n<=m), which is implemented in nat.scm. So we enter (use "NatLtToLe") followed by (use "case1"). The second case is analogously proven. Here we need the theorem NatNotLtToLe: all n,m((n<m -> F) -> m<=n) to get n<=m from the assumption m<n -> F.

Acknowledgements

A lot of thanks to Helmut Schwichtenberg for proofreading this chapter and to the Istituto Nazionale di Alta Matematica "F. Severi" for supporting me financially.
Any remaining inaccuracies in this chapter are my own responsibility. I am grateful for correcting any kind of mistakes.

References

[1] U. Berger, K. Miyamoto, H. Schwichtenberg, and M. Seisenberger. Minlog — a tool for program extraction supporting algebras and coalgebras. In *CALCO*, pp. 393–399 (2011).
[2] K. Miyamoto. The Minlog System. URL http://www.mathematik. uni-muenchen.de/~logik/minlog/index.php#installation (2017). [Online; accessed 20-November-2017].

[3] H. Schwichtenberg and S. S. Wainer, *Proofs and Computations*. Perspectives in Logic, Association for Symbolic Logic and Cambridge University Press (2012).

[4] F. Wiesnet. Kostruktive Analysis mit exakten reellen Zahlen. Master's thesis, Ludwig-Maximilian University of Munich (2017).

List of Contributors

Josef Berger obtained a Diploma (1997) and a PhD (2002) in Mathematics at the LMU Munich. His PhD studies on infinitesimal approaches to stochastic analysis were advised by Horst Osswald. Since 2016, he has been Privatdozent (private researcher) at the LMU Munich. He is working in constructive mathematics. Recently, he is investigating constructive approaches to convex optimization.

Laura Crosilla is a Teaching Fellow in Logic and Philosophy of Mathematics at Birmingham University. She graduated in philosophy from the University of Florence and obtained her PhD in mathematical logic from the University of Leeds. She has held postdoctoral positions at the Ludwig-Maximilian University Munich, the University of Florence, and the University of Leeds. She was recipient of a grant awarded by the John Templeton Foundation for a project on Infinity in Mathematics. She recently completed a second PhD in the philosophy of mathematics at Leeds University. Her research spans from the philosophy of mathematics to mathematical logic. She has worked on constructive set theory, proof theory, and constructive type theory. Her main research interest at present is the philosophy of mathematics, with particular focus on the philosophy of constructive mathematics and predicativity.

Hajime Ishihara is Professor for Mathematical Logic at Japan Advanced Institute of Science and Technology (JAIST). After completing his doctorate in Information Science at Tokyo Institute of Technology, he was appointed Associate at Hiroshima University and Associate Professor at JAIST. Apart from constructive mathematics and mathematical logic at large, his principal research interests are constructive analysis and topology, constructive set theory, constructive reverse mathematics, and proof theory.

Klaus Mainzer is Emeritus of Excellence at the Technical University of Munich (TUM). After studies of mathematics, physics, and philosophy at the University of Münster, he was professor for the foundations and history of exact sciences and vice-president at the University of Constance, professor for philosophy of science and director of the institute of interdisciplinary informatics at the University of Augsburg, and professor for philosophy of science, director of the Carl von Linde Academy and founding director of the Munich Center for Technology in Society (MCTS) at TUM. His principal research interests are about constructive and computational foundations of mathematics, science, and philosophy with a special focus on AI-technology and its societal impact.

Kenji Miyamoto received a PhD in Mathematics from the University of Munich in 2013. He joined the University of Innsbruck as research assistant in 2016. He was a researcher at fortiss GmbH in Munich and software developer at software companies Hagenberg Software GmbH in Austria and Hitachi Software Engineering Co. Ltd. in Japan. His principal research interests are the application of proof theory in constructive mathematics and implementation of the proof assistant Minlog.

Masahiko Sato is Professor Emeritus at the Graduate School of Informatics, Kyoto University. He graduated from the Department of Mathematics, University of Tokyo in 1971 and received his PhD in mathematics from Kyoto University in 1977. He was Professor at Tohoku Unversity from 1986 to 1995, and Professor at Kyoto University from 1995 to 2012.

Peter Schuster is Associate Professor for Mathematical Logic at the University of Verona. After both doctorate and habilitation in mathematics he was Privatdozent at the University of Munich, and Lecturer at the University of Leeds. Apart from constructive mathematics at large, his principal research interests are about Hilbert's programme in abstract mathematics, especially the computational content of classical proofs in algebra and related fields in which transfinite methods such as Zorn's Lemma are invoked.

Helmut Schwichtenberg is an Emeritus Professor of Mathematics at Ludwig-Maximilians-Universität München. After both doctorate and habilitation at the Institut für Mathematische Logik der Fakultät für Mathematik, Universität Münster he was Wissenschaftlicher Rat und Professor at Universität Heidelberg, before becoming Professor (Ordinarius) at

Ludwig-Maximilians-Universität München. His principal research interests are proof theory, lambda calculus, recursion theory, and applications of logic to computer science.

Gregor Svindland is Ákademischer Rat at LMU Munich. He studied mathematics at HU Berlin, before doing his PhD at LMU Munich. After a year as Senior Researcher at EPF Lausanne, Gregor rejoined the stochastics/financial mathematics group at LMU Munich as assistant professor (Juniorprofessor) before taking his current position. His research interests lie mainly in mathematical models of risk and uncertainty, stochastics combined with convex analysis, and recently a constructive approach to convex optimization (joint project with Josef Berger).

Stanley Wainer is an Emeritus Professor of the University of Leeds, a one-time Head of its School of Mathematics, and a past-President of the British Logic Colloquium. His research interests lie in the area of proof theory, and in the sub-recursive hierarchies of bounding functions obtained via ordinal analysis. His co-authored monograph with long-term collaborator Helmut Schwichtenberg, entitled "Proofs and Computations", was published in 2012 by Cambridge University Press.

Franziskus Wiesnet is a PhD student at the Joint Doctoral School in Mathematics at the University of Trento and the University of Verona. He obtained both bachelor's and master's degrees in mathematics from the University of Munich. He holds a Marie Skłodowska-Curie fellowship of the Istituto Nazionale di Alta Matematica "F. Severi". His current research interests include algorithms for the signed digit code of real numbers, and set theory..

Printed in the United States
By Bookmasters